Techniques for Prospecting: Prospect or Perish

Financial Advisor Series   Walt J. Woerheide, Editor

Sales Skills Techniques

Techniques for Exploring Personal Markets

Techniques for Meeting Client Needs

Techniques for Prospecting: Prospect or Perish

Product Essentials

Essentials of Annuities

Essentials of Business Insurance

Essentials of Disability Income Insurance

Essentials of Employee Benefits

Essentials of Life Insurance Products

Essentials of Long-Term Care Insurance

*Essentials of Multiline Insurance Products**

Planning Foundations

Foundations of Estate Planning

Foundations of Financial Planning: An Overview

Foundations of Financial Planning: The Process

*Foundations of Investment Planning**

Foundations of Retirement Planning

Foundations of Senior Planning

* Courses under development

Financial Advisor Series

Techniques for Prospecting:
Prospect or Perish

I. David Cohen
Richard A. Dulisse
Kirk S. Okumura
Glenn E. Stevick, Jr.

Lynn Hayes, Editor

The American College Press/*Bryn Mawr, Pennsylvania*

This publication is designed to provide accurate and authoritative information about the subject covered. While every precaution has been taken in the preparation of this material, the authors and The American College® assume no liability for damages resulting from the use of the information contained in this publication. The American College is not engaged in rendering legal, accounting, or other professional advice. If legal or other expert advice is required, the services of an appropriate professional should be sought.

Library of Congress Control Number 2006932177
ISBN 193281938X

Printed in the United States of America

To Bruce Worsham, who developed the Financial Advisor Series of LUTC and FSS courses and who has served The American College for more than 35 years to create quality educational materials for the enrichment of financial services professionals.

In our role as authors he taught us never to compromise the integrity of our work and challenged us always to strive for perfection. We thank him.

RAD
KSO
GES

Contents

Preface

The mission of this book is to develop your professionalism as a financial advisor who counsels prospects and clients about the need for insurance and financial planning. We intend to do this by teaching you the eight-step selling/planning process. The text covers the first two steps of the process, focusing on prospecting, marketing, and approaching prospects. It explains the skills, techniques, and knowledge relating to gaining appointments to discuss financial matters with prospects. To acquire an understanding of the prospecting process, you need to read the entire book. This will give you both an overview of the whole process and an in-depth look at the skills and techniques necessary to become successful in prospecting in the financial services industry.

Although much of the text material will be new to you, some will, no doubt, refresh knowledge you acquired in the past. In either case, all of the text material is both valuable and necessary if you aspire to be a successful financial advisor. The benefits you gain from studying the text material will be directly proportional to the effort you expend. So read each chapter carefully and answer the essay and multiple-choice review questions for the chapter (preferably before looking in the back of the book for the answers); to do less would be to deprive yourself of the unique opportunity to become familiar with the selling/planning process and all that it entails.

The book includes numerous educational features designed to help you focus your study of the selling/planning process as it relates to prospecting. Among the features found in each chapter of the book are

- learning objectives
- a chapter outline, examples, figures, and lists
- key terms and concepts
- review questions (essay format)
- self-test questions (multiple-choice format)

Features located in the back of the book are a(n)

- glossary
- answers to questions section
- index

Finally, all of the individuals noted on the acknowledgments page made this a better book, and we are grateful. In spite of the help of all these fine folks, however, some errors have undoubtedly been successful in eluding our eyes. For these the authors are solely responsible.

Richard A. Dulisse
Kirk S. Okumura
Glenn E. Stevick, Jr.

The American College

The American College® is an independent, nonprofit, accredited institution founded in 1927. Through a continuum of education program, The College offers professional certification and graduate-degree distance education to men and women seeking career growth in financial services.

At the beginning end of the continuum, the Center for Financial Advisor Education at The American College offers the LUTC Fellow (LUTCF) professional designation jointly with the National Association of Insurance and Financial Advisors (NAIFA) and the Financial Services Specialist (FSS) professional designation. The Center's curriculum is designed to introduce students to the technical side of financial services while at the same time providing them with the requisite sales training skills.

In the middle of the continuum, the Solomon S. Huebner School® of The American College administers the Chartered Life Underwriter (CLU®), the Chartered Financial Consultant (ChFC®), the Chartered Advisor for Senior Living (CASL™), the Registered Health Underwriter (RHU®), the Registered Employee Benefits Consultant (REBC®), and the Chartered Leadership Fellow® (CLF®) professional designation programs. In addition, the Huebner School also administers The College's CFP Board-registered education program, the CFP Certification Curriculum, for those individuals interested in pursuing CFP® certification.

Finally, at the advanced end of the continuum, the Richard D. Irwin Graduate School® of The American College offers the master of science in financial services (MSFS) degree; the master of science in management (MSM) degree with an emphasis in leadership; the Chartered Advisory in Philanthropy (CAP™) professional designation; the Graduate Financial Planning Track (another CFP Board-registered education program), and several graduate-level certificates that concentrate on specific subject areas. The National Association of Estate Planners & Councils has named The College as the provider of the education required to earn its prestigious Accredited Estate Planner (AEP) designation.

The American College is accredited by the Commission on Higher Education of the Middle States Association of Colleges and Schools, 3624 Market Street, Philadelphia, PA 19104; telephone number: (215) 662-5606. The Commission on Higher Education is an institutional accrediting agency recognized by the U.S. Secretary of Education and the Commission on Recognition of Postsecondary Accreditation.

The American College does not discriminate on the basis of race, religion, sex, handicap, or national and ethnic origin in its admissions policies, educational programs and activities, or employment policies.

The American College is located at 270 S. Bryn Mawr Avenue, Bryn Mawr, PA 19010. The toll-free telephone number of the Office of Student Services is (888) AMERCOL (263-7265), the fax number is (610) 526-1465, and http://www.theamericancollege.edu is the home page address. The College welcomes visitors to its 35-acre campus during regular business hours, 8:00 a.m. to 5:30 p.m., Monday through Friday.

Acknowledgments

This book was written by I. David Cohen, Richard A. Dulisse, Kirk S. Okumura, and Glenn E. Stevick, of the faculty at The American College.

The College wishes to thank Larry Barton, president and CEO of The American College, and David Woods, CEO of the National Association of Insurance and Financial Advisors (NAIFA), for their roles in creating the LUTCF and Financial Services Specialist (FSS) designations for which this book is an elective. Also, The College wishes to thank Walt J. Woerheide, executive vice president and dean of academic affairs at The American College, for providing continued support and encouragement in writing this book.

For her valuable contribution to the development of this book, appreciation is extended to Lynn Hayes for editing the textbook and for graphic/production expertise. Appreciation is also extended to Todd Denton for editing the workbook and to Lynn Hayes and Todd Denton for editing the moderator guide.

Special thanks is extended to Gail Goodman and Shawn Hopper for their contributions of material for this textbook.

To all of these individuals, without whom this book would not have been possible, The College expresses its sincere appreciation and gratitude.

I. David Cohen
Richard A. Dulisse
Kirk S. Okumura
Glenn E. Stevick, Jr.

About the Authors

I. David Cohen, LUTCF, CLU, ChFC, began his service in the financial services industry in 1958. He has been a member of the Million Dollar Round Table since 1961, achieving Court and Top of the Table status, and currently is a Life Member. Mr. Cohen is past president of the Columbus Chapter, National Association of Insurance and Financial Advisors (NAIFA Columbus) and the Society of Financial Service Professionals (SFSP). He is a member of the Hall of Fame of both organizations as well as a recipient of the national Paul S. Mills Scholarship award from the SFSP. In 2001, Mr. Cohen received the prestigious Edmund L. Zalinski Award for lifetime achievement in education and community service at the NAIFA convention. In 2003 Mr. Cohen received the Golden Apple Award from the Columbus Chapter SFSP for his career-long service in teaching and mentoring. In 2004, in honor of this lifetime service, NAIFA Columbus retitled its annual recognition award to the I. David Cohen Lifetime Achievement Award.

Mr. Cohen is past president of the Mutual Benefit Life Agents Association and has served as National Trustee of LUTC. He has moderated LUTC classes for 26 years and has taught numerous CLU and ChFC classes. This textbook was derived from his book, *Prospect or Perish: A Success Guide for Financial Services Professionals.* He has co-authored two other books: *Alligator Proofing Your Estate* and *Get What You Want.*

Richard A. Dulisse, LUTCF, CLU, ChFC, CASL, CFP, RHU, REBC, is an author/editor and assistant professor of financial planning at The American College. His responsibilities at The College include writing and preparing text materials for the LUTCF and FSS programs. He also teaches insurance and financial planning courses at The College.

Mr. Dulisse is author of *Essentials of Disability Income Insurance* and *Foundations of Senior Planning,* co-author of *Essentials of Long-Term Care Insurance* and *Foundations of Financial Planning: The Process,* and a contributing author to *Financial Decisions for Retirement.* In addition, he edited *Ethics for the Financial Services Professional* and made extensive contributions to *Essentials of Annuities.* All of these books are published by The American College. Mr. Dulisse also contributes articles to *Advisor Today,* the national magazine distributed to members of NAIFA.

Before joining The College, Mr. Dulisse worked in the life insurance industry from 1979 through 2001. His experience includes 5 years as a life insurance agent, initially with Metropolitan Life and then with New York Life. At New York Life, he also served as a sales manager before becoming a training manager in 1985. As a training manager, he helped implement the company's training curriculum to teach agents product knowledge and selling skills.

Mr. Dulisse earned a BSoc.S degree, cum laude, from The Pennsylvania State University. He also holds both the MSM and the MSFS degrees awarded by The College.

Kirk S. Okumura is an author/editor at The American College. His responsibilities at The College include writing and preparing text materials for the LUTCF and FSS programs.

Mr. Okumura is co-author of *Essentials of Long-Term Care Insurance, Techniques for Exploring Personal Markets* and *Foundations of Financial Planning: The Process*; he is author of *Essentials of Employee Benefits.* All of these books are published by The American College. Mr. Okumura also writes articles for *Advisor Today*, the national magazine distributed to members of NAIFA.

Before joining The College, Mr. Okumura worked for State Farm Insurance Company as a supervisor in a regional life/health office and as a trainer in the Pennsylvania regional office's agency training area.

Mr. Okumura earned a BS degree from The Pennsylvania State University.

Glenn E. Stevick, Jr., LUTCF, CLU, ChFC, is an author/editor and assistant professor of insurance at The American College. His responsibilities at The College include writing and preparing text materials for the LUTCF and FSS programs. He also has taught insurance and financial planning courses at The College.

Mr. Stevick is co-author of *Essentials of Long-Term Care Insurance, Techniques for Exploring Personal Markets, Essentials of Life Insurance Products,* and *Foundations of Financial Planning: The Process*; he is author of *Techniques for Meeting Client Needs* and *Essentials of Business Insurance.* All of these books are published by The American College. Mr. Stevick also writes articles for *Advisor Today*, the national magazine distributed to members of NAIFA.

Before joining The College, Mr. Stevick worked for New York Life as a training supervisor for 15 years in its South Jersey office. He also served as an agent with New York Life for more than 2 years. Prior to his insurance industry experience, Mr. Stevick taught psychology at the college level and worked in various educational and mental health programs.

Mr. Stevick earned his BA degree from Villanova University and his MA degree from Duquesne University. Mr. Stevick is a 22-year member of NAIFA and a 17-year member of the Society of Financial Service Professionals.

Special Notes to Advisors

Text Materials Disclaimer

This publication is designed to provide accurate and authoritative information about the subject covered. While every precaution has been taken in the preparation of this material to ensure that it is both accurate and up-to-date, it is still possible that some errors eluded detection. Moreover, some material may become inaccurate and/or outdated either because it is time sensitive or because new legislation will make it so. Still other material may be viewed as inaccurate because your company's products and procedures are different from those described in the book. Therefore, the authors and The American College assume no liability for damages resulting from the use of the information contained in this book. The American College is not engaged in rendering legal, accounting, or other professional advice. If legal or other expert advice is required, the services of an appropriate professional should be sought.

Caution Regarding Use of Illustrations

Any illustrations, fact finders, sales ideas, techniques and/or approaches contained in this book are not to be used with the public unless you have obtained approval from your company. Your company's general support of The American College's programs for training and educational purposes does not constitute blanket approval of any illustrations, fact finders, sales ideas, techniques, and/or approaches presented in this book unless so communicated in writing by your company.

Use of the Term Financial Advisor or Advisor

Use of the term "financial advisor" as it appears in this book is intended as the generic reference to professional members of our reading audience. It is used interchangeably with the term "advisor" so as to avoid unnecessary redundancy. Financial advisor takes the place of the following terms:

Account Executive	Financial Planner	Planner
Agent	Financial Planning	Practitioner
Associate	Professional	Property & Casualty Agent
Broker (stock or	Financial Services	Registered Investment
insurance)	Professional	Advisor
Employee Benefit	Health Underwriter	Registered Representative
Specialist	Insurance Professional	Retirement Planner
Estate Planner	Life Insurance Agent	Senior Advisor
Financial Consultant	Life Underwriter	Tax Advisor
	Producer	

Answers to the Questions in the Book

The answers to all essay and multiple-choice questions in this book are based on the text materials as written.

About the Financial Advisor Series

The mission of The American College is to raise the level of professionalism of its students and, by extension, the financial services industry as a whole. As an educational product of The College, the Financial Advisor Series shares in this mission. Because knowledge is the key to professionalism, a thorough and comprehensive reading of each book in the series will help the practitioner-advisor to better service his or her clients—a task made all the more difficult because the typical client is becoming more financially sophisticated every day and demands that his or her financial advisor be knowledgeable about the latest products and planning methodologies. By providing practitioner-advisors in the financial services industry with up-to-date, authoritative information about various marketing and sales techniques, product knowledge, and planning considerations, the books of the Financial Advisor Series will enable many practitioner-advisors to continue their studies so as to develop and maintain a high level of professional competence.

When all books in the Financial Advisor Series are completed, the series will encompass 16 titles spread across three separate subseries, each with a special focus. The first subseries, *Sales Skills Techniques,* will focus on enhancing the practitioner-advisor's marketing and sales skills but will also cover some product knowledge and planning considerations. The second subseries, *Product Essentials,* will focus on product knowledge but will also delve into marketing and sales skills, as well as planning considerations in many of its books. The third subseries, *Planning Foundations,* will focus on various planning considerations and processes that form the foundation for a successful career as a financial services professional. When appropriate, many of its books will also touch upon product knowledge and sales and marketing skills. Current and forthcoming titles are listed earlier in this book.

Overview of the Book

Techniques for Prospecting: Prospect or Perish examines methods to successfully identify, select, and approach prospects for financial products and services.

Chapter 1 explains prospecting within the context of the overall selling/planning process, and it explores ways to overcome the hurdles to successful prospecting. It also looks at the psychology of the buying process from the prospect's perspective and examines what motivates people to buy.

The focus of chapter 2 is marketing. The chapter defines target marketing and its advantages, and it and explains the procedure to identify niche markets using target marketing techniques, including segmenting markets and positioning the advisor's personal brand and products.

Chapter 3 looks at how to select prospecting sources and effective prospecting methods so that the advisor can identify names and contact information for specific prospects from the advisor's target markets.

Chapter 4 discusses ways the advisor can create awareness of his or her personal brand through prestige-building techniques. It also discusses how the advisor can use preapproach activities effectively to educate and motivate prospects regarding the emotional and financial needs that the advisor's products can meet.

Techniques for approaching prospects to get an appointment are discussed in chapter 5. The chapter explores ways to awaken prospects' interest in the advisor's products and services. There is a brief look at face-to-face approaches, followed by an in-depth look at telephone approaches that can be used in the personal and business markets.

Chapter 6 concentrates on goal setting and time management. It explains the principles of goal setting, the difference between goal setting and problem solving, and the concept of structure and its application to meet goals. The chapter concludes with a discussion of techniques for effective time management.

Chapter 7 covers prospecting through service. It enumerates the laws of extraordinary service, describes ways to deliver it, and outlines a procedure by which the advisor can segment his or her client base into different categories of clients, each of which will receive a different level of service, to enable the advisor to be most productive and profitable.

Finally, professional practice management is the subject of chapter 8. This includes mentoring relationships, strategic alliances, and developing and managing a personal board of directors. The chapter also reviews the use of contact management systems, business planning and organization concepts, and issues relating to compliance, ethics and professionalism.

Techniques for Prospecting:
Prospect or Perish

1

The Psychology of Prospecting

Learning Objectives

An understanding of the material in this chapter should enable the student to

1-1. Describe the differences between marketing, selling, and prospecting.

1-2. Explain how prospecting fits into the overall selling/planning process.

1-3. Explain how to overcome the five hurdles to successful prospecting.

1-4. Discuss how the psychology of the prospect affects how an advisor prospects.

Chapter Outline

COURSE OVERVIEW

You have heard the old adage, "All dressed up but no place to go." The equivalent of it for financial advisors is, "Fully credentialed but no one to see." Even if you are the most qualified advisor with the best products and services, if you do not have prospective clients to see, you will starve. Prospecting is to financial advisors as blood is to the human body. It is arguably the lifeblood of your practice; the better you can do it, the greater the likelihood of achieving your goals.

> **Prospecting is to financial advisors as blood is to the human body.**

As critical as prospecting is to a financial advising practice, many advisors struggle to do it well. Consequently, this book is devoted to unleashing the prospecting machine within you.

Chapter 1 explains the context of prospecting within the overall selling/planning process. Then it looks at the psychology of prospecting from the advisor's point of view. It also examines the psychology of the buying process from the prospect's point of view, discussing some of its effects on prospecting. This discussion provides the foundation for the principles that follow.

The rest of the book explains the prospecting process and the marketing activities required to prospect effectively and efficiently. Target marketing—what it is and how to do it—is the focus of chapter 2. Chapters 3 and 4 discuss various prospecting techniques and preapproach methods, which often overlap. Chapter 5 leads you through the process of approaching prospects to set appointments. In addition, it addresses the task of qualifying prospects. Goal setting and time management are discussed in chapter 6. Chapter 7 examines how service is tied into prospecting. The text closes with a discussion of practice management and ethics.

THE PROSPECTING ENIGMA

What do you think when you hear the term *prospecting*? Does it bring negative images or feelings to mind? Most people have their own personal definition of prospecting, just as they do for other terms that are thrown about too loosely in the financial services industry. That is why it is important to define prospecting, differentiating it from marketing and selling and to understand how prospecting fits in the overall selling/planning process.

Defining Marketing, Selling, and Prospecting

In order to define prospecting, we will first define two other related terms: marketing and selling. Then, we will define prospecting and describe the characteristics of a qualified prospect.

Marketing

marketing

Many people mistakenly equate marketing and selling. But they are not the same. The American Marketing Association defines *marketing* as "an organizational function and a set of processes for creating, communicating, and delivering value to customers and for managing customer relationships in ways that benefit the organization and its stakeholders."[1] In the financial services industry, this means that marketing is the planning and implementation of a process that involves

1. identifying specific consumer needs and wants that your products and services can meet
2. defining groups of consumers who have those needs and wants
3. creating or customizing a solution that meets those needs and wants effectively
4. creating and articulating messages that raise consumer awareness of these needs and wants
5. positioning your ability in the consumers' minds to meet those needs and wants
6. identifying and contacting specific consumers with whom to meet
7. motivating consumers to buy the product or service that will satisfy those needs and wants from you
8. exchanging the product or service for something of value (money)
9. managing postsale relationships

Selling

selling

In comparison, *selling* is the process of motivating, guiding, and asking the prospect to buy—to become a buyer. (Selling involves activities 7 and 8 in the above list.) As you can see, selling is a part of marketing, but marketing is much more than selling. As marketing expert Dr. Philip Kotler notes, "[T]he most important part of marketing is not selling. Selling is only the tip of the marketing iceberg."[2] Selling should be the natural result of the effective completion of the other activities associated with marketing such as building product and/or brand awareness or attaining an audience with the prospective customer. Without these other marketing activities, selling would be like trying to move a sailboat without any wind.

prospecting

> **Prospecting is the continual activity of exploring for and prequalifying new people to meet with and talk to concerning your business.**

prospect

qualified prospect

Prospecting

Last but not least, *prospecting* is the continual activity of identifying, approaching, and prequalifying new people to meet with and talk to concerning your business. Prospecting is the marketing activity designed to identify individuals, couples, businesses, and so on to call for an appointment. It is the marketing activity that narrows an advisor's focus from the broader market to the selling of products and services to individual prospects.

Prospecting starts with a large pool of people or businesses, known as prospects. A *prospect*, for the purposes of this discussion, is a person for whom you have a name, contact information, and a suspicion that he or she might benefit from your products and services. (Many would define such a person as a suspect rather than a prospect. The difference between a suspect and a prospect is subject to interpretation. Table 1-1 shows one such interpretation. For simplicity, we will use the term prospect to refer generally to both.) From this pool, you want to identify those persons who have a high probability of being qualified prospects. A *qualified prospect* is someone who

- needs and values your product and/or service
- can afford it
- is suitable for it (if applicable to the product or service)
- can be approached by you on a favorable basis

Obviously, you will want to focus your efforts on identifying and working with prospects who have a greater probability of meeting the above criteria. Thus, understanding the four criteria will enable you to identify what you need to know and perhaps determine how to obtain it sooner rather than later. Let's look at these four criteria in greater detail.

Needs and Values It. If a person does not need your products or services, no matter how much he or she wants them, that person is not a qualified

TABLE 1-1 The Prospect-Client Continuum	
Suspect	Person who could become a prospect
Prospect	Potential buyer identified by the advisor
Qualified prospect	Prospect who needs and values a product or service, has the ability to pay, is suitable for the product or service, and can be approached by the advisor
Buyer (account owner, policyowner, and so forth)	Qualified prospect who has bought a product or service at least once
Client	Repeat buyer who refers you to others

prospect. A sale made in this situation has a great chance of being non-compliant and unethical.

Need is important, but value is critical. Value determines want, and the reality is that people will most often buy what they want before what they need. For example, almost everyone with dependents needs life insurance, but does that mean they have it, or they have enough? Wanting is the key, and people want that which they perceive adds value to their lives.

> **Need is important, but value is critical.**

✓ ***Can Afford It.*** A person cannot be a qualified prospect unless you are confident that he or she can afford what you offer. The issue of affordability often can be resolved by implementing a spending plan (a budget). However, this works only if the person has an adequate amount of discretionary expenditures that can be reduced and wants what you offer badly enough to make the necessary sacrifice.

Example:	If you expect someone to give up cable TV (an immediate gratification) so that he or she can pay long-term care insurance premiums (not so immediately gratifying to most of us), there is a chance that the policy may not stay in force for very long.

∫ ***Is Suitable for It.*** Unfortunately, there are people who need and want the products and services you provide but cannot meet the necessary financial and/or health requirements. Working with prospects who will probably satisfy any financial and/or health conditions will increase your effectiveness and efficiency.

Example:	People with low risk tolerance and/or a short investment time horizon may not be suitable for stocks, certain mutual funds, or other financial instruments. Likewise, someone with a serious health condition may not qualify for life, health, disability income, or long-term care insurance.

↙ ***Can Be Approached.*** In lines of business where relationships drive most activity, this criterion is critical. A person may meet all the other requirements, but if you cannot approach him or her on a favorable basis, then he or she is not a qualified prospect for you right now. Rapport is crucial, especially in a business that gets personal. Financial advisors get about as personal with their clients as anyone can. In fact, the only people who may get any closer are usually health care professionals. That said, do

Working with people who have a high probability of being qualified prospects is the key to success for financial advisors.

not throw the unapproachable prospect's name away—you may find a way to approach him or her on a favorable basis at some point in the future.

Working with people who have a high probability of being qualified prospects is the key to success for financial advisors. Selling and planning are much easier when you are dealing with someone who needs and wants your help, can afford it, is suitable for it, and is approachable.

Unfortunately, such prospects generally do not come to advisors; advisors typically must identify and approach them for appointments using the appropriate prospecting methods. From this perspective, selling and planning are easy. But prospecting? That is difficult.

The next section will explore further the relationship between prospecting and selling by reviewing the selling/planning process.

Selling/Planning Process

selling/planning process

The *selling/planning process* is the process that advisors use to identify and meet with prospects, help them create and implement plans to address their financial needs, and provide ongoing service, which leads to repeat sales and referrals (see figure 1-1). The phrase "selling/planning" is used to recognize that advisors need sales skills, regardless of their product—insurance, mutual funds, or financial plans. All financial advising involves communicating, motivating, and persuading—elements of successful selling.

The first two steps of the eight-step process—identify the prospect and approach the prospect—implement the marketing activities related to prospecting. Steps 3 through 8 comprise the actual (financial) planning process. These six steps involve personal interaction between you and a prospect or client (a distinction previously described). The use of a circle in figure 1-1 illustrates that if you complete the eight-step process effectively, servicing the plan in step 8 will lead the buyer back for other products and/or services—and, in addition, referrals to new prospects.

Although this textbook will focus primarily on the topics associated with prospecting (steps 1, 2, and 8), it is important to be familiar with the underlying philosophy and the other steps of the selling/planning process. The client-focused philosophy (more in a minute) pervades every step of the process. Furthermore, all eight steps are integral parts of the whole. It is necessary to understand, and eventually implement, the entire process to gain maximum benefit. (See *Techniques for Meeting Client Needs* or *Foundations of Financial Planning: The Process* for a detailed treatment of steps 3 through 7 of the selling/planning process.) Following is an overview of the individual steps of the selling/planning process and a brief discussion of the fundamental selling/planning philosophy.

Eight Steps of the Process

The discussion that follows explains the eight steps of the selling/planning process.

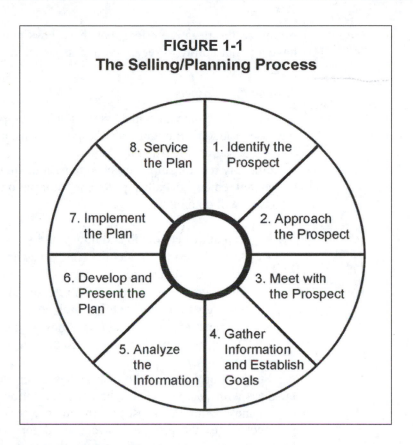

FIGURE 1-1
The Selling/Planning Process

(Wheel diagram showing eight segments:)

1. Identify the Prospect
2. Approach the Prospect
3. Meet with the Prospect
4. Gather Information and Establish Goals
5. Analyze the Information
6. Develop and Present the Plan
7. Implement the Plan
8. Service the Plan

1. Identify the Prospect

Effective selling and planning begin with getting in front of the "right" prospects. But who are the right prospects? How do you identify them effectively and efficiently? The answer to these questions lies in basic target marketing strategies and the selection of prospecting techniques and pre-approach methods that are especially effective for your desired target markets.

The recent enactment of federal and state Do Not Call Laws aimed at placing restrictions on telemarketers has changed the landscape of prospecting. Some lament that these laws will hurt the financial services industry. But the change is perhaps good news for advisors who use prospecting techniques that generate prospects who already have a favorable view of the advisor such as referrals and seminars.

2. Approach the Prospect

> **There is only one objective to this step: to set the appointment.**

There is only one objective to this step: to set the appointment. The approach may be done via the telephone or face-to-face. When using a telephone approach, you will need to adhere to the federal and any state Do Not Call Laws that apply. Fortunately, there are prospecting methods that can minimize the negative effect that these laws and the Do Not Call Registry have on your ability to set appointments, as alluded to earlier.

3. Meet the Prospect

This is the step in which you establish rapport, explain your business purpose, ask some thought-provoking questions, and listen, listen, and listen. The importance of listening cannot be overstated; it is essential when building any relationship. You need to earn the prospect's trust and obtain his or her agreement to gather pertinent information about him or her. In this step, you will do the following:

- Explain what you do and how you do it.
- Ask questions to identify the prospect's needs, goals, problems, priorities, and attitudes.
- Motivate the prospect to take action to meet those needs, achieve those goals, and solve those problems.
- Demonstrate, in general terms, how you can help the prospect.
- Obtain agreement to proceed to the information-gathering step.

Furthermore, just as the prospect is determining whether or not to do business with you, you should gather information to assess whether or not to do business with the prospect. In other words, you need to qualify the prospect. Does the prospect have a need for your products and services? More important, is the prospect motivated to meet that need? Can the prospect afford the necessary products and services? Finally, is the prospect the type of person with whom you want to work?

4. Gather Information and Establish Goals

It is critical to gather detailed, relevant information about the prospect's personal and financial situation. Your objective is to collect sufficient information to make suitable and sound recommendations to the prospect. Use an effective and compliant fact finder designed for this purpose.

To be thorough, give prospects a list of financial documents to bring with them to the interview such as a will, budget, balance sheet, cash flow statement, insurance plans, benefit statements, most recent Social Security statement, mutual fund account statements, and so on. You need to know as much as you can about each prospect's financial situation.

Before you leave this step of the process, confirm your understanding of the prospect's goals and priorities. Also, establish agreement on planning assumptions (interest rates, inflation rates, investment return rates, risk tolerance, and so on). Finally, obtain a dollar commitment so you will know how much

> **Your recommendations must be based on the prospect's ability to pay or they probably will not be implemented.**

money is available to implement a plan. Your recommendations must be based on the prospect's ability to pay or they probably will not be implemented.

The key skills in the fact-finding process are as follows:

- *questioning.* Ask the appropriate closed- and open-ended questions that will give you an overall sense of the prospect's current financial situation, attitudes, goals, values, and needs. Ask confirming questions to clarify.
- *listening.* Actively listen, rephrase, and reflect to ensure that you and the prospect are on the same page.
- *taking notes.* Jot down assumptions, goals, needs, priorities, and feelings.
- *summarizing.* Review what the prospect has told you, and confirm that you have clearly understood him or her.
- *acting.* Agree on a proposed next action.

5. Analyze the Information

Once you have gathered information and established goals and planning assumptions, you are ready to analyze the information. Your analysis will include activities such as

- preparing and analyzing appropriate financial statements
- projecting the outcomes of the current plan (if the prospect does nothing different)
- identifying obstacles to desired goals
- analyzing the prospect's current and desired situations as they relate to the major planning areas: insurance planning and risk management, employee benefits planning, investment planning, income tax planning, retirement planning, and estate planning
- determining viable alternatives and their outcomes
- analyzing alternatives using possible less favorable assumptions (higher tax rates, lower rates of return, and so on)
- quantifying the opportunity costs of each viable alternative
- comparing the costs of implementing solutions with the client's ability to pay

6. Develop and Present the Plan

After you have analyzed the prospect's situation, you will be ready to develop a plan. In addition to summarizing the client's situation and the findings of your analysis, the plan should include recommended actions. If you provide alternate solutions, you should include a synopsis of their relevant advantages and disadvantages and a projection of anticipated costs and outcomes.

In between developing the plan and presenting it you should spend some time preparing your presentation (unless this is a one-interview planning session). Anticipate questions, objections, and concerns.

Now you are ready to present the plan. During the presentation you will do the following:

- Review your prospect's needs, wants, and desires.
- Restate the situation to confirm your understanding of it.
- Identify the solutions and any additional alternatives that will cover the needs.
- Present your recommendations.

7. Implement the Plan

If you have conducted fact finding properly and completely and your recommendations are based on information from the fact finder, implementing the plan should simply be the logical next step in working together. That does not mean the prospect will not have some concerns or misunderstandings. It does not mean that the plan will not have drawbacks. But if you have done your job uncovering the prospect's attitudes and values, as well as his or her financial situation, you will be prepared to explain how your recommendations will enable the prospect to achieve his or her goals.

Once the prospect has agreed to your recommendations, help him or her acquire the necessary products and services. For those of you who sell financial products, implementation includes completing all of the required forms and applications and, in some instances, delivering an insurance policy.

Of course, not every prospect will implement your recommendations. Do not be discouraged. It may take even the best advisor several attempts to close. Sometimes prospects postpone purchases, and the advisor is forced to make several appointments and compensate for a lapse of time between the presentation and implementation. As frustrating as this will be to you, there are recognized techniques that you can use to motivate your prospect to follow your recommendations.

8. Service the Plan

This is the step in which you turn buyers into lifetime clients. Service cements the relationship with a buyer, giving you the opportunity to make additional sales and obtain referrals. Some service is reactive—the buyer or client initiates it by requesting a needed change, such as an increase in coverage. In these situations, the buyer or client should expect to receive excellent service. What differentiates one advisor from another, however, is the proactive element of his or her service strategy. Many people buy a

> If you've done your job uncovering prospects' attitudes, values, and financial situation, you will be prepared to explain how your recommendations will enable them to achieve their goals.

> What differentiates one advisor from another is the proactive element of his or her service strategy.

product and never hear from the advisor again. Proactive servicing strategies, such as monitoring the plan through periodic reviews and relationship-building activities enable an advisor to stay in touch with clients. It is this high-contact service that builds clientele. You need to communicate what reactive and proactive services you will provide and then make them happen!

Client-Focused Selling/Planning Philosophy

As mentioned earlier, all selling involves communicating, motivating, and persuading. Selling skills are inherently neutral. You can use them to help people create a financial plan to achieve their dreams or to manipulate people to buy a product that is unsuitable for them. Thus, the philosophy underlying the selling/planning process is critical.

The selling/planning process described in this textbook is based on a client-focused approach to selling and planning. The objective is to cultivate a mutually beneficial, long-term relationship with a *client*, a repeat buyer who will refer you to others. (Note: For our purposes, a person who pays an annual retainer, asset management fee, and so on is a repeat buyer.) In other words, the end result is an ongoing relationship that benefits both parties. The initial sale is an intermediate rather than final step. Thus, the product or service reflects the prospect's needs, values, and personal situation.

client

Obviously, manipulative, high-pressure strategies are incompatible with the client-focused philosophy, which utilizes consultative and financial planning strategies. Creating solutions that reflect overall client goals, values, and needs requires a careful gathering and analysis of very personal information and feelings. As a result, good communication skills such as asking probing open-ended questions, listening carefully, and confirming your understanding are invaluable. Moreover, motivation and persuasion are employed with the prospect's best interest in mind—not to generate commissions or fees. Thus, everyone benefits. Clients buy valuable products that meet their needs. Advisors receive repeat business and referrals. Consequently, the reputation of financial advisors in general is enhanced.

> **Creating solutions that reflect overall client goals, values, and needs requires a careful gathering and analysis of very personal information and feelings.**

The client-focused philosophy pervades the entire process, beginning with prospecting, arguably the most difficult part. The next section examines the major hurdles to effective prospecting and how to overcome them.

OVERCOMING THE FIVE HURDLES TO SUCCESSFUL PROSPECTING

Prospecting is the enigma that confronts every financial advisor. It is also the number one reason that salespeople fail or leave their respective business. They simply cannot master the art of prospecting. In the financial services industry, the failure of advisors adversely affects both the companies and the clients, as well as the lives and families of the advisors themselves.

> **The reason so many advisors run out of prospects is that they have not made prospecting a habit.**

The reason so many advisors run out of prospects is that they have not made prospecting a habit. Doing so involves much more than simply memorizing the techniques. If it were all technique, everyone would be doing it—and succeeding at it.

So why do advisors struggle to prospect? Everyone has his or her own personal reasons or psychological hurdles. But all the reasons probably fall into one or more of the five primary hurdles to successful prospecting:

- lack of conviction
- fear of rejection
- not knowing exactly what to do or how to do it
- lack of motivation
- the inherent difficulties of prospecting

Everyone has reasons why prospecting is difficult. Recognizing the reasons enables advisors to identify the root causes and begin to address them. Make a list of your reasons.

Assume that your list is similar to the issues we just identified. Let us examine each one a little more closely and consider some action steps you can take to overcome them and jump-start your prospecting engine!

Hurdle #1: Lack of Conviction

Most advisors would not start asking people to meet with them if they knew nothing about their products and services. Some can describe in detail the feature of every product they sell. But they do not understand the related buyer benefits that their products and services bring to their clients. The result is a lack of conviction that their products and services are not simply important, but critical to people's well-being. They have not grasped the true value that their products and services deliver to their clients.

In contrast, advisors who are armed with an understanding of how each product makes clients' lives better are filled with confidence and conviction. They believe that they really can make a difference in people's lives. Their excitement is infectious, and people are drawn to them. Their conviction pushes them forward and, because of a sense of mission, they have the courage to initiate a conversation about their products and services.

The following are four ways to create a sense of conviction about what you do and what you sell.

Identify the Value of Your Products

Have you ever encountered a new acquaintance at a social or business gathering and when you ask the person about what he or she does, the individual trivializes it? Responses such as these may sound familiar:

- "I'm just an accountant."
- "All I do is sell insurance."
- "I'm only a broker."

Advisors who respond in this manner need to take the time to understand the difference between *what they sell* (their product) and the *value it delivers* (why people would want it). Specifically, they need to think about ways their product makes a difference in people's lives. See table 1-2 for a few examples.

TABLE 1-2
Examples of Products and the Value They Deliver

If You Are	What You Offer (Product/Service)	Why People Want It (Value)
An accountant	Advice and services on taxes and cash management	Individuals and/or businesses want to make the best use of their earned capital and achieve the maximum savings and benefits under current tax law.
An insurance representative	Insurance products for life, home, and other valuable assets	Individuals and families want to protect their assets in the event of an accident or other type of loss. Your products also provide the foundation for a solid financial plan that creates long-term financial security for your clients and their children.
A financial planner	Advice on managing assets, investments, and other financial-related issues	People want security and peace of mind. They want to know their hard-earned dollars will provide security and comfort for themselves and their children.

Take a look at the portfolio of products and/or services you offer. What excites you about these products or services? Perhaps you do comprehensive financial planning and your particular software program is truly outstanding. You feel it can help people identify, set, and achieve their financial goals. If this is important to you, share that sense of importance with others. Create an excitement for the company you represent and your particular services. It will go a long way toward overcoming any prospecting reluctance you may have. Moreover, it will help you develop a clear sense of *how your products and services improve people's lives.*

> Create an excitement for the company you represent and your particular services.

Talk to Your Clients

Want to gain a better understanding of how your products and/or services help people? Ask the people you are helping now. You can learn quite a bit simply by talking to your clients. Ask them how your products and/or services are meeting their needs or enriching their lives. Find out how you can improve and if there are any concerns or issues they might like to discuss. You may be surprised at how much praise you will hear, merely by asking if there is anything you can do better. Take any suggestions seriously, and make certain you implement them.

Talk to Others in Your Business

Seek out possible mentors and others whose success you would like to emulate, and ask them what they like about their work. Beyond the financial rewards, what do they receive? Why did they pick this business? Why not something else? Ask them to share stories of their favorite clients (minus the names, of course) and about their relationships. The essence of their experiences will probably go much deeper than insurance and financial services. Understand and capture their passion!

Define Brand "You"

brand

Although the products you sell may have different features and benefits from those of your competitors, do not overlook the importance of your personal brand or, as business guru Tom Peters calls it, "a BRAND CALLED YOU" [all caps his].[3] A *brand*, according to the American Marketing Association, is "a name, term, design, symbol, or any other feature that identifies one seller's good or service as distinct from those of other sellers."[4]

For financial advisors, the identifying features are typically the prospect's perception of the advisor's knowledge, skills, ideas, and personality that differentiate him or her from competing advisors. How is the service the advisor provides unique and set apart from the competition? What advice can the advisor offer on critical, possibly life-changing issues? What kinds of positive client experiences can the advisor create? Answer these questions; then make sure everything you do reflects brand "You."

> **If you are excited about what you offer people and how you can help them, you will have greater success as a financial advisor.**

Understanding the value of your products and services is vitally important. If you are excited about what you offer people and how you can help them, you will have greater success as a financial advisor. Furthermore, you will find that as your enthusiasm builds, it will have a tremendous impact on your capacity to overcome other obstacles that hinder your prospecting efforts.

Hurdle #2: Fear of Rejection

No one likes to face rejection. It is an emotionally hurtful experience, and people go to great lengths to protect themselves from it. Unfortunately, for financial advisors who generally must seek out prospects with whom to meet, rejection is an occupational hazard.

When it comes to rejection, remember that it is your product or service that is being rejected, not you. Do not take rejection personally.

Fear of rejection is a manifestation of insecurity. Most advisors are insecure about something. All people have their issues, whether the issues are their appearance, how much money they earn, or just an overall lack of faith in themselves to accomplish and attain what they desire. Thus, there is a measure of detachment that advisors must adopt if they are to overcome any inherent insecurity. When it comes to rejection, remember that it is your product or service that is being rejected, not you. Do not take rejection personally.

Of course, that does not mean that an advisor's personality (likability) and rapport with others are not important. An advisor can offer the most beneficial product in the world, but if he or she is not particularly pleasant to be around, the advisor generally will not succeed. So, while advisors should be cognizant of how their personality is perceived by others, they must be careful not to be overly sensitive—a delicate balancing act, to be sure. The fact is, people most likely are rejecting the advisor's products and not the advisor. Therefore, the goal is for advisors to acknowledge their own insecurity and put it in perspective.

That said, let us address fear in general and get it out in the open. What about prospecting frightens you? Do you have any of the following feelings?

- "I am imposing on people."
- "No one wants to buy from me."
- "What if I say something wrong?"
- "What if they say no? Will they not like me? I can't risk being rejected."
- "I am sure they already have all they need."

Fear is normal. Thus, a key question to explore is, "What can you do to help you overcome your fear?" Until you answer this question, you can know all the best prospecting techniques in the world, but you will be too afraid to apply them.

What Is the Worst That Could Happen?

One strategy is to ask, "What is the worst that could happen?" Consider the following, for example:

- If you send a letter, perhaps enclose a copy of a helpful article on financial planning, and then follow up with a phone call, what's the worst that could happen?
- If you telephone a person, tell him or her about your services, and request an appointment, what's the worst that could happen?
- If you approach a person at a social gathering and gently encourage the person to open up to you a little in regard to his or her financial and insurance concerns, what's the worst that could happen?

Probably, the worst thing that will happen is that the person will politely decline your offer. Is that really such a terrible thing?

Believe You Can Do It; Do It; Keep Doing It

We are all afraid of rejection, of embarrassment, of suffering or pain. We are only human. But think about other things you have accomplished in life—things that made you afraid when you first approached them. Have you ever, for example, done any of the following?

- spoken before an audience
- bought a house
- taken on an exciting new job
- jumped out of an airplane
- learned to ride a bicycle
- mastered a foreign language

In accomplishing any of the above, you had to overcome your fear. This meant you had to do three things: First, you needed to believe that you would succeed. Then you needed to do it. Finally, you needed to keep doing it.

Example: Imagine you are a child again. You want to get up on that big bicycle without training wheels, but you are afraid. You are afraid it will topple over and take you with it. No matter how many other children and adults you have seen riding their bikes without training wheels, you know, you insist, it just will *not* work for you. It is a belief that many children have despite evidence to the contrary all around them.

The fears stemmed from a faulty belief system. Do you have these same fears today? Of course not. Today, you operate with different beliefs. What changed first, your belief or your ability to ride the bicycle? At some point, you had to believe enough that you could do it before you got on the bicycle. But all the belief in the world would have meant nothing if you had never acted on it. More than likely, you fell a few times. But after many attempts, you finally were able to ride on two wheels.

Once mastered, riding a bike was like walking, jumping, and climbing. It became second nature. So, too, will prospecting.

You will always be prospecting. It will become second nature. Prospecting, like any skill, eventually comes down to believing that you can do it, doing it, and then doing it over and over again.

Hurdle #3: Not Knowing Exactly What to Do or How to Do It

> **Until you determine one or two prospecting approaches that you find effective, you will always be in a fog.**

You may ask a successful financial advisor, "How do you know what to do?" Your follow-up question: "How do I do it?" There are dozens of ways to prospect; you must make a choice—you must have the goal to keep it simple. Until you determine one or two prospecting approaches that you find effective, you will always be in a fog.

This gets into the "techniques" side of the issue. Chapters 3 and 4 will review many different prospecting techniques. The key, though, is to select just a few techniques and concentrate on doing them well. Choose techniques that are relevant to the kinds of prospects you want to see and that emphasize your strengths and personality, if possible.

Hurdle #4: Lack of Motivation

Everyone has times when he or she feels unmotivated. Unfortunately, a lack of motivation to a financial advisor is like kryptonite is to Superman; it eventually will be fatal. Staying motivated begins with creating a business plan that outlines clear goals and objectives, and implementing an action plan for achieving them. Having a vision provides a sense of purpose that can keep you going during times of low motivation.

Ultimately, even with a plan, motivation comes down to deciding to be a doer and then taking action. You may have heard the following expression: There are three types of people in the world—those who make things happen, those who watch things happen, and those who wonder what happened. Which one are you? Are you a doer, making things happen?

Traits of Doers

There are many ways to observe and define the characteristics of doers. For the purposes of this discussion, there are 10 distinguishing traits of "doers."

1. Doers Know Where They Are Going. First and foremost, doers have a vision and a sense of mission. They do not run just in any direction. They have a sense of the future they desire, and they set goals that will enable them to achieve their vision.

Doers evaluate their activities in light of their vision, giving priority to the activities that further their mission and ignoring those that do not. Their desired future anchors their focus. Doers make adjustments when confronted with unforeseen obstacles and distractions by evaluating the situation in conjunction with their vision and goals.

Example: Sailing is a good analogy. When you are sailing, you typically have your eye on a destination or a point up ahead. But you rarely move in a perfectly straight direction; you are probably at least slightly off course some of the time. While you are steering the boat, you are continually anticipating and responding to your environment—the waves, the water currents, and the wind. You are always compensating to make the necessary adjustments. Nevertheless, as you move—even though you may be off course—you never lose sight of your ultimate destination. You never ignore where you are going.

The visioning process comes down to asking, "What do I want?" Author and speaker Stephen Covey advises to "begin with the end in mind."[5] In other words, long-term vision should shape your intermediate and short-term goals and actions. Chapter 6 will discuss this subject in greater detail.

2. Doers Are Passionate. Doers want to succeed and make a difference in people's lives. They have discovered that the key to success is to look inside themselves and become passionate about their goals. Advisors who are doers want to realize their goals. They care so deeply about achieving them that they rise above the bad days and the fierce competition. They look past the difficult periods, learn from their mistakes and adverse experiences, and focus on their goals.

Passion is what energizes the spirits of doers, gets them up early in the morning, and frees them from being prisoners of the moment. Doers are focused on the big picture. It does not matter if they are tired, if their houses are messy, or if their cars break down. Doers are passionate people and not excuse makers—they are constantly moving forward. They are proactive, taking responsibility for the results they produce, both good and bad.

> **Doers are passionate people and not excuse makers—they are constantly moving forward.**

Picasso, Passion, and Profit

A young man asked Picasso to draw the man's portrait. The artist completed the drawing, which took 10 minutes, and gave it to the young man. Picasso then said, "That will be $2,500."

"What?" the man gasped. "But it took you only 10 minutes to do this drawing! You expect $2,500 for 10 minutes' worth of work?"

"Not 10 minutes," Picasso replied, "30 years."

Picasso had begun developing his talent long before the visitor ever set foot into his world. Was Picasso an overnight success? Of course not. He had a passion for his work, and he had paid the price. It was his time to be rewarded for his hard work—to reap the benefits.

Many people think that successful entrepreneurs are passionate only about earning money. According to Steve Mariotti, however, founder and president of the National Foundation for Teaching Entrepreneurship, most successful entrepreneurs are *not* focused on the money. Instead, they are passionate about their work, what they do, and how they are serving people. Furthermore, once they reach some level of financial security or independence, they are passionate about what they are able to do with their lives as a result of their success. A great example of this is the partnership between Bill Gates and Warren Buffet formed in June 2006. The two richest men in the world designated between them an astounding $68 billion—that's billion with a "b"—to philanthropic causes.[6]

3. Doers Believe. What the mind can conceive and believe, it can achieve. This is a powerful statement. Perhaps you have also heard this one: Whether you believe you can or you can't, you're probably right. In other words, people who believe they can succeed will succeed. People who believe they cannot succeed will fail. In either case, the inward belief becomes a self-fulfilling prophecy.

Every religious tradition gives testimony to the power of an individual's beliefs. Thus, it should not be a surprise to find that people's beliefs play a great role in determining their futures. Beliefs affect daily life, they have an impact on the choices people make and the results they produce. This is more than Pollyanna-type positive thinking—this is the power of the mind!

For example, doers believe that the world plays no favorites. The world does not care whether or not people get what they want. The only reason the odds may favor doers is that doers have the right attitude and approach to achieving their objectives.

Ask yourself, "What are my beliefs?" Specifically, what do you believe you can and cannot do? What outside influences do you believe will prevent you from reaching your goals? How might your perceived personal limitations and outside barriers be more a state of mind than a state of reality?

Doers either work around or through personal limitations and external obstacles. They find a way.

> "When you tell me I can't do something, you may be right, but I don't believe it. Nothing builds a fire under me more than if I'm told I can't do something. Maybe I can't—but I'm sure going to try."

Example:

Life insurance agent Ben Feldman lacked the physical attributes, silver tongue, and charismatic presence often (mistakenly) associated with successful people. In fact, he was told at the beginning of his career that he did not look the part of a life insurance salesperson.

Reflecting later on this assessment, Ben wrote, "When you tell me I can't do something, you may be right, *but I don't believe it.* [italics his] Nothing builds a fire under me more than if I'm told I *can't* do something. Maybe I can't—but I'm sure going to try."[7]

Despite his "personal limitations" and the fact that he sold in East Liverpool, a small Ohio town of 20,000 people,[8] the legendary Ben Feldman sold more than $1.5 *billion* in volume between 1941 and 1993.[9] Although few others believed he could succeed, Ben chose to believe differently. He was convinced he could achieve success—and he did!

---

4. Doers Create Plans. Doers believe, but that does not mean they are unaware of the challenges and obstacles to their goal. Indeed, they see a worthy goal realistically, with both the challenges and the opportunities it presents. They understand where they are and where they want to be. Then they create a plan and navigate the route that will enable them to achieve their goal.

Most of the time, things do not turn out exactly as planned. It is important to take a "roadmap" approach. If you are driving somewhere, even across town, you might decide the best way to get there is on the major freeway. As you head for the entrance ramp, however, you discover that the ramp is closed. What do you do now? Do you give up on reaching your destination and go back home? Of course not. You look at a map and determine an alternative route. All this time, what is on your mind? Your ultimate destination.

5. Doers Do. But it takes more than a plan. Doers do. Wishers only wish, "Wouldn't it be nice," "If only," or "One of these days. . . ." This kind of unproductive thinking occurs not only in individuals but also in organizations. Many companies and organizations engage in strategic planning. It is a long, drawn-out, and expensive process that ultimately results in a plan, usually printed on reams of paper and presented in a heavy binder.

But when this plan is completed, what do you think happens? In many cases, the binder becomes an expensive bookend on the CEO's shelf. It sits there collecting dust. Employees do not follow the plan because they are mired in meeting their day-to-day demands. Obviously, then, it takes more than merely having a plan—the plan must be implemented.

> People who decide that they really want to do something do not wait until certain events occur—they just do it.

People who decide that they really want to do something do not wait until certain events occur—they just do it. The proof is in their actions, not in what they say. Doers believe the best time to act is now. There is no "someday"—there is only today.

6. Doers Are People of Strong Character. One of the greatest qualities that anyone can offer to someone else is strong character. Character is what gives individuals the power to make the right choices. The course of our lives is determined by our choices, and our choices are determined by the depth of our character.

People of strong character have integrity, which means that they are true to themselves and to everyone else. Integrity means that there is value in a person's promise or commitment because the person follows through. You might find that people of integrity make promises sparingly. But when they do make them, they keep them. You can have the most valuable skills on earth, but if you cannot be trusted, what good are you to anyone?

Some people may feel that integrity will prevent them from making money. In fact, there is a common misperception that millionaires are dishonest and crooked. Perhaps that is because of the media reports on the misdeeds of a few bad apples. Author Thomas Stanley has spent 20 years observing and writing about the habits of our nation's millionaires. Stanley notes in his book *The Millionaire Mind*, that a majority of millionaires (who live frugally, drive American-made cars, and enjoy long-term marriages) regard honesty and integrity as a major part of their success.

7. Doers Have Energy. How many of you are tired? It's been a long day; you've been at work or dealing with your kids. You are exhausted. How many of you are *always* tired? How many are just living from day to day, focusing on the time that you can get back into bed from the moment you get out of it?

Your body is like a battery. Did you know that right now, every cell in your body is generating power? The food you eat is the fuel. The challenge is to find ways to tap into each cell and generate even more energy. Ask yourself, "What did I eat for breakfast? Did I eat breakfast at all? What about lunch and dinner? What do I eat every day? How about exercising?"

Your body is the only tool you have. From your brain to your feet, it needs constant care. If you do not care for your body and treat it with respect, it will not perform the way it should. Imagine, there are people who take better care of their cars than their bodies. Eating right, exercising regularly (several times a week), reading good books (to exercise your mind) are all practices toward maintaining good health. There is plenty of information on how to live a healthier life. When you are healthy and taking care of yourself, you have the energy you need to achieve your goals.

8. Doers Communicate. What would you say is the most important communication skill? Of all the communication skills we have, the most important is our capacity to listen. The adage, "God gave us two ears and one mouth; thus, we should listen twice as much as we speak," is quite true. The term *listen* refers to more than hearing someone; it means understanding someone.

When people speak to us, give us their perspective, and share their concerns, frustrations, and joys, the most important thing we can do for them is to understand them. When there is conflict, it is not nearly as important for two people to come to an agreement as it is for them to understand each other.

When there is conflict, it is not nearly as important for two people to come to an agreement as it is for them to understand each other.

How many of you can recall having a conversation in which you were thinking about what you were going to say while the other person was speaking? Everyone has done it; it is a bad habit for a lot of people. When another person is speaking, it is time to set your own agenda aside and put yourself in the other person's shoes. Then confirm that you have understood by giving the person your interpretation of what he or she said. When you have demonstrated that you understand the other person, you open the door for him or her to listen to you.

How important is listening? If parents made the habit of listening more often to their children, it could dramatically reduce drug abuse[10] and perhaps other ills that plague our youth. Countless relationships would heal and begin to thrive. Financial services advisors would greatly increase their commissions. In situations of conflict, communication is not about winning an argument. (When was the last time you actually won an argument?) It is about understanding one another with empathy, with love, and with compassion.

9. Doers Take Responsibility. Taking responsibility for your life is liberating. Doers assume full responsibility for the results they produce. It is often said that the moment John F. Kennedy became a true leader was during the Bay of Pigs crisis. As president, he ordered a military operation that failed utterly. When he had to face the American people, he did not hesitate or place blame on anyone else. His answer was simple: "I am your president. I am responsible."

> **In your life, you will make mistakes and you will have successes. Assume responsibility for each.**

In your life, you will make mistakes and you will have successes. Assume responsibility for each.

10. Doers Are Courageous. To succeed as a financial advisor requires taking risks. Setting foot into a realm where the outcomes are unpredictable is frightening. Fear calls for courage.

Courage is defined as "mental or moral strength to venture, persevere, and withstand danger, fear, or difficulty." When you have courage, you will persevere and overcome the obstacles you face.

Are you afraid to call someone you do not know very well and ask him or her to meet with you? That is normal. But when you have passion, vision, and integrity, you will look past the fears and obstacles and act courageously.

Hurdle #5: The Inherent Difficulties of Prospecting

Perhaps none of the previous four hurdles are stumbling blocks for you. You recognize the value of what you are selling, you do not fear rejection, you know what to do and how to do it, and you are most definitely motivated. Then why is prospecting so difficult? Maybe the hurdle is that prospecting is inherently difficult. Certainly, if it were easy, everyone would be good at it.

Prospecting is an activity that requires skill and technique just like playing golf, basketball, chess, or a musical instrument. You can learn the prerequisite skills fairly easily. But to excel takes hard work, dedication, and practice.

To be sure, there may be advisors who have a natural ability to prospect. But nothing replaces a commitment to hard work and practice. Michael Jordan arguably is the greatest basketball player ever. Yes, he had a tremendous amount of natural ability. But the reason for his dominance was his relentless pursuit of perfection demonstrated by the quantity and quality of time he spent practicing, working on his weaknesses.[11] The same holds true with prospecting.

To pinpoint your prospecting weaknesses, you must focus on the marketing activities that prospecting involves:

- identifying specific needs and wants your products and services can meet
- defining groups of customers who have those needs and wants (target marketing)
- creating or customizing a solution that will meet those needs and wants effectively (defining a value proposition)
- creating and articulating messages that raise prospect awareness of these needs and wants
- positioning in the prospects' minds your ability to meet those needs and wants (preapproaching and creating awareness)
- identifying and contacting individual prospects with whom to meet (setting appointments)

Assess yourself honestly. How good are you at generating appointments that do not result in cancellations and no-shows? If you are experiencing many cancellations, this indicates your prospecting activities are deficient. What part of the overall process of prospecting do you find the most difficult? If this seems rather vague, then analyze how you achieve the marketing objectives listed above. Describe the activities related to each objective, and grade your performance in completing them. Furthermore, evaluate the activities themselves for effectiveness and relevance. Pay special attention to how **value proposition** well you have defined a distinct target market and your corresponding *value proposition*, a compelling reason to conduct business with you, to that target market (the second and third objectives). Most business enterprises fail to accomplish these two objectives.[12] Chapter 2 will examine these two topics in more depth.

Once you have identified your weaknesses, focus on improving them until they become strengths. In this way you will overcome the inherent difficulties in prospecting.

UNDERSTANDING THE PROSPECT

> It is important to be able to articulate the value of any product you offer and what you personally can achieve for your clients.

It is assumed at this point that you have identified your hurdles to prospecting and are acting to overcome them. Thus, you have (or are committed to obtaining) an in-depth understanding of and a missionary zeal for how your products and services enrich people's lives (hurdle number one). Again, it is important to be able to articulate the value of any product you offer and what you personally can achieve for your clients. The next step is to know your prospects.

Knowing your prospects means understanding how prospects make buying decisions, what motivates them to buy, and what motivates them to purchase from you instead of your competition. Then you can discover and customize specific ways to identify prospects, create awareness of their needs, pre-approach them, and approach them. Ultimately, you will apply what you learn from the discussion below of these topics to specific groups of prospects and to individual prospects (discussed in later chapters).

How Prospects Make Buying Decisions

buying process

Like the selling/planning process, the buying process provides a map for understanding the sale—this time from the prospect's perspective. Marketing expert Philip Kotler divides the *buying process* into five stages:[13]

- *problem recognition.* The prospect becomes aware of a problem or need.
- *information search.* The prospect searches for relevant information to quantify the need and the possible solutions.
- *evaluation of alternatives.* The prospect compares the alternatives using various criteria.
- *purchase decision.* The prospect selects a product or service.
- *postpurchase behavior.* The buyer is either satisfied or dissatisfied. Satisfaction leads to future purchases and referrals.

The following discussion explains these five stages more fully.

Problem Recognition

> It may take a combination of triggers before a prospect recognizes that he or she has a need.

Most of your prospects will be in the problem-recognition stage when you contact them. Problem recognition for financial services products does not involve a physiological response, like hunger or thirst. Instead, problem recognition is triggered by some external stimulus. It could be an experience, event, exposure to relevant information, and so forth. Triggers often result in an emotional response. It may take a combination of triggers before a prospect recognizes that he or she has a need and begins searching for information. But sometimes an event is so powerful that it pushes people to act. For example,

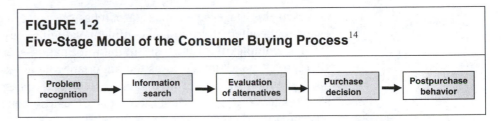

FIGURE 1-2
Five-Stage Model of the Consumer Buying Process[14]

estate planning advisors had advocated the importance of a living will long before the Terry Schiavo controversy. But Ms. Schiavo's situation served as the compelling force that motivated many people finally to create one, resulting in a substantial increase in the number of individuals who were seeking information and drafting legal documents to address end-of-life issues.[15]

Information Search

A prospect in the information-search stage will proactively seek relevant information, or he or she might be open to an advisor's offer to discuss the particular matter with him or her. Sources for information other than an advisor include friends, relatives, books, newspapers, journals, broadcast media, the Internet, and so forth. During this phase, the prospect is looking for guidance regarding aspects of his or her financial need such as the following:

- What is the nature of the need?
- How do I quantify it?
- What are my options for meeting it?
- What have others in my situation done?
- Who can provide the solution?

Evaluation of Alternatives

After the prospect has gained an understanding of the alternatives, he or she will compare them using various criteria that will be different for every prospect. Possible criteria are as follows:

- price
- relevant benefits (for example, the ability to receive premiums back if long-term care benefits are never used)
- performance (quality)
- convenience (ease of obtaining)
- company's reputation
- advisor's reputation

Prospects often never reach this stage because they do not feel confident that they have fully understood the need and possible alternatives. Many who do reach this stage never make a purchase because they cannot adequately evaluate the alternatives to feel confident enough to buy.

The complexity of the tasks in the information-search and evaluation-of-alternatives stages underscores the important role advisors play to help prospects realize and protect their dreams. However, prospects' natural distrust for strangers (you) who want their money creates unquestionably the greatest obstacle to addressing their needs.

Purchase Decision

After evaluating the alternatives, the prospect makes a decision to buy. For financial services products, the purchase decision signals that the buyer's confidence in his or her purchase has exceeded (for that moment) the potential regret of being wrong.

Regret is a function of the perceived (real or imagined) potential consequences of the known alternatives.[16] More often than not, it comes down to cost and value. The buyer perceives at the moment of purchase that the benefits of acting now outweigh the potential regret associated with paying too much, receiving too little in value, or a combination of the two.

> In general, the greater the potential financial and emotional consequences or benefits relative to the prospect's situation, the higher the regret associated with being wrong.

In general, the greater the potential financial and emotional consequences or benefits relative to the prospect's situation, the higher the regret associated with being wrong.

Example: A prospect who balks at the cost of a comprehensive financial plan feels that the potential benefits of creating such a plan do not outweigh the perceived risk of paying $1,500 for a plan that he or she could find impractical or unworkable.

Postpurchase Behavior

Many financial advisors overlook the prospecting possibilities in the postpurchase-behavior stage. It is in this stage that three significant buyer behaviors can occur: First, the buyer may purchase other products (or more of the same one) from the advisor. As noted earlier, a repeat buyer includes a renewal of annual retainer and asset management fees. Second, the buyer may refer prospects to you. Third—the objective from the onset—the buyer may become a client and do both.

What Motivates Prospects to Buy

Understanding what motivates prospects to buy financial products and services may also shed some light on why they would agree to meet with advisors in the first place. Motivations involve psychology—the body of knowledge that attempts to explain how people's thoughts, feelings, and emotions affect their behavior. The following is a brief synopsis of human motivations for meeting needs, based on the work of Abraham Maslow.

Maslow's Hierarchy of Needs

Maslow's hierarchy of needs

Abraham Maslow's theory proposes that human motivations for meeting needs can be best described by a hierarchy known as *Maslow's hierarchy of needs*. Maslow identified five primary needs that every individual strives to satisfy and the order in which those needs arise. The five needs, in ascending order of importance, are shown in figure 1-3.

Physiological. Physiological needs are for food, water, air, shelter, and clothing. People who want any of these things are first preoccupied with attaining them. Only after these basic needs are met can the individual consider satisfying higher-level needs. You have only to remember stories of families caught in a war zone without shelter and food to visualize this level of need.

Security. Security needs are for safety, stability, and the absence of pain or illness. Again, people with these needs become preoccupied with satisfying them. Many workers require medical, unemployment and retirement benefits to help satisfy needs in this level of the hierarchy.

Affiliation. Affiliation needs are for love, affection, and a feeling of friendship and belonging. When the physiological and security needs are satisfied,

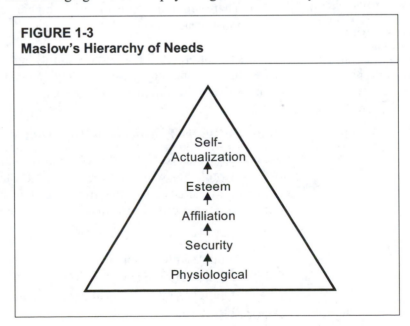

FIGURE 1-3
Maslow's Hierarchy of Needs

Self-Actualization

Esteem

Affiliation

Security

Physiological

the social or affiliation needs arise to motivate individual behavior. These needs are most evident in a person's conduct as it relates to job satisfaction, work ethic, team participation, family ties, and general well-being.

Esteem. Esteem needs include personal feelings of self-worth, recognition, and respect from others. People with esteem needs want others to accept them for what they are and to view them as capable. The needs are fulfilled when the individual receives recognition and feedback from others regarding his or her competence and ability.

Self-Actualization. Self-actualization needs are the highest in Maslow's ranking. They are the needs for self-fulfillment and the realization of personal potential. People who are striving for self-actualization usually accept themselves and others, are superior at problem solving, are more detached, and have a desire for privacy. One irony is that people driven by ambition often willingly sacrifice everything to achieve their quest.

Maslow added four basic assumptions to this hierarchy of needs:

- Satisfying a need is not enough. Another one will always arise. People are always trying to satisfy some need.
- An individual's needs are usually complex, many times involving more than one level. These needs drive behavior and personality.
- The most basic needs must be satisfied before higher-level ones are strong enough to drive behavior.
- There are more ways to satisfy higher-level needs than to satisfy the most basic ones.

> **Maslow's hierarchy of needs provides the key to understanding prospects' buying behavior and what might motivate them to meet with you.**

This is important. Review these points again because you have just read the key to understanding prospects' buying behavior and, consequently, what might motivate them to meet with you.

What Motivates Prospects to Buy from You

Ask your clients, "Why did you choose me over others who do what I do?" You probably will not hear, "You had the right product to meet my needs," or "Your price was the lowest." Although the logic and price buyers are out there, your clients most likely will respond with answers such as

- "I like you."
- "You're a professional."
- "I feel comfortable with you."
- "I believe you understand my needs and care about me."
- "I trust you."

As mentioned earlier, one of the greatest barriers that prevents prospects from buying your products and services is that they do not know you and, therefore, do not like and trust you. Ask yourself this question: Have you ever done business with another person or place of business because you *liked them better* than other, even cheaper, alternatives? Would you rather pay an honest mechanic a little more than a mechanic you are not sure you can trust? Your answer to both questions is probably, "Yes." The reason is simple. When you like another person, you enjoy being around him or her. If that person became unlikable, it would push you away from him or her. Similarly, when professional advice is required, you want to know you can trust the professional giving it—especially when you have little or no expertise yourself.

Financial services products seem complex to the average prospect. The process of evaluating needs and alternatives and then making decisions requires more expertise than most prospects have. In addition, these financial decisions often involve large outlays of money. Thus, there tends to be a very high threshold that an advisor must surpass in order for the prospect's confidence in the advisor to exceed the prospect's fear of regret. Furthermore, personal finances are just that—personal. Not many people would entrust such personal matters to someone they do not like any more than they would to someone they do not trust.

Think about it. All other things being equal, with whom would you entrust your $100,000 inheritance or $4,000 a year Roth IRA contribution? A total stranger? A stranger whom your friends recommend to you? Someone you already know? Someone you already know and like? Someone you already know, like, and trust?

Why people buy from you is greatly influenced by your personal brand—that is, your reputation. That means that it is important to ask, "What do you want your personal brand to be? What reputation do you want to precede you?" Your personal brand should reflect your personal traits, experiences, hobbies, interests, and so on. We will discuss these aspects more in chapter 2. But there are some elements of reputation, related to personal conduct, that are universally applicable to all advisors. Without these core qualities, advisors will find it difficult to gain the trust and respect of those to whom they market. Consequently, it will affect their sales and/or retention. Let us review a few of those core qualities.

1. Be Punctual

Show up on time. Do not make people wait. Lateness can be interpreted as disrespect. When you make a habit of not showing up on time, you send one or more of these messages:

- I am lazy.
- I am disorganized.

- I have no regard for your time.
- I have no respect for you or me.

Punctuality is tied to many character traits. When you are punctual, you reflect other essential traits of strong character and likability, as follows:

- *humility*. You are not perceived as arrogant, selfish, or conceited. You do not see yourself as the center of the universe, and you keep your ego in check.
- *empathy*. Because you respect other people and their time, you place yourself in their shoes. You are constantly working to see things from other people's point of view.
- *caring*. You care about others.

2. Be Honest

> **What good is your word if you cannot be trusted and cannot be relied upon?**

Honesty is about saying what you mean and meaning what you say. It is about truthfulness, character, and integrity—qualities of a doer discussed earlier. If you are not true to yourself and everyone with whom you come in contact, everything else you offer is worthless. What good is your word if you cannot be trusted and cannot be relied upon?

3. Be Polite

Treat others as you would like to be treated. Always say "Please" and "Thank you." Be positive, laugh often, and be upbeat. Act likable. Be the kind of person that others enjoy being around.

4. Be Responsive

Reply immediately to all telephone calls, voice mails, pages, faxes, and e-mails. In the minds of your clients, friends, and prospects, a prompt reply is a welcomed gesture. Responsiveness is a strong character trait that you should always emulate.

Impact on Prospecting

How prospects make buying decisions, why they buy, and why they buy from you should have a clear impact on the following marketing activities associated with prospecting:

- identifying prospects
- creating awareness (preapproaches)
- approaching prospects
- servicing buyers and clients

What follows is a brief summary of a few of the more important applications from the preceding material. The rest of this text will develop these concepts more fully.

Identifying Prospects

Most of your prospects will be in the problem-recognition stage of the buying process. If you identify the problem-recognition triggers, you will have a profile for prospects who need your products. Some of these triggers have a greater likelihood of affecting certain groups of people who share some common characteristics. This is the basis for identifying target markets (the main topic of chapter 2) and identifying specific prospects to contact (discussed in chapter 3). For instance, parents of young children have a great need for life insurance. There are several sources of such prospects: local elementary schools (public or private), day care centers, youth sports leagues, and so on. Some prospecting methods will work better for you than others. Therefore, select methods based on your target market and your particular strengths.

Creating Awareness

As mentioned repeatedly, people typically entrust their financial matters to advisors they know, like, and trust. This means that prospects need to know who you are, have a favorable impression of you, and have some basic level of trust and respect for you. Therefore, defining and promoting your personal brand is mission critical for successful prospecting and selling. Furthermore, awareness strategies that feature more personal interaction between you and prospects are more effective than Yellow Pages advertisements and billboards.

Likewise, most financial products are not demand products. Not everyone who needs a particular financial product experiences a life event that causes him or her to acknowledge a possible financial need. In addition, even when life events occur apart from other triggers, they typically do not cause problem recognition. Thus, advisors need to create and implement strategies that help trigger prospects' recognition of their needs. Without recognition, according to the buying process, prospects are not going to buy.

Prospects' motives to meet with you are generally the same as the ones that provoke them to buy. Thus, messages designed to increase awareness of financial needs should incorporate relevant facts and appropriate motivations, using Maslow's hierarchy of needs as a guide.

> **Advisors need to create and implement strategies that trigger prospects' recognition of their needs. Without recognition, prospects are not going to buy.**

Example: An affluent prospect is generally not going to respond to appeals that satisfy his or her physiological and/or security needs. Therefore, your awareness message should focus on this prospect's affiliation, esteem, and/or self-actualization needs.

Figure 1-4 compares the advertising message for the 2007 Hyundai Accent and the message for the 2007 Mercedes E350. Note the different needs that each message addresses.

FIGURE 1-4 Sample of Message Differentiation	
Advertising Message	**Maslow's Need(s)**
Hyundai Accent	
"Peace of mind you can afford. The 2007 Accent delivers the safety equipment, interior roominess and power you seek, at the price you are looking for. . . ."[17]	Security, physiological (affordability) Key words/phrases: peace of mind, afford, safety, at the price you are looking for
Mercedes E350	
"Offering unprecedented luxury, the newly redesigned E350 exhilarates with elegance. Hand-polished Burl Walnut wood accents distinguish a spacious interior while engineered precision and classic Mercedes-Benz design come together in the most elegant manner."[18]	Esteem Key words/phrases: unprecedented luxury, elegance, hand-polished, spacious, elegant

Understanding the information-search stage can enable you to identify where and how to create awareness of the products and services that can solve prospects' problems and promote your personal brand. Specifically, you can identify the sources of information to which prospects turn and the types of information they seek. Most likely, they are looking for third-party information from unbiased sources, like a neighbor, a family member, or a website. If you can identify those sources for a particular group of prospects, you may be able to establish a relationship (if the source is a person) or a presence (if it is some form of media). If neither is possible, then find ways to incorporate the information source in your marketing efforts. This could include quoting or referring to the information source in your communications and meetings with that group of prospects. Or it could mean referring such prospects to the information source, provided the source is credible.

Example: Suppose you learn that some members of a group of prospects with common characteristics and needs have read Lee Eisenberg's *The Number: A Completely Different Way to Think about the Rest of Your Life*. You could use the book to market yourself and your products in any of the following ways:

- Quote from Eisenberg's book in your communications to this group of prospects.
- Use the book as a basis for a book club.
- Create a seminar around the book.
- Send it as a gift to prospects who are members of this group.

Financial advisors must create awareness to prospect effectively. They need to select appropriate and relevant methods and strategies to generate problem recognition, product awareness, and appreciation of their personal brand. You will study how to create awareness and preapproach prospects in chapter 4.

Approaching Prospects √

The ways that you approach prospects to ask for appointments must complement each prospect's behavior and motivation. Obviously, your approaches will be more effective when you contact prospects who are familiar with you and find your personal brand appealing, such as prospects who have been referred or introduced to you by a client or someone highly respected in the community.

Appointment-setting success really begins with the source of your prospects. However, the content and mechanics of your approach are also critical. Your approaches should reflect the most probable motives the prospect would have for conducting business with you. Your mechanics should reflect your professionalism and personal brand as a trusted advisor. Because every contact with prospects early on is crucial, it is necessary to have a well-thought-out plan and, of course, to practice. Planning, practicing, and executing approaches is the topic of chapter 5.

Servicing Buyers and Clients

> You need to provide the type of service that is not merely efficient and good; you must implement a service strategy that cultivates relationships.

Finally, because referrals are integral to effective prospecting, you need to provide the type of service that is not merely efficient and good; you must implement a service strategy that cultivates relationships. Furthermore, trust determines whether or not people will become clients. Your customer service strategy must build your reputation and establish your personal brand, the fundamental elements of which must be integrity and ethics. These topics are explored in chapters 7 and 8.

CONCLUSION

This chapter explained the relationship between marketing, selling, and prospecting. It also described the role that prospecting plays in the eight-step selling/planning process designed to create clients—repeat buyers who refer you to others. Understanding these relationships can help you define the necessary tasks for effective and efficient prospecting, thus generating more clients.

The chapter looked at the psychology of prospecting, beginning with an exploration of five psychological hurdles to prospecting. Because prospecting is difficult for most advisors, it is vitally important that each advisor identify his or her individual hurdles and devise ways to overcome them.

Finally, the chapter examined the psychology of the prospect—how a prospect makes buying decisions, why the prospect buys, and why the prospect buys from a particular advisor. The discussion concluded with a look at the implications of this psychology on the marketing activities associated with prospecting, which will form the basis for the chapters that follow.

> As you persevere, prospecting will become a life-long habit, a daily routine, and a means to achieve your goals rather than an end unto itself.

The primary objective of this book is to enable you to develop prospecting skills, as you have developed other skills such as riding a bicycle. When you were first learning to ride, you had to overcome your fear and concentrate on each small step in the overall process. Now you can ride a bicycle confidently, and your focus is on where you are going. In the same way, prospecting requires you to overcome your fears and focus on each particular aspect of the process. But as you persevere, prospecting will become a life-long habit, a daily routine, and a means to achieve your goals rather than an end unto itself.

CHAPTER ONE REVIEW

Key Terms and Concepts are explained in the Glossary. Answers to the Review Questions and Self-Test Questions are found in the back of the textbook in the Answers to Questions section.

Key Terms and Concepts

marketing	client
selling	brand
prospecting	value proposition
prospect	buying process
qualified prospect	Maslow's hierarchy of
selling/planning process	needs

Review Questions

1-1. Describe how prospecting directly affects selling and planning success.

1-2. Discuss the relevance of service to clients within the context of prospecting. What does service have to do with prospecting?

1-3.   Explain the relationship between marketing, selling, and prospecting.

1-4.   List the four criteria for a qualified prospect and explain their importance.

1-5.   Describe two ways to overcome a lack of conviction.

1-6.   Explain two ways to overcome the fear of rejection.

1-7.   List the 10 characteristics of a doer.

1-8.   Identify and briefly describe the five stages of the buying process.

1-9.   Define the different needs in Maslow's hierarchy of needs.

1-10.   Describe two ways to apply information on prospect behavior to the four marketing activities associated with prospecting.

Self-Test Questions

Instructions: Read chapter 1 first, then answer the following questions to test your knowledge. There are 10 questions; circle the correct answer, then check your answers with the answer key in the back of the book.

1-1.   Which of the following actions is consistent with the client-focused or consultative approach to selling and planning?

    (A)   Exaggerate the prospect's need in order to create interest in an initial meeting.
    (B)   Talk 90 percent of the time during an initial meeting.
    √ (C)   Conduct comprehensive fact finding during the information-gathering step.
    (D)   During the presentation of the plan, explain every feature and corresponding benefit of a product.

1-2.   Devante does not like calling prospects on the telephone because he feels bad when they say they do not want to meet with him. Which of the five hurdles to successful prospecting does Devante need to overcome?

    (A)   lack of motivation
    √ (B)   fear of rejection
    (C)   not knowing what to do or how to do it
    (D)   lack of product knowledge

1-3. Which of the following financial advisors will Pedra have a higher probability of retaining to assist her with her financial plan?

    (A)   Kendra, who cold-called Pedra yesterday
    (B)   Alisha, whom Pedra met last week at a wedding
    (C)   Esther, whose office is down the street from Pedra's employer
√ (D)   Bob, who was highly recommended by Pedra's accountant, whom she trusts

1-4. Which of the following persons meets the definition of a client for Joe Advisor?

√ I.   Mary, who has four policies with Joe and has sent four friends to see him
  II. Barry, who purchased a policy from Joe 20 years ago

    (A)   I only
    (B)   II only
    (C)   Both I and II
    (D)   Neither I nor II

1-5. Barclay and Cara are in their early 30s and have three children under age 7. They live paycheck to paycheck. To which of the following needs from Maslow's hierarchy would you appeal in your communication with them?

√ I.   Esteem
  II. Self-actualization

    (A)   I only
    (B)   II only
    (C)   Both I and II
    (D)   Neither I nor II

1-6. Kelly lacks conviction. Which of the following actions will help her overcome this hurdle?

  I.   Selecting 10 of her best clients and asking them how her products and services are enriching their lives
  II. Identifying and articulating the knowledge, skills, ideas, and personal characteristics that make her unique

    (A)   I only
    (B)   II only
√ (C)   Both I and II
    (D)   Neither I nor II

1-7. All the following are characteristics of a qualified prospect EXCEPT:

  (A)   He or she lives in another state.
  (B)   He or she has some disposable income.
  (C)   He or she is suitable for your products and services.
  (D)   He or she has called you for an appointment.

1-8. All the following can be attributed to doers EXCEPT:

  (A)   They are passionate about what they do.
  (B)   They have great public speaking skills.
  (C)   They have a plan.
  (D)   They have integrity.

1-9. All the following are stages in the buying process EXCEPT

  (A)   problem recognition
  (B)   evaluation of alternatives
  (C)   prepurchase behavior
  (D)   information search

1-10. All the following prospects are conceivably in the information-search stage of the buying process EXCEPT

  (A)   Roy, who is reading a few books related to his need
  (B)   Joy, who recently experienced a life event relevant to her need
  (C)   Malloy, who has visited Internet sites that provide facts about his need
  (D)   Troy, who has discussed his need with a few friends with the same need

NOTES

1. American Marketing Association: "Marketing Definitions," accessed July 26, 2006, from www.marketingpower.com/content4620.php.
2. Philip Kotler, *Marketing Management,* 11th ed. (Upper Saddle River, NJ: Prentice Hall, 2003), p. 9.
3. Tom Peters, *The Circle of Innovation: You Can't Shrink Your Way to Greatness* (New York, NY: Alfred A. Knopf, Inc., 1997), p. 194.
4. American Marketing Association: "Dictionary of Marketing Terms," accessed August 21, 2006, from http://www.marketingpower.com/mg-dictionary.php?Searched=1&SearchFor=brand.
5. Stephen R. Covey, *Seven Habits of Highly Effective People: Powerful Lessons in Personal Change* (New York, NY: Simon and Schuster, 1989).
6. "The New Powers in Giving," *The Economist,* July 1, 2006, pp. 63–65.
7. Andrew H. Thompson with Lee Rosler, *The Feldman Method* (Chicago, IL: Dearborn Financial, 1989), 12.
8. Ibid, pp. 3–8.

9. Wikipedia: "Ben Feldman (insurance salesman)," accessed August 8, 2006, from en.wikipedia. org/wiki/Ben_Feldman%28Insurance_Salesman%29.

10. University of Illinois at Urbana–Champaign: "Parents need to listen to their teens before the teens will listen to them," accessed August 8, 2006, from www.news.uiuc.edu/news/ 04/0318teens.html.

11. Art Thiel, "Michael Jordan: Modern-Day Icon," accessed August 11, 2006, from www.nba. com/jordan/jordan_icon_thiel.html. Jordan's head coach, Phil Jackson, said, "The thing about Michael is he takes nothing about his game for granted. . . . When he first came into the league in 1984, he was primarily a penetrator. His outside shooting wasn't up to pro standards. So he put in his gym time in the offseason, shooting hundreds of shots each day. Eventually, he became a deadly three-point shooter."

12. Kotler, p. 4.

13. Ibid, pp. 204–209.

14. Ibid, p. 204.

15. John Schwarts and James Estrin, "Many Still Seek Final Say on Ending Life," *The New York Times*, June 17, 2005, late edition—final.

16. Ron S. Dembo and Andrew Freeman, *The Rules of Risk: A Guide for Investors* (New York, NY: John Wiley & Sons, Inc., 1998) p. 80.

17. Hyundai Motor America, "2007 Accent," accessed August 22, 2006 from www.hyundaiusa. com/vehicle/accent/accent.aspx.

18. Mercedes Benz, "The 2007 E350 Sedan," accessed August 22, 2006 from www.mbusa.com/ models/main.do?modelCode=3520W.

2

Getting Started with Target Marketing

Learning Objectives

An understanding of the material in this chapter should enable the student to

---

2-1. Define target marketing and identify its advantages.

2-2. Describe the role that financial needs and personal traits play in the "define the product" step of the target marketing process.

2-3. Explain how the advisor can identify market segments from his or her natural markets by analyzing his or her personal background and history, and past production.

2-4. Detail the process for creating an ideal client profile.

2-5. Describe the three activities related to selecting a target market from the available market segment options.

2-6. Explain how an advisor can position his or her personal brand and products.

Chapter Outline

INTRODUCTION

Obviously, finding groups of qualified prospects (markets) will make prospecting more effective and efficient. That is, you will produce more prospects with a lesser outlay of time, energy, and money. Over the years, the most successful strategy for achieving this objective has been target marketing.

Note that the focus of this textbook is on prospecting, not target marketing. Thus, the goals of this chapter are to define target marketing and its advantages and to describe a simple procedure to identify niche markets using basic target marketing concepts. These concepts include segmenting markets and positioning your personal brand and products.

DEFINITION AND ADVANTAGES

Before delving into how to target markets, let us look at what target marketing is and how it is beneficial to the financial advisor.

Target Marketing Defined

mass marketing

general market

> Long-term success requires moving to a target marketing strategy at some point.

target marketing

A good starting place to understand target marketing is to describe its opposite—*mass marketing*. An advisor who uses a mass marketing strategy aims his or her products and services to a general market without accounting for differences in the characteristics or needs of the various groups within it. The *general market* is everyone within the advisor's territory as defined by geography, whether voluntarily selected by the advisor or assigned by a company, and restricted to the jurisdictions in which the advisor is licensed to practice.

For new advisors, appealing to the general market may be a necessary evil, essential to short-term survival. But long-term success requires moving to a target marketing strategy at some point (the sooner, the better).

In contrast to mass marketing, *target marketing* aims products and services at well-defined target or niche markets. (Although there are some technical differences between a target market and a niche market, they are

target market
(niche market)

too subtle for the purposes of this text. Hence, throughout the text, the two terms are used interchangeably.) A *target market (niche market)* is a group of prospects with common needs and characteristics that make them distinct from nonmembers of the group. The group is large enough to provide a continual flow of prospects and, ideally, has a communication network through which to share information. Furthermore, the common characteristics are actionable, meaning that they provide a practical way to distinguish members from nonmembers and to create unique marketing messages and approaches that produce sales.

Regardless of the procedure advisors use to identify target markets, the objective is to pinpoint groups of qualified prospects who have needs and characteristics that are similar among members and distinct from nonmembers. Thus, the process must be effective in determining the groups of people who generally need and value your products and/or services, can afford them, are suitable for them (if applicable), and whom you can approach on a favorable basis.

Recall from chapter 1 that one of the hurdles to prospecting effectively is the fear of rejection. Combine that with the fact that prospects would prefer to work with advisors they know and trust, and you can see why it is so important to choose target markets in which you can approach prospects on a favorable basis. For this reason, the recommended process that follows limits the target market search to natural markets, which are groups of prospects that the advisor has a natural affinity for or access to. This process involves completing the following tasks:

- defining your product
- identifying and segmenting your natural markets
- selecting a target market
- developing a position strategy for the selected target market

Of course, advisors may use this process to identify target markets that are *not* a subset of their natural markets. But for advisors new to the business or to the target marketing process, this approach is the best place to start.

Before discussing the steps of the process in depth, let us review some of the advantages of target marketing.

Advantages of Target Marketing

Think of a target market as a town. If you have ever moved to an unfamiliar town or city, you can relate to the initial anxiety of finding your way to stores, doctors, schools, and so forth. The first time you drove to a particular location, you paid attention almost solely to the landmarks and street names included in the directions you were given. This preoccupation may have caused you to miss the shortcuts, your favorite stores, the park, and so on. But the more you drove around, the more you became acquainted with the area, and the more your comfort level increased, allowing you to see things you had missed before.

Likewise, the more that you work with prospects from a well-defined target market, the more accustomed and, consequently, the more comfortable you will be with diagnosing prospects' needs and prescribing solutions. As with the new town (or city) experience, a greater familiarity will allow you to observe more and provide better recommendations. You will become an expert in the issues and concerns surrounding your chosen niche markets. Accordingly, several advantages will result:

- enhanced referability
- increased customer loyalty
- less competition
- higher profits
- greater sense of satisfaction for the work you do

Example: Nancy sells long-term care insurance. Early in her career, she built some relationships with the more reputable long-term care facilities and home health care agencies in the area. She spent time learning about the services the facilities and home health care agencies provided and their related costs. As a result, she understands how claim payments can be used. This understanding gives her insight into coverage amounts and additional benefits and riders to recommend. She also has cultivated relationships with local chapters of the Alzheimer's Association and Area Agency on Aging, gaining an appreciation for the emotional and financial challenges facing claimants and their loved ones.

Enhanced Referability

If you had to have heart surgery, who would you like to perform the surgery—a general surgeon or a heart surgeon? Similarly, prospects will feel more comfortable with you if they perceive that you have unique insight into their situation and needs. This will result in enhanced referability.

How important are referrals? Remember that the biggest barrier that prevents most prospects from buying from you is that they do not know and, consequently, do not trust you. But a prospect will have some level of trust in you if someone he or she knows and/or respects recommends you.

✓ *Increased Loyalty*

Concentrating on one or a few target markets enables you to tailor your general postsale service strategies to not only meet your clients' needs, but also to do so in a manner that builds deeper relationships. All things being equal, deeper relationships engender increased loyalty.

Example:	Al's clients are parents with young children, many from his daughter's preschool. Al sends a quarterly newsletter to his clients. Because he focuses on a target market with common needs, he is able to select articles that have almost universal appeal to that market. His newsletter includes one article that discusses practical information relevant to his clients' situation but has little to do with his products and services. For example, one month he included an article on toddler car seats, written by a local consumer advocate. Another article will discuss something relevant to the products and services Al provides as a financial advisor such as education planning. His clients appreciate both types of information. Al's newsletters help cement his relationship with his clients.

✓ *Less Competition*

> If you become a dominant player in serving the needs of your target market, it will deter competitors from intruding.

With enhanced referability and increased loyalty, there will be less competition for the same audience. Indeed, if you become a dominant player in serving the needs of your target market, it will deter competitors from intruding. By the same token, if you discover a market segment in which a local advisor is the well-regarded expert in your discipline, you would be wise to think twice before also targeting that market.

✓ *Higher Profitability*

Target marketing typically results in lower acquisition costs. By narrowing your focus, you can spend less on advertising and awareness campaigns. Consider the cost savings of mailing a marketing piece to a small segment of your geographic territory as opposed to every household, many of which do not need your products and services. In addition, your ability to tailor messages to your target markets will mean a greater probability that your promotion and awareness methods will be more effective. Thus, you will be able to reduce the number of contacts you need to set an appointment and appointments you need to make a sale. The result is a lower cost to acquire new clients, which means higher profitability.

Greater Sense of Satisfaction for the Work You Do

If you select a target market using the process that follows, you will be working with prospects and clients with whom you have a great deal in common. The commonalities will lead to more satisfying relationships. Furthermore, because of the other benefits described above, you will generate clients with greater ease, freeing you to pursue other interests. All of these benefits will lead to a greater sense of satisfaction for the work that you do.

The four steps in the target marketing process, discussed in detail in the upcoming sections, are as follows:

- Define the product.
- Identify your natural markets.
- Select a target market (or markets).
- Position your personal brand and products.

DEFINE THE PRODUCT

The first step in the target marketing process is to define the product. The temptation here is to think immediately in terms of specific financial products: life insurance, mutual funds, health insurance, and so on. But such a limited view of products is inadequate and inaccurate. The advisor must define products in terms of the financial needs of the target market that he or she is able to meet and his or her personal traits.

Financial Needs

People may buy the same product for a variety of different reasons. For example, millions of people each day put a dollar or two into the vending machine and drink an ice-cold Coke. Although the outcome is the same (they buy a Coke), their reasons for buying are probably different—quenching their thirst, satisfying their need for caffeine, and so forth. Similarly, people may buy the same financial product to meet different needs. A stock mutual fund can be used to invest for retirement, future education costs, a down payment on a house, a walk-away fund, a business start-up, and so on.

Recall from chapter 1 that relevant triggers are required for a buyer to become aware of his or her need and decide to do something about it. Which advisor will be more successful—the one who attempts to create a general awareness of mutual funds or the one who focuses on creating awareness about relevant financial needs of a particular group of prospects? The answer is obvious. Thus, it is important for an advisor to define his or her products in terms of the needs they meet and, specifically, the needs for which the advisor is qualified to address.

> **An advisor who knows his or her strengths and weaknesses will be able to target markets whose needs match the advisor's strengths.**

In addition, the advisor should assess his or her proficiency in the necessary skills and knowledge related to each need identified. Both a self-evaluation and an objective third-party assessment from a manager or colleague, if possible, are recommended. An advisor who knows his or her strengths and weaknesses will be able to target markets whose needs match the advisor's strengths.

Example 1:   Ryan is a first-year life insurance agent. He has identified certain financial needs for which he believes he has some knowledge, skills, and ideas. He has completed a self-assessment (rating himself on a scale of 1 to 10, with 1 being the lowest proficiency and 10 being the highest). In addition, his general agent has evaluated him. Their assessment is as follows:

Financial Needs	Self	GA
Income replacement (for parents)	8	7
Education funding	7	5
Final expenses	6	6
Retirement income (with annuities)	3	1

Example 2:   Leila is a 5-year life insurance agent. She has identified the financial needs below. She, too, has completed a self-assessment and has had her manager complete one.

Financial Needs	Self	Mgr
Income replacement (for parents)	10	10
Education funding	7	7
Final expenses	9	9
Business continuation	7	5
Estate planning	8	8
Rewarding key employees (nonqualified deferred-compensation plans)	5	4
Retirement income (with annuities)	3	2

Personal Traits

Because you should target markets in which you can approach prospects on a favorable basis, your personal traits are also important to define. *Personal traits* are your personality, experiences, hobbies, and so forth that make you unique. They have much to do with your "likability" and are assets for creating marketing opportunities. Personal traits have everything to do

personal traits

with personal branding, a topic we will discuss later in this chapter. Take some time to ask yourself, "Who am I? What do I like? About what am I passionate?" As you answer these questions, consider the following.

First, examine your personality. Admittedly, there are some "bad" personality traits out there, and each advisor should be aware of his or her own and find ways to mitigate or change them. (If you are not sure what bad personality traits you may have, ask a sibling or a spouse; they seem to have an eye for your flaws!) But the focus here is the way you are uniquely wired to approach a task or relationship. For example, do you execute the financial advising task as a teacher? Are you an extrovert? Do you approach tasks and people with an analytical mind?

Second, identify any significant life experiences that have given you a deep perspective into life. For example, if you have gone through a divorce, you have an understanding of both the financial and emotional challenges in that situation.

Third, list previous employment experiences. Were you a nurse, an engineer, a teacher? Were you in the armed forces? Do you have an appreciation and knowledge of the financial needs and other aspects of life for members of these occupations?

Fourth, note any hobbies, skills, and interests you have. Perhaps you are into gardening or computers. Maybe you have a passion for certain causes.

The following example is a brief description of a hypothetical advisor's personal traits. Your summary will reflect your unique characteristics.

Example: *Personality:* analytical, introverted, opinionated, perfectionist, methodical, teacher, risk and debt averse

Significant life experiences: have three children, father died in a car accident when I was 19, both grandmothers in nursing homes

Previous employment: math teacher, convenience store clerk

Hobbies, skills, and interests: playing guitar, volleyball, volunteering at ministry for at-risk children and families, writing, computers, mentoring, coaching little league baseball, helping people make wise financial decisions

---

IDENTIFY YOUR NATURAL MARKETS

The next step is to identify your natural markets. To complete this step, you will first need to know what a natural market is. Second, you will need to

[handwritten margin note: identify who you are to develop your approach]

complete two tasks that will enable you to identify your natural markets. Third, from these natural markets, you will need to select one or a few target markets. Let us begin with a definition of a natural market.

Definition of a Natural Market

natural market

A *natural market* is a group of people to whom you have a natural affinity or access because of similar values, lifestyles, experiences, attitudes, and so on.

affinity

Affinity is the natural gravitation you have toward certain people because of shared interests, hobbies, background, lifestyle, or stage of life. Natural markets based on affinity include such groups of people as your friends, family, and acquaintances.

Make a list of your friends. How many of them are roughly your age, share the same lifestyle and values, enjoy the same hobbies or activities, or are in the same stage of life (for example, parenting young children)? Your friends tend to mirror you in these key areas. You like them; you have a certain comfort level with them because of these commonalities. It should not surprise you, therefore, that clients who most resemble your friends in terms of attitudes, values, interests, and so on are the people with whom you find it easiest to create a relationship.

access

Natural markets, however, are not limited to groups of people you know. They also can be formed by groups of people to whom you have access for reasons other than direct personal knowledge or acquaintance. *Access* means that you can approach a group and gain entry with greater ease because you have something in common. This translates into a personal understanding of their needs and the way they think.

Example 1:   Tammy was a lawyer before she became a financial advisor. She has greater access to attorneys not only because of the other lawyers she knows, but also because of her understanding of what it is like to be an attorney.

Example 2:   Teaching was Jack's passion before he started his financial planning practice. His natural market includes teachers because his personal experience gives him insight into the unique challenges teachers face and the financial needs they might have.

Example 3:   Samantha has young children, one in day care and two in elementary school. Her natural markets include the parents of her children's friends, the teachers at her children's elementary school, and the day care workers. If she becomes acquainted with the owners of the day care center, she can gain access to any of their friends who are also business owners. The possibilities are many.

Long ago, advisors identified their natural markets only after years of selling. They just kept selling wherever it led—to anyone who would buy. After a few years, these advisors eventually found their natural markets through pure trial and error and without any planning and reflection. This was the lesson of experience. But the lost opportunity cost in gaining that experience was high. Had they recognized their natural markets earlier, they would have found qualified prospects with less effort and expense and generated clients with fewer interviews. Each of those clients may have resulted in referrals of more people like them—people to whom the advisors related easily because of commonalities.

You can either allow your natural markets to develop or you can take steps to identify them. Identifying them is not time consuming and will result in greater long-term success and satisfaction.

There are two important factors in identifying your natural markets— your personal background and your personal production.

Personal Background and History

Your personal background and history have a lot to do with defining your natural markets.

Your personal background and history have a lot to do with defining your natural markets. For example, if you know a number of people in your town, you have an advantage. If you are particularly young, you may have trouble getting older prospects to respond well to you. If you have only limited business experience, it may be difficult at first for you to penetrate the business markets that may available to others with strong business backgrounds. It is important to look at yourself realistically.

Perhaps the most serious mismatch occurs if you do not place the same values on money, family, and other critical life values as your prospects do. This often arises because of cultural, religious, and economic differences between advisors and the markets they are trying to penetrate. Imagine an advisor dressed in a three-piece Armani suit, wearing a Rolex watch, and driving a shiny new Lamborghini, meeting with a farmer at the farmer's dairy farm. Can you say, "Clashing values?"

It is essential, therefore, to identify your natural markets. To do this, list the occupations with which you feel you work well. In addition, list social organizations and associations with which you have had a measure of success. Also describe the types of people with whom you enjoy working. Finally, consider any businesses that you know fairly well (for example, entrepreneurs and corporations). Use the "Natural Markets Checklist" in figure 2-1 as a guide.

FIGURE 2-1
Natural Markets Checklist

Occupations I have worked well with in the past

Professionals, industrial

Social organizations/associations in which I have been successful

Networking - BNI, Coffees, After hours
trade Shows

Types of people whose company I enjoy and with whom I enjoy working

Networkers, Professionals, Nat Mkt

Businesses I know fairly well, especially how they operate

Professionals, Industrial

Organizations, businesses, and associations where I am recognized by most people

---

Next, you must cross-reference this list with your analysis of your past personal production, which is explained in the next section.

Past Personal Production

Your current clients can indicate the path of least resistance to future sales success. Thus, it is important to analyze your past personal production.

One way to do this is to list 20 of your best current clients and divide them into market segments using the process that will be described shortly. Some advisors also create an ideal client profile, which is a natural byproduct of identifying natural markets. The ideal client profile will also be discussed.

Identify Your Top 20 Clients

Ask yourself , "Who are the 20 clients I enjoy working with the most?"

Print a list of your current clients (if you are newer to the business, use your most promising prospect list, although the results will not be as reliable) and ask yourself, "Who are the 20 clients I enjoy working with the most?" This does not necessarily mean your biggest commissions or fees, but those clients with whom you have the greatest rapport and whose company you enjoy the most.

Select Relevant Market Segmentation Variables

market segment

Next, you will select relevant market segmentation variables, which you will use to divide your natural markets into market segments. A *market segment* is a potential target market.

Segmentation Variables. There are generally four types of segmentation variables that marketing experts use to divide a market: geographic, demographic, psychographic, and behavioristic.[1] The segmentation process utilizes a number of variables from one or more of these four categories.

geographic variables

Geographic Variables. These variables segment a market by using political divisions such as states, counties, cities, boroughs, and so forth, or by territories delineated by neighborhoods, regions, miles, and so on. For most financial advisors, geography is used to define their territory, mainly because of licensing requirements (advisors must be licensed in each state in which they practice)—although with technological advances, especially cell phones and broadband Internet, more and more advisors are setting up virtual practices that market to a very small high-net-worth niche market (such as dentists, doctors, or professional athletes) located in several states or even the entire country.

Example:	Rachel lives in Burlington County, New Jersey. One of the characteristics of her target market is that many individuals in that market are preretirees. Rachel knows that there are several over-age-55 communities in the county. Geography is a meaningful segmentation variable for her.

✓**demographic variables**

Demographic Variables. Partially because they are easier to measure (through a census or marketing data companies), demographic variables are the most commonly used segmentation variable. Even more important is that the differences in prospects' needs and wants are often linked to these variables.[2] For example, interest in Medicare supplement insurance occurs primarily around age 65 because that is when such insurance is an issue for people. Along with age, demographic segmentation includes variables such as gender, education, ethnicity, occupation, income, size of family, marital status, religion, generational cohort, (baby boom, generation X, silent generation, and so forth), and family situation (single, married with kids and a single income, married with no kids and two incomes, single parent, empty nester(s), divorced, and so on).

✓**psychographic variables**

Psychographic Variables. Advisors use psychographic variables to divide a market by lifestyle and attitudes. These variables include things like leisure activities, values, personality, interests, and hobbies. For example, you may have a few clients who are running enthusiasts, play in a softball league, participate in a bowling league, and so forth.

✓**behavioristic variables**

Behavioristic Variables. Finally, behavioristic variables group people by their buying and usage behaviors. This type of segmentation categorizes people according to when they buy—life events (birth of a child, marriage, divorce), type of user (do-it-yourselfer, collaborator, delegator), brand loyalty, benefits sought (convenience, price, quality), and so forth. For instance, perhaps you find that upcoming nuptials triggered a few of your best clients to seek your financial advice. It could be a coincidence but, then again, it may be a great target marketing opportunity.

Common segmentation variables used to segment natural markets appear below.

Example: ***Common Market Segmentation Variables Used to Segment Natural Markets***

Geography: neighborhood, location of employer

Demography: age, income, investable assets, occupation or profession, ethnicity, employer, family situation (married, kids, and so on)

Psychography: type of lifestyle (modest, luxurious, and so forth), attitudes and values, community interests, social organizations, religious affiliations, clubs, alumni organizations, fraternity, sorority

Behavior: financial reasons for buying (income protection, children's education, asset protection, and so on), emotional reasons for buying (desire for security, sense of responsibility, and so forth), and referral activity (how many, the quality of referrals, and so on)

Selecting Relevant Variables. To select variables, you need to think in terms of the criteria for a qualified prospect as well as the distinguishing characteristics that identify members of the target market.

Thus, you will need to analyze the financial needs you identified in the product-definition step for possible segmentation variables that indicate (1) a need and a desire to meet that need and (2) the ability to afford and qualify for your products. (See figure 2-2.) Let us assume that by segmenting your natural markets that you already can approach prospects on a favorable basis. Note that in many cases, you will probably not be able to find segmentation variables that will help you define *all* the criteria of a qualified prospect.

FIGURE 2-2
Possible Segmentation Variables

	Criteria
Need/want/value	Age Sex Family status Products or services they bought from you Reasons for product purchase Values and attitudes Total commissions, fees, and so forth they represent Referrals Occupation or profession
Ability to pay	Income Lifestyle Assets
Suitability	Income Lifestyle Assets
Approachability	Occupation or profession Employer Lifestyle Personal traits Ethnicity Social organizations Hobbies Referrals

> **If there are no distinguishing characteristics that enable you to identify a market segment, then you do not have a market segment.**

One or more of the variables should reflect possible ways to distinguish members from nonmembers and to access the market segment. Such characteristics might include occupation or profession, employer, activities (working out at the same gym or joining the same club, for example), neighborhood, and so forth. If there are no distinguishing characteristics that enable you to identify and access the market segment, then you do not have a market segment.

Segment the Market

After you have selected the segmentation variables you will use, begin to segment your market. Make a chart with columns for the client's name and to record information corresponding with each of the demographic variables you chose. Enter the information for each client. (See figure 2-3 for an example of an abbreviated form).

FIGURE 2-3
Segmentation Chart

Client	Product	Financial Need	Inc and Assets	Life	Family Status	Org	Emp
1	Term	Inc rep Edu Debt	70K 100K	Mod	Married 3 children	Kiwanis Ltl league	Fed Ex Elec Bout
2	Annuity	Ret plan	150K 500K	Extra	Married	Kiwanis	SMS SMS
3	Mut Fd	Gen inv	50K 35K	Mod	Single		City of Phil
4	Term	Inc rep Div dec	85K 110K	Mod	Divorced 2 children	Chamb comm Ltl league	Bugs R Us (Self-emp)
5	Term	Inc rep Debt	65K 120K	Mod	Married 2 children	Ltl league	EDS

A spreadsheet application will expedite the process; the sort function will make the analysis much easier. For more powerful analysis, it is worth considering a database or contact manager (with customizable fields). (Note: OpenOffice is an open source office suite that includes a word processor, spreadsheet, database, and presentation applications. As of this writing, it can be downloaded free. Visit www.openoffice.org for information.)

Create Market Segment Profiles

Note the common characteristics of clients who bought a product for the same reasons.

Once you have gathered the information, look for any trends (exclude exceptions or extremes). Seek out commonalities. Begin by grouping according to product and need. Note the common characteristics of clients who bought a product for the same reasons. For example, in figure 2-3, the clients who have an income replacement need for term life insurance (clients 1, 4, and 5) all have roughly the same income and assets. Two of them are married. All three of them have children, are involved with little league, and have a moderate lifestyle.

Although figure 2-3 does not show these behavioral variables, consider how much each of the top 20 clients means to you in commissions, fees, and so on. Also consider the quantity and quality of referrals—you will probably have a great deal of referability within a market segment that includes these clients.

When you have completed this task, you should have market segments from your natural markets that are good potential target markets. Develop profiles for each of them, identifying their relevant distinguishing characteristics. Add subjective characteristics that you feel are missing and for which you have no hard data. For example, you may sense that most of your clients in the "Responsible Parents" target market are planners. Although you have no real data or no way to observe or measure this without interacting with them, you feel intuitively that this is true. (See figure 2-4.)

Cross-reference the results of your personal background and history and your past personal production activities. Ask the following questions:

- Which occupations and businesses appear in both analyses?
- Which businesses, organizations, and associations in which you are recognized show up in the analysis of your past personal production?
- Which types of people whose company you enjoy and with whom you enjoy working do you find in both analyses?
- If businesses, organizations, and associations in the "Natural Markets Checklist" are not represented in the analysis of your past personal production, why not? What obstacles have prevented you from penetrating these potential target markets? How difficult would it be?

Any overlap in the two analyses points to market segments that deserve strong consideration as possible target markets.

Create an Ideal Client Profile

ideal client profile

Some successful advisors go one step further and create an *ideal client profile*, which describes the type of prospects with whom the advisor prefers to work. The ideal client profile is slightly different from a target market

FIGURE 2-4
Example of Market Segments

Responsible Parents

- Between ages 25 and 40
- Average income: $70,000
- Average investable assets: $100,000
- Children in little league
- Looking to insure income at death
- Budget conscious
- Sensible spenders
- Planners

Financially Savvy Singles

- Ages 23 to 30
- Average income: $50,000
- Average investable assets: $25,000
- Thinking about retirement
- Responsible spenders
- Looking for affirmation in their decisions
- Civic-minded

Maxed-out Empty Nesters

- Ages 50 to 65
- Dual income
- Average income: $150,000
- Average investable assets: $600,000
- Partner who stayed home with children has re-entered workforce
- Serious about retirement, maxing out tax-advantaged plans
- Budget conscious

profile in that it concentrates more on personality traits and attitudes than on demographics. The ideal client profile typically tells you the type of person you think you would enjoy working with, focusing less on but not totally ignoring the need for your product. It follows that ideal clients are a subset of each of your target markets. They are the type of people in each target market with whom you would prefer to work.

Although you can create a client profile without analyzing your natural markets, using the segmentation process gives you a more grounded profile. For example, without any regard to experience, you may desire clients who have $1 million in investable assets. Whether or not that is practical, however, will depend on your most recent past experience. If your current client base has average investable assets of $100,000, the $1 million figure is probably not a reasonable immediate or short-term aspiration.

The process to create an ideal client profile is very much the same, whether you choose to segment your natural markets or not. Make a chart with two columns. The first column heading should be "What I Want." The second should be "What I Don't Want." Then think about your 20 best and worst clients, and note the demographics and attributes that correspond with each, placing them in the appropriate column. For example, if some of your worst experiences were with very analytical prospects, that characteristic should appear in the "What I Don't Want" column. The following is an example of characteristics and attributes of one advisor's best and worst clients.

(handwritten note: Wants & Don't Wants)

Example:

What I Want	**What I Don't Want**
• Has household income of $70,000 or more • Has investable assets of $100,000 or more • Desires face-to-face interaction • Values family • Is thinking about retirement and/or education planning	• Very analytical • Unreliable • Price shopper • Do-it-yourselfer

You can now create a profile from the lists of traits you want your ideal client to have and those you do not want him or her to have. As you create the profile, convert the negative traits into positive ones. For instance, if you do not want a do-it-yourselfer, you want a delegator or someone who seeks advice. Using the two columns in the example above, you can describe the ideal client as shown below.

(handwritten note: Ideal Client list)

Example: My ideal client is someone who

- has a household income of $70,000 or more
- has investable assets of $100,000 or more
- wants a face-to-face relationship
- desires advice from a competent professional about retirement and education planning
- is family-oriented
- appreciates the value of good advice and is willing to pay for it
- is reliable

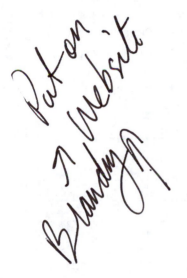

> **The most obvious application for your ideal client profile is to focus your prospecting activity.**

The most obvious application for your ideal client profile is to focus your prospecting activity. Compare your new or current prospects with the profile to determine whether you should continue to pursue these prospects, eliminating those who do not match the profile. In addition, you should periodically repeat the profiling process to see if your ideal client has changed. You may discover, for instance, that you are now working predominantly with prospects who have investable assets of $150,000 or more instead of the $100,000 or more you originally listed on the profile.

Another application is to determine who your "A" clients are—that is, the clients to whom you provide your top level of service. This will enable you to create a service strategy that caters to your top clients (see chapter 7).

Additional worthwhile applications (some of which are mentioned later in this textbook) are as follows:

- Use the profile to guide you in your marketing strategies and materials.
- Post your ideal client profile on your website and in your office.
- Give the profile to friends, family, current clients, and centers of influence. (These are influential people who know you, have a favorable impression of you, and agree to introduce or recommend you to others.) That way, these people will know the type of client with whom you want to work.

Ideal Client Profile on Your Website

Assuming you have a website (captive agents may not have the option of creating one), you may find it helpful to place your ideal client profile, or relevant excerpts, on the site.

One financial planning firm has on its website both the financial status and personal characteristics of its ideal clients, describing them, in part, as follows:

- Recognize the connection between accumulated assets and the freedom to pursue their life vision
- Are successful in their careers
- Participate in charitable activities and foundations[3]

SELECT A TARGET MARKET

The next step is to narrow your target market options and select one or a few to test and pursue. The following activities are recommended for completing this step:

- creating selection criteria
- conducting market research
- choosing an approach

Creating Selection Criteria

> Some market segments make better target markets than others.

Some market segments make better target markets than others. Take the top five or so market segments from your natural markets and compare and prioritize them using criteria of your choosing. Such criteria will depend largely on your product mix (product here includes services).

You can group your criteria into three main categories[4] (you may add your own, of course):

- fit of resources to segment's needs
- level of income
- level of competitiveness

Assess each market segment and give it a score. You may use a chart like the one in figure 2-5.

FIGURE 2-5
Selection Criteria

	Market Segment				
	1	2	3	4	5
Fit of resources to segment's needs					
Products (financial needs I can meet)					
Personal traits					
Subtotal					
Level of income					
Potential income					
Referability					
Size of market segment					
Subtotal					
Level of competitiveness					
The number of competitors is few					
The competition is not firmly established					
Subtotal					
Total					

Market Segments
1—Responsible parents of Caln Elementary School students
2—Financially savvy singles from Exton and Downingtown
3—Maxed-out empty nesters who live at Hershey Mills over-55 community
4—Established women business professionals in Chester County
5—High-life DINKs who own luxury cars

Score low to high—0 to 5

Consider the details for each of the categories, as explained below.

Fit of Resources to Segment's Needs

Products. Assess how well your resources match your market segment's needs. The best way to do this is to create another chart; this time, the first column lists the needs your product(s) meet. Constrain your list to the needs for which you are knowledgeable and skilled to handle. Then create columns for each of the market segments you identified. Finally, for each need and feature, rank the relevance to each segment, as illustrated in figure 2-6. Factor into your assessment the priority that market segment members would place on meeting that need as well. How high a priority is it in light of other needs, wants, and desires?

In addition to the chart, consider your product's affordability. Price your product. For example, if you sell life insurance, calculate the cost of a policy

---

FIGURE 2-6
Matching Resources to Needs

	Market Segment				
Life Insurance	**1**	**2**	**3**	**4**	**5**
Income replacement	5	1	2	4	1
Education funding	5	0	0	3	0
Debt repayment	5	0	2	5	4
Estate planning	1	1	4	4	4
Philanthropy	0	0	0	3	1
Estate creation	4	1	4	4	1
Maintaining responsibility and preparedness	5	1	1	4	1
Long-Term Care Insurance					
Asset protection for heirs	1	1	4	5	1
Preservation of dignity and assurance of choice (avoidance of Medicaid)	1	1	5	4	4
Security that loved ones are not burdened	4	4	5	5	5
Avoidance of Medicaid 5-year lookback period	1	1	3	3	1
Maintaining responsibility and preparedness	1	1	4	5	3

Market Segments

    1—Responsible parents
    2—Financially savvy singles
    3—Maxed-out empty nesters
    4—Established women business professionals
    5—High-life DINKs

Relevance of need or feature, scored low to high—0 to 5

with a typical coverage amount that you expect members of a particular market segment need. Then compare the cost to the market segment's average income. How reasonable is it?

Personal Traits. Finally, how well do your personal traits match each market segment? Remember, your personality will have a lot to do with your ability to approach members of the market segment on a favorable basis. If you have created an ideal client profile, how closely does the market segment match the profile? (Obviously, it will not be a 100 percent match.) What is your basis for a favorable approach? For example, are members in the market segment your friends; do they belong to the same country club? How favorable is your basis to approach them relative to other market segments? You could construct a chart similar to the one you made to evaluate the fit of your product.

Level of Income

You will want to quantify the possible lifetime profitability that each market segment can produce. Lifetime profitability includes income from product purchases (the initial purchase and repeat purchases of the same or other products) and from referrals. It also accounts for the size of the segment (think in terms of sales volume).

Income from Product Purchases. Calculate the price for a likely product purchase, and estimate the income you would receive for all *probable* purchases over an individual client's lifetime. Do not simply total up the prices for all *possible* products. Remember the affordability factor. A $1,500 premium per month for multiple insurance policies for a family with a gross income of $5,000 per month is not reasonable and thus not probable. Estimate a probable average amount and then figure the income that amount produces for you (if you are dealing with a commission).

Referrals. How many referrals do you think a client in this market segment will provide if he or she is satisfied with your service? To gauge your referability, determine the formal or informal communication systems that clients in this segment have and how much they use these systems. The presence of a communication system, whether formal (a meeting, a newsletter, and so on) or informal (neighborhood relationships), increases your potential referability. Estimate the income potential from referrals by assuming that a percentage of the referrals will produce the same average income as just calculated above. Multiply the number of referrals by the amount of average income produced. These are crude estimates for sure. But all you need is a general idea.

Size of Market Segment. Demographics are important. How large is the market segment? Also, how fast is it growing? If you are evaluating the market segment of Hispanic families in Chester County, Pennsylvania, for instance, you would find that the population increased from 8,652 in 1990 to 16,126 in 2000. And it is anticipated to be 25,787 by 2010.[5] That is a growing market!

Level of Competitiveness

> As sure as a picnic will attract ants, a good target market will attract competition.

As sure as a picnic will attract ants, a good target market will attract competition. How competitive are you? To determine your competitiveness you will need to evaluate your competition. Specifically:

- Who are your competitors? How many are there?
- How well are your competitors established in the market?
- How highly rated are your competitors within each market segment?

If the competition already has an established presence within one of your possible target markets, you must evaluate your potential to be a formidable rival or to carve out a specialized niche within that segment. Determine whether or not the competition has a good reputation or if there are other unexploited opportunities. All things being equal, you should choose target markets that are underserved or in which the competition does not yet have a solid presence.

Conducting Market Research

You may not know enough about the market segments you have identified to evaluate them accurately. As a result, you will need to conduct market research. (Note: It may be appropriate to have done some investigatory work prior to this point.) Regardless of when you initiate market research, the principles are the same; only the purpose and questions are different.

Initial Research

A good place to begin your research is on the Internet or at the local library. Incidentally, if you do not have access to broadband Internet service (cable, DSL, and so on), purchase it. The Internet is a treasure trove of valuable information (see the box titled "Market Research Aids" for some examples). The local chamber of commerce is also an appropriate place to start if you are targeting business owners.

Your main goal is to gather information regarding the selection criteria you previously developed. The information does not have to be precise; a very rough estimate will do fine.

Example:	Cecil wants to target realtors in Chester County, Pennsylvania. Using an Internet search engine, he searches for a realtor association in that county and discovers

the Suburban West REALTORS® Association at www.suburbanwestrealtors.com.

At this website, Cecil can find the following marketing information: There are 4,400 members in the association. Some members are not realtors but businesspeople who provide products and services related to the real estate industry.

Cecil determines that the target market is large enough to warrant his time and effort. The existence of the website gives him reason to believe there is a communication network. Also, he has access to a database of names, numbers, and addresses of real estate agents for market research and, ultimately, for prospecting.

Use your initial research to learn about market segments and to prioritize them. Because of the time commitment required to survey a market segment, you will want to invest that time only in the top two or three for now. Note that the more information you can gather about any of the market segments

---

Market Research Aids

An excellent resource is *The Lifestyle Market Analysis*, which gives demographic and psychographic information for 210 designated market areas all over the United States. One report provides demographic information about designated market areas such as the percentage of the population that has a certain occupation, household income, children, and so forth. Another report gives psychographic information about designated market areas such as the percentage of the population who invest, participate in certain sports, have certain hobbies and interests, and so on. You may be able to find a copy of *The Lifestyle Market Analysis* at your public library. For more information, see www.srds.com.

In addition, there are many Internet resources:

Free Sites

U.S. Census Bureau	www.census.gov
Statistical Abstract of the U.S.	www.census.gov/prod/www/statistical-abstract-us.html
County and City Data Book	www.census.gov/statab/www/ccdb.html
The Social Statistics Briefing Room	www.whitehouse.gov/fsbr/ssbr.html
State Chamber of Commerce	(varies by state)

Search Engines (Free)

Google	www.google.com
Hotbot	www.hotbot.com

Pay Sites

The American Marketing Association	www.marketingpower.org
ESRI Business Information Solutions	www.esri.com
American Demographics	www.adage.com

you have identified, the fewer questions you will need to ask in your survey. Your initial research can also guide you in developing appropriate questions for the survey.

Market Survey

Select your best potential target markets from the market segments you have chosen, and prepare to conduct a market survey.

Identifying Survey Participants. Before you construct the survey, you must identify whom you will survey. This will affect what questions you ask. Thus, clearly identify your prospective survey participants.

At first glance, this may seem obvious. You want to survey people from the market segments you identified in the previous step. In particular, you want to survey current clients who are members of one of the identified market segments, beginning with your best clients. But other professionals who currently target this market segment and potential centers of influence are good people to survey, too. Although this will mean crafting two distinct surveys, it will be worth your time.

Selecting Questions. Once you know whom you will interview, construct your survey. Keep in mind that the primary purpose of your market survey is to gather information related to the selection criteria that you were unable to obtain from your initial research efforts. In addition, you will want to confirm or discover distinguishing and relevant characteristics common to members of the potential target market. As a result, construct a survey that enables you to

| **The primary purpose of your market survey is to gather enough information to enable you to decide if the market is worth penetrating.** |

- confirm or discover relevant distinguishing characteristics of members of the market segment
- define clearly the most important problems and goals common to most members, especially financial problems and goals
- understand how members perceive the financial products and/or services you provide
- determine the income potential of the market segment
- evaluate the competition's marketing efforts and success within the target market

Make your survey thorough but concise; you do not want to overwhelm your participants. Keep the number of questions under 10 if you can. Figure 2-7 contains some sample questions from which you can build your own survey.

| **Make your market survey thorough but concise; you do not want to overwhelm your participants.** |

Conducting the Survey. Ask several members of each market segment to participate in your survey, assuring them that your purpose is not to sell them

---

FIGURE 2-7
Market Survey Sample Questions

Characteristics

- What attitudes and interests do you share with others in your market?
- If an outsider went to a convention for (name of market segment), what impressions would he or she get from meeting members?
- Is there something special or interesting that members in your market have in common? Why is this common to your market?

Fit of Resources to Segment's Needs

- What are the common problems and concerns people in your situation face? Which are the most important and pressing?
- If you could have a day's fee paid that would allow you to hire a good financial advisor to solve a problem that (name of market segment) face, what would you have that person do? Why?
- How do you feel about (type of product or service)? How important do you think it is? Why?
- What determines how you buy (product or service)? Price? Quality? Brand name? Customer service? Convenience? Other?
- What do you think is an affordable price for (your product or service)?
- Who or where would you go to find information about (product or service)?
- What do you expect from (your product)? What are you looking for in a (your profession)?

Level of Income

- How would you determine from which company to purchase (name of product or service)?
- What do (name of market segment) read to keep up with the latest information?
- What industries, associations, or clubs support (name of market segment)?
- How often do you get together with other people in (name of market segment), and what do you do when that happens?
- How often do you tell friends, family, business associates, and so on about good products and excellent service?

Level of Competitiveness

- Who comes to mind when (name of market segment) think of (your products and services)?
- How do you know about (name of a competitor)? Does (competitor) send you mail or e-mail? Does (competitor) advertise? Where?
- What reputation does (competitor) have among (name of market segment)?

---

something but to find out more about them and the market segment to which they belong. Here are some additional thoughts and tips:

- Be polite and professional. You are probably dealing with a possible prospect and/or center of influence. Therefore, the survey is an opportunity to create a favorable impression of yourself.

- If you are interviewing a current client or noncompeting professional, consider treating him or her to lunch or a cup of coffee.
- Be observant when interviewing members from your target market. How does the person dress (conservatively, casually, ostentatiously)? What kind of car does the person drive (luxury, minivan, truck, compact)? Where does the person live?
- Listen carefully, especially during small talk. What does the person value? What nonfinancial needs does he or she have? To which organizations does the person belong?

The Benefits. Surveys provide a number of benefits. First, you obtain information about the market segment directly from its members or someone who knows them very well. Consequently, you are able to make your own judgments rather than relying solely on statistics and data from others. Second, because you develop the questions and possibly interview people yourself, the information you acquire is specific and relevant to your marketing efforts. Moreover, the members of the market segment you interview may notice your efforts, which will let them know that you can help solve their problems. This alone can build your prestige and create awareness of you in your role as an advisor. In fact, it is an excellent first step to prospecting within your target market.

Selecting a Marketing Coverage Strategy

marketing coverage strategies

The market research is complete. The market segments have been evaluated according to the selected criteria. The next phase of the selection process is to choose from one of several *marketing coverage strategies*—that is, "alternative approaches that a company can use to select and target markets."[6]

Five Basic Marketing Coverage Strategies

There are five basic marketing coverage strategies:

- single-market concentration
- selective specialization
- product specialization
- market specialization
- full-market coverage

Figure 2-8 graphically illustrates these five types of market coverage.

Single-Segment Concentration. The first strategy is the narrowest. It involves marketing one product to a single market segment. It may be either a rather large market segment or a small but very lucrative one (doctors, dentists, professional athletes, and so forth).

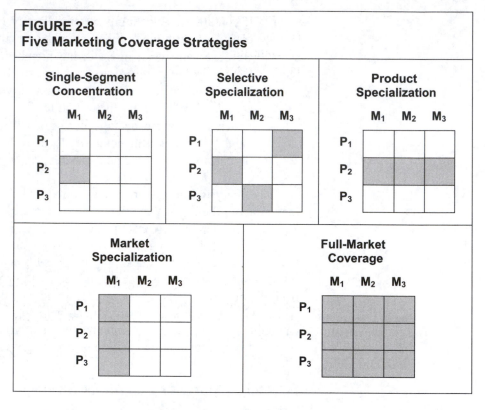

FIGURE 2-8
Five Marketing Coverage Strategies

This strategy is typical in situations in which the value of a single client is thousands or even tens of thousands of dollars. Thus, advisors who target high-net-worth clients would employ this market coverage—for example, an advisor who focuses on estate planning for physicians. If the market segment is very small, however, there is the risk to the advisor of having all his or her eggs in one basket.

> **The risk to the advisor of a very small market segment is having all his or her eggs in one basket.**

Selective Specialization. The second coverage strategy involves marketing a few different products to multiple target markets. This strategy diversifies an advisor's market portfolio both in terms of its products and target markets. Advisors who sell more than one product can pursue this coverage strategy effectively.

A good idea is to select target markets that are related to one another. For example, real estate agents, title insurance agents, home inspectors, and mortgage brokers are related market segments; in fact, if there is a local real estate agents association, the others may very well be members. Although the needs and characteristics are different between market segments, there is a common point of access (if there is an association as described) that affords an opportunity to use the selective specialization coverage strategy if desired.

Marketing to related markets

Product Specialization. Advisors who use the product specialization strategy market a product to multiple target markets. As with the selective specialization approach, it is ideal to target market segments that are related to one another in some way.

Market Specialization. Market specialization is like the single-segment concentration strategy in that the advisor specializes in one market's needs. It allows the advisor to devote his or her undivided attention to a particular target market and to become an expert in that market's financial needs. The difference is that the advisor sells multiple products to that market. Obviously, the advisor should choose a niche market that needs and values the products that the advisor sells.

Full-Market Coverage. Finally, there is full-market coverage, in which all products are marketed to all segments. This is the least efficient strategy for individual financial advisors but is a viable strategy for multi-advisor practices.

Advisors who sell property and casualty insurance along with other insurance and financial products can use a hybrid version of the full-market coverage. That is, they can use a full-market coverage to sell auto and homeowners insurance but then use one of the other coverage strategies, most likely selective specialization, for their other products.

Considering the Options

Your current practice, specifically the number of products you sell, will dictate your initial options for choosing a market coverage strategy. It is important to weigh the advantages and disadvantages in each situation. In addition, you need to take into consideration your strengths and weaknesses as well as your vision for the practice. For example, some advisors like to focus on one product; others enjoy the variety of mastering several products.

Advisors who enjoy managing people could consider a wider product and market range because they can always hire others to work with them. In contrast, some advisors prefer to work alone. Thus, fewer products and/or markets would probably be more suitable. The bottom line is that in considering the options, you should gravitate toward strategies that leverage your strengths and avoid your weaknesses.

As you can see, the two decisions—selecting target markets and a market coverage strategy—are interrelated. If you have identified three markets related to one another, that could influence you to choose a selective specialization coverage strategy. At any rate, the result to this point is that you have determined a coverage strategy and, therefore, one or a few target markets. Now you need to come up with a strategy to position your personal brand and products within each target market.

> **In considering your market coverage options, you should gravitate toward strategies that leverage your strengths and avoid your weaknesses.**

POSITION YOUR PERSONAL BRAND AND PRODUCTS

As most of you are aware, Charles Barkley played in the National Basketball Association in the 1980s and 1990s. Undersized at 6' 6", Barkley was a dominant rebounder. The key to his success was the masterful way he established the right position underneath the basket. Other players were taller, stronger, and could jump higher. It did not matter. Where Barkley gave away such competitive advantages as height, strength, and leaping ability, he made up for them in positioning.

In an ideal world, people would choose you as their advisor simply because of your superior expertise and skill. But the real world does not work that way. The truth is that people often base choices about products and services more on perception than on real quality and performance. Like Barkley's dominance as a rebounder in the NBA, the key to success is being well positioned in your target market. The right positioning matters more than having the best products, the best fact finder, and even the most expertise, as objectionable as that might sound.

> **The truth is that people often base choices about products and services more on perception than on real quality and performance.**

This section defines positioning. Then it looks at how positioning applies to your personal brand and products.

Positioning Defined

positioning

CREATE In Your Prospects minds

According to management expert Philip Kotler, *positioning* is "the act of designing the company's offering and image to occupy a distinctive place in the mind of the target market."[7] Thus, the objective of positioning is to create in the minds of your prospects a predetermined favorable perception of both your personal brand (image) and your product (offering). Consequently, effective positioning results in a *value proposition*, a compelling reason offered to prospects to buy financial products and to buy them from you.[8]

value proposition

Positioning Your Personal Brand

As discussed in chapter 1, your financial plans, insurance policies, and mutual funds cannot set you apart from the competition. If an insurance policy includes a much coveted feature, you can be sure that other advisors will soon offer it to their clients. But one thing is virtually impossible for your competitors to copy—you! You are *the* differentiator—your personal brand. Your personal brand is an amalgamation of the qualities, characteristics, personal experiences, and skills that make you who you are. It is critical to identify the relevant, unique aspects of your personal brand and position them appropriately in your target market. The process includes the following steps:

> **One thing is virtually impossible for your competitors to copy—you!**

- Identify a relevant, unique position.
- Put it in writing.

- Test it.
- Establish your position.
- Monitor and protect it.

Identify a Relevant, Unique Position

Your relevant, unique position is determined by analyzing all of the information that you have gathered about your target market to this point. Specifically, organize and analyze the following:

- what the target market needs and values
- your resources, skills, and so forth that position you well in the target market
- how your competition has positioned itself in the target market

Look at the intersection of the characteristics of your personal brand (your life experiences, skills, hobbies, interests, and so on) and what the target market needs and values *most*. Then examine how your competitors are positioned to address that same intersection of personal characteristics and target market needs. Identify their weak and nonexistent positions, and choose the one you are best suited to address. This should all sound familiar because it is the process to select a target market (see pages 2.19 to 2.23). In other words, by completing the target marketing process as described, you will have done most of the work to identify a relevant, unique position.

What remains is to select the aspects of your personal brand that you will emphasize from the possible positions you have identified. There are at least two aspects to which you should give your attention: deliverables and public image.

deliverables

what you Promise !

Deliverables. *Deliverables* are the products and service experiences you promise and provide to your prospects and clients. As you carve a position in the target market, look for ways that you can use your skills and background to offer a unique approach to or perspective of the planning process (see examples 1 and 2, respectively). Consider any complementary products and services you can provide because of your personal resources and skills (see example 3). Just because they are complementary does not mean you should not consider charging for them (assuming you are permitted to do so).

Example 1:	Marge O'Hara is a financial planner with a background in psychology and counseling. Her target market is young women business owners. She has found that they are open to a coach or mentor. Marge decides to position herself as a life coach.

Example 2:   Pat Smith is a former human resources employee of General Motors and thus understands the company's employee benefits package. Given the number of GM employees in the area, Pat decides to position himself as a financial advisor who specializes in retirement planning for GM employees.

Example 3:   Henry Jones is a former English teacher and guidance counselor with many contacts who are teachers and college admissions counselors. Parents in his target market have asked for help in assessing their children's educational progress, completing financial aid applications, and finding scholarships. Henry is considering offering information about and access to resources to assist parents with these issues. Thus, he will position himself as a "more holistic" education planner.

public image

Perception

Guard your character; your reputation is priceless.

Public Image. *Public image* is how people—your target market in particular—perceive you. It is composed of all the characteristics that make you recognizable and memorable. As much as you hope it is your financial expertise that causes people to connect with you, more often it is your personal traits and/or the reputation of your character that attracts people to you.

In fact, Authors Peter Montoya and Tim Vandehey contend that you should not market your abilities as a financial advisor as much as you should "market yourself by highlighting a set of characteristics—your military experience, your travels, your family's local history, your previous career racing cars, whatever—that makes you stand out from the crowd, and then hammer those characteristics home over a period of time through consistent marketing efforts."[9]

Obviously, how much you should use your personal traits to position your personal brand will depend on your target market. Some target markets will not respond to such an approach. But others will find personalization of their financial advisor quite appealing.

Regardless of whether you buy into Montoya and Vandehey's mantra of personal marketing, one thing about your public image that is mission critical is the reputation of your character. As discussed in chapter 1, character qualities such as punctuality, honesty, politeness, and responsiveness should be a part of every advisor's brand. Although your character is not something you can easily advertise and publicize, it is crucial to your long-term success. Guard your character; your reputation is priceless.

Put It in Writing

The culmination of your analysis of personal characteristics and target market needs should result in two written statements: a positioning statement and a value proposition. If you are targeting two or more markets that have very different needs (for example, parents of young children and seniors—not that such a selection of target markets is recommended), you may have to create positioning statements and value propositions for each.

Positioning Statement. The positioning statement consists of one short paragraph. It is a private declaration of how you want your target market to perceive you—not necessarily how members do right now. The statement should guide all that you do—from prospecting methods, advertisements, and planning approaches to postsale service. It is the basis for any value proposition you make to prospects in your target market.

Your positioning statement should answer the following questions:[10]

- Who are you?
- What business are you in?
- For whom?
- What do you do?
- With whom do you compete?
- What will differentiate you from them?

> **Your positioning statement should be reflected in all that you do—from prospecting methods, advertisements, and planning approaches to postsale service.**

Example 1:	Marge is a former marriage counselor turned financial planner who works with young women business owners to implement business and personal financial plans. Unlike other financial planners, Marge coaches her clients to hone their skills for making financial decisions that enable them to achieve their life goals.
Example 2:	Pat is a financial advisor who works with General Motors employees to develop and implement retirement plans. Unlike other mutual fund brokers, Pat understands the GM benefits package and can help GM employees structure optimal retirement plans.
Example 3:	Henry is a financial services representative who helps middle-class parents accumulate assets for their retirement and children's education and protect their families' financial futures with life and disability income insurance. Unlike other life insurance and mutual fund providers, Henry offers information and access to resources that will help parents manage their children's educational progress and plan the logistics related to applications to schools, for financial aid, and for scholarships.

> **Write your positioning statement in terms of the tangible results you commit to deliver.**

Value Proposition. From your positioning statement, you can create your value proposition. Write it in terms of the tangible results you commit to deliver to prospects and clients. Unlike the positioning statement, the value proposition is for external communication to your prospects and other interested parties. In addition, it is not aspirational but based on what you can deliver currently.

Ideally, you can boil your value proposition down to a few sentences, articulating the relevant benefits that fall into such categories as

- improving the prospect's financial position through increased income, decreased expenses and taxes, or both
- optimizing the prospect's productivity
- creating emotional satisfaction or relief
- increasing the prospect's competitive advantage (for business owners)

Example 1:	Marge is a financial planner who teaches women business owners the skills for making sound financial decisions that help them achieve their life goals. Marge's commitment to her clients is a plan and the necessary coaching to generate the income they need to achieve their current and future life goals.
Example 2:	Pat has developed his financial planning process to serve the needs of GM employees. His 16 years as a human resources employee at GM give him extensive knowledge of GM's benefits package. Pat can help his clients take full advantage of their employee benefits and design integrated retirement plans that will mean a smoother and more enjoyable ride during the best years of their life.
Example 3:	Henry turns saving for both retirement and children's education from an either-or to a both-and proposition. He also helps clients insure these plans against the death or disability of an income earner. Furthermore, Henry's former experience as a high school guidance counselor can mean less stress in executing the logistics of applying to schools, for financial aid, and even for scholarships. With Henry, clients can fund, protect, and facilitate their children's educational futures with less stress and more confidence.

Take time now to define any complementary deliverables (ancillary products and services) you offer, even though they are technically not part of the value proposition. These products and services are the equivalent of the stylish in-store café, hospitable seating arrangements, spacious aisles, and spotless restrooms that contribute to the distinct shopping experience offered by clothing retailer Nordstrom's. None of these amenities are the sole reason customers shop at Nordstrom's; nor are your ancillary products and services *the* reason people choose you. They do, however, reinforce the customer-centric brand you are cultivating. Again, the focus is on relevance to the target market and differentiation, when possible.

Example:	Henry is considering offering the following complementary services to his clients and prospects:

- assisting with completion of financial aid online
- sponsoring a website with evaluations and aids, giving parents information about what their children should have mastered at various grade levels, and providing supplementary activities to strengthen their children's abilities
- teaching adult education classes sponsored by the school district that guide parents through the process of preparing their children and planning for postsecondary education
- offering seminars to high school juniors and seniors on writing essays for college applications
- providing a link with SAT preparation tutors
- sponsoring seminars taught by financial aid officers for parents of middle school and high school students

Test It

Before you start pitching your value proposition and any complementary deliverables in earnest, evaluate them. Then ask a few of your best clients and centers of influence from this market segment for their feedback. Ask such questions as

- Does my position reflect the most important needs and values of the target market?
- Do I offer meaningful differences from what the competition offers?

- Will the target market be attracted to my value proposition (position statement plus meaningful differences)?
- Are the elements of any complementary deliverables cost effective in terms of time, energy, and money?
- Most important, can I deliver?

After feedback and evaluation, make any adjustments to your value proposition and complementary offerings.

Establish Your Position

Establishing your position involves two basic principles: creating awareness of your personal brand and value proposition and delivering on your promises.

Members of your target market need to know who you are and what you do. This is all about creating awareness of your personal brand and thus your value proposition. In chapter 4, we will look at the most common, basic methods of creating awareness: social mobility, personal brochures, websites, newsletters, and advertising. Start with these foundational methods and innovate. The possibilities are endless, limited only by your creativity (and compliance, of course). In addition, you must make sure selected methods are consistent with your target market's expectations and needs as well as your personal brand.

Above all else, the most important aspect of establishing your position is to deliver what you promised in your value proposition. As the old adage goes, "Say what you'll do and do what you say."

> **Above all else, the most import aspect of establishing your position is to deliver what you promised in your value proposition.**

and Consistent and repititive [handwritten marginalia]

Example: What comes to mind when you think of a Honda? For many car owners, Honda means quality. It is positioned as a reliable, durable, gas-efficient, well-engineered, driving machine.

It was not always that way. If you had asked that question in the 1970s before the gas shortage, the answer might have been, "cheap, low quality, and tinny." This poor reputation existed because of the prior bad experience American consumers had with Japanese products. But when the gas crisis hit in the 1970s, the American consumer began to value fuel economy, something the Honda Civic offered that American cars did not. People began to buy Hondas, and to their surprise—and much to the chagrin of American automakers—they found the cars to be very reliable and well-engineered.[11]

*track
adjust survey*

For the past quarter of a century, Honda has forged its hold on a quality position in the automobile industry by delivering reliability, durability, fuel economy, and design consistently as advertised.

Monitor and Protect Your Position

Finally, if you have success in the target market, you will need to monitor and protect your position. As emphasized several times, success will attract competitors. Vigilantly monitoring your target market and competition will enable you to make necessary changes to your value proposition, which will keep you in a dominant position.

Example:	When the first powerful spreadsheet application Lotus 123 came out, it dominated the market. That did not deter Microsoft from attempting to unseat the champion. Now that the dust has settled on this epic duel, Microsoft's Excel is the clear winner.
	There were a variety of factors in Lotus's demise, but they all stemmed from one or all of the following: ignoring environmental changes (the change in operating systems from clumsy DOS to the current what-you-see-is-what-you-get [WYSIWYG] environment of Windows), overlooking the strategy of "insignificant" competition, and/or formulating a deficient strategy for maintaining its position.
	Microsoft, on the other hand, is doing what it can to maintain its position. Although its core position has not changed (an integrated suite of computer applications commonly used in an office or home), its successive iterations add features that continue to differentiate Excel from its competition.

You must keep one eye on your target market. By cultivating relationships in the target market, you will have access to the information you need to make your position more responsive to the market.

Keep your other eye on the competition. Assess their movements within the target market and around it. Although they may not try a head-on attack, they may try to outflank you. This is what Microsoft did effectively to dethrone Lotus 123. It changed the operating system standard and integrated Excel with its word processing program, Microsoft Word. Whereas Lotus

When you have an understanding of any changes in your target market and in your competition, you are ready to make good strategic choices to protect your position.

123 was paying attention to the spreadsheet market, Microsoft saw the spreadsheet as part of an integrated office suite. Had Lotus realized this sooner, it could have created its office suite before it had lost its market share. Your monitoring and analysis of your competitors will not be as complex as in the Lotus 123 example, but it is still important to sense your competition's priorities and intentions.

When you have an understanding of any changes in your target market and in your competition, you are ready to make good strategic choices to protect your position. As you develop your strategy, consider the following:

- Keep your name in front of the target market.
- Build a reputation as an expert advisor within your target market.
- Cultivate relationships with key members of your target market.
- Form strategic alliances with other key noncompeting advisors who serve the same target market.
- Look for new, relevant, and meaningful ways to distinguish yourself from the competition.

Personal Branding Is for You!

The next time you are dealing with a service provider or salesperson who is extraordinary, ask yourself, "What did this person do that distinguished himself or herself from the competition?" That is the power of personal branding. Consider the following experience of one of the authors of this textbook:

A number of years ago while dining at an upscale restaurant in Columbus, Ohio, I was fortunate to be waited on by Maxine. Although the wait staff at this restaurant is known for its excellent service, Maxine differentiated herself from the other servers and, for the record, from any other server I have ever known.

Because my wife and I had made reservations, Maxine knew us by name. In passing, or so it seemed, she asked me for my business card, which I gave her. I was shocked, however, when she handed me *her* business card. A restaurant server with a business card? Highly unusual!

Curious, I asked her how many business cards she has collected. She told me that she has a book with more than 1,500 of them; when people come to the restaurant again, she knows each person by name.

Maxine provided excellent service all evening. At some point, she asked if she could take a picture of my wife and me. A few days later, I received a thank you note with the photograph enclosed. Wow! Maxine turned eating into a memorable experience.

Since that experience, I have eaten at this restaurant dozens of times—and guess which server I ask for? And I am not the only one. It probably does not surprise you that Maxine is the top wage earner at the restaurant. Maxine's service is not only excellent, it is also memorable because it stands out from the crowd—that is what personal branding is all about!

If it works for a restaurant server, why not for financial advisors?

What will you do to distinguish yourself from your competition? How will you be memorable?

Positioning Your Products

Personal branding is not enough. Advisors can stand out from the competition. Their target market may think highly of them. But advisors still need to help prospects see the need for their products and services. Just as advisors must provide a compelling reason for choosing them as advisors, they must also provide a compelling reason for purchasing their products. Why do prospects need a financial plan? Why do they need life insurance? Why do they need mutual funds?

Logic and Emotion

At the heart of positioning is persuasion. Advisors want to persuade people to have a certain perspective—in this case, the need for products or services. According to Aristotle in his classic *The Art of Rhetoric,* there are three elements to a person's argument (persuasive reasoning): ethos, logos, and pathos. Ethos refers to the person's character. Is the person credible? Knowledgeable? Trustworthy? Logos refers to the facts of an argument. Is the argument logical? Finally, pathos refers to the emotional appeal of the argument. How does it make the listener feel? We have already dealt with ethos in explaining personal branding. Let us look at the other two aspects: logic (logos) and emotion (pathos).

The Logical Aspect. To quote the immortal Sergeant Friday, "Just the facts, ma'am." All persuasion must be rooted in the truthful presentation or reflection of the facts. It is important to identify the relevant facts that pertain to the prominent financial needs of the target market. Examine the financial needs you listed in the first step of the target marketing process. What are the facts concerning household income at the death of a wage earner in your target market? What types of current expenditures financed by that income and important to your target market are affected? For example, if you are target marketing parents of children in a private school, the inability to afford tuition could be a relevant logical outcome of the death of a wage earner.

Statistics are often used to present a logical appeal. Be careful, however, in your use of statistics. To safeguard your credibility, use only statistics that have a reliable source. If a book or website quotes a study, it is advisable to find the original source if you can. Also, be careful not to overuse statistics. They will detract from your appeal.

> Be careful not to overuse statistics. They will detract from your appeal.

The Emotional Aspect. Recall that a buyer responds to external stimuli that trigger a sense of need. Triggers are usually based on an emotional response. Therefore, the positioning of your products must appeal to your prospects' emotions. In selecting reasons for purchasing the products, advisors need to examine the key emotional reasons to buy that flow from the financial need that was created. Maslow's hierarchy of needs (discussed in chapter 1) is especially helpful in this process.

Creating a Position

Once you have identified the key logical and emotional reasons for buying your products and services, you should create a way to communicate this position to prospects in the target market. As life insurance great Ben Feldman puts it, "What otherwise would be dull, complex actuarial legalisms are transformed into attractive 'packages of money' that anyone can understand. Packages that solve a problem. Packages that do a specific job; a job that the prospect needs done."[12]

According to Feldman, positioning a product should accomplish at least three objectives:[13]

- First, it should help prospects see their financial needs clearly. This takes a few well-placed facts that stir up prospects' emotions. Consider creative ways to pique prospects' curiosity. Imagine, for example, that you are a prospect who received a two-foot canoe paddle in the mail from an advisor who has been trying to set an appointment with you. How would you react to the note, "Will your current financial strategies leave you up the creek without one of these?"[14] You might be a little curious!

- Second, prospects should have a general sense of the cost of doing nothing compared to the cost of addressing their financial needs. In other words, you must help them see that "a stitch in time saves nine," that it is more costly to do nothing than it is to tackle the problem now while they can.

- Third, the positioning of products should motivate prospects to take action. It should be a trigger to move them from awareness of their need to searching for information and being willing to solve their problems. This is accomplished by statements and questions that cause prospects to evaluate the harsh reality of the status quo.

| *Example:* | Ben Feldman targeted business owners who had cash flow problems related to their businesses and/or estates. Here is one of the ways he positioned his products: |

"You spent 30 years putting this company together. I've never known anybody who had a lease on life; do you? No? Then it's only a question of time until you walk out and Uncle Sam walks in. Know what he wants? Money! And he has a way of getting it. First. Not last. Could you, right now, give me 30 percent of everything you own—in cash— that's the least Uncle Sam will take—without hurting a little bit? Do you think it would affect your credit position? Your working position?"[15]

These are the facts he used to accomplish his objectives: Everyone dies. The federal government assesses an estate tax at a taxpayer's death. The business owner will have to pay an estate tax. The funds for that tax must come from the current cash flow or a liquidation of assets.

The emotional appeals he uses are as follows: The business owner has poured his or her life into the company. It would hurt the company if the business owner had to give the federal government 30 percent of what he or she owns. The owner would hate to have the company go under because of his or her death.

Granted, you will probably not be able to accomplish all of these objectives in one mailer or advertisement. The important thing at this point is to identify what you must communicate to accomplish your objectives. Remember, too, what works for one advisor and his or her target market may not work for you. As with everything else, your position must reflect facts and feelings that are relevant and important to your target market. Finally, expect this creative process to take some time. Feldman was known to work for 6 hours on a 30-second statement.[16]

CONCLUSION

Successful, cost-effective prospecting utilizes the target marketing process. This process identifies target markets—groups of people in which members have common characteristics and needs that distinguish them from nonmembers. The groups are sufficiently large and, ideally, have a communication network.

The first step of the process is to define products in terms of the financial needs the advisor is able to meet based on the products he or she sells, the advisor's skills and knowledge, and his or her personal traits.

The second step requires the advisor to identify and segment his or her natural markets. Natural markets are groups of people to whom the advisor has an affinity or access. The advisor can determine his or her natural markets by completing a natural markets checklist, analyzing his or her past personal production, and examining the results of these tasks for overlap. The analysis of the advisor's past personal production utilizes the basic target marketing principle of market segmentation, dividing a market by geographic, demographic, psychographic, and behavioristic variables in an attempt to identify market segments that can be potential target markets. The end result is a few potential target markets from the advisor's natural markets and, if desired, an ideal client profile.

Selecting a target market is the third step. In this step, the advisor creates selection criteria that help measure the fit of resources to a market segment's needs, level of income, and level of competitiveness, in addition to any other criteria the advisor wishes to use. Next, the advisor conducts research to evaluate each potential target market in regard to the selected criteria. Initial resources through which to conduct research include the Internet, the local chamber of commerce, and the library. At some point, the advisor will want to interview members of the potential target markets, as well as other advisors and professionals who serve these target markets. Then the advisor decides on a marketing coverage strategy and determines which and how many target markets to pursue.

The last step in the target marketing process is positioning the advisor's personal brand and products. The advisor identifies his or her unique position, a selection of relevant deliverables and public image characteristics of the advisor's personal brand, and creates a positioning statement that will guide all marketing efforts. In addition, the advisor develops a value proposition based on the positioning statement, which encapsulates the deliverables associated with working with him or her. Finally, the advisor considers how to position his or her products, combining logic and emotion to develop powerful statements and questions that generate prospects' need for the advisor's products. The positioning of a personal brand and products is important in creating awareness, the subject of chapter 4.

CHAPTER TWO REVIEW

Key Terms and Concepts are explained in the Glossary. Answers to the Review Questions and Self-Test Questions are found in the back of the textbook in the Answers to Questions section.

Key Terms and Concepts

mass marketing	access
general market	market segment
target marketing	geographic variables
target market (niche market)	demographic variables
personal traits	psychographic variables
natural market	behavioristic variables
affinity	ideal client profile

marketing coverage strategies   deliverables
positioning   public image
value proposition

Review Questions

2-1. Explain the difference between target marketing and mass marketing.

2-2. Target marketing enables the advisor to gain expertise in one or a few market segments. Identify five of the advantages that result from such a focused approach.

2-3. Explain why it is important for an advisor to define his or her products in terms of the prospect's financial needs and the advisor's personal traits.

2-4. Give an example of (a) affinity and (b) access as they relate to natural markets.

2-5. Provide a detailed explanation of the four types of segmentation variables used in market segmentation.

2-6. Explain the selection criteria for choosing a target market.

2-7. Describe the five basic market coverage strategies.

2-8. Discuss the two aspects of persuasion—logic and emotion—as they relate to positioning a product.

2-9. Identify and explain three objectives that positioning a product should accomplish.

Self-Test Questions

Instructions: Read chapter 2 first, then answer the following questions to test your knowledge. There are 10 questions; circle the correct answer, then check your answers with the answer key in the back of the book.

2-1. Which of the following statements regarding target marketing is correct?

(A) In target marketing, an advisor aims his or her products to a general market.

(B) Target marketing ignores the differences between groups in a general market.

(C) Target marketing is essential only to the advisor's short-term survival.

(D) The objective of target market is to identify groups of qualified prospects with common characteristics and needs.

2-2.   Which of the following statements regarding defining the product is correct?

(A)   The advisor should list all the products he or she could potentially sell.
(B)   The advisor should assess his or her proficiency in the necessary skills and knowledge related to each prospect need identified.
(C)   The advisor should list all of the financial needs the product can meet, regardless of the advisor's expertise.
(D)   The advisor should ignore any of his or her bad personality traits; they are irrelevant.

2-3.   Which of the following market coverage strategies entails marketing one product to a single market segment?

(A)   product specialization
(B)   single-segment concentration
(C)   selective specialization
(D)   full-market coverage

2-4.   Which of the following statements regarding positioning a personal brand is correct?

(A)   It requires a lot of advertising to the general market.
(B)   It relies heavily on the financial plans, insurance policies, and so forth that an advisor sells to differentiate himself or herself from the competition.
(C)   It is the advisor's attempt to occupy a distinct place in the mind of the target market.
(D)   An advisor should include a positioning statement on his or her business card.

2-5.   A natural market is a group of people to whom the advisor has which of the following?

I.   A natural affinity
II.   Access for reasons other than direct personal knowledge or acquaintance

(A)   I only
(B)   II only
(C)   Both I and II
(D)   Neither I nor II

2-6.   An advisor might conduct a market survey for which of the following purposes?

I.   To determine how members communicate with one another
II.   To confirm or discover relevant distinguishing characteristics of the market segment

(A)   I only
(B)   II only
(C)   Both I and II
(D)   Neither I nor II

2-7.   Which of the following statements regarding an ideal client profile is (are) correct?

I.   The advisor must analyze his or her natural markets to create such a profile.
II.   The advisor should not share his or her ideal client profile with prospects or clients.

(A)   I only
(B)   II only
(C)   Both I and II
(D)   Neither I nor II

2-8.   All the following are advantages of target marketing EXCEPT:

(A)   It enhances an advisor's referability.
(B)   It leads to increased customer loyalty.
(C)   It eliminates having to call prospects for appointments.
(D)   It results in higher profits.

2-9.   All the following are types of segmentation variables EXCEPT

(A)   psychiatric
(B)   demographic
(C)   behavioristic
(D)   geographic

2-10.   All the following are criteria for evaluating the income potential of a market segment EXCEPT

(A)   referrals
(B)   size of market segment
(C)   practice model
(D)   income from product purchases

NOTES

1.   Marian Burk Wood, *The Marketing Plan: A Handbook* (Upper Saddle River, NJ: Pearson Education, Inc., 2003), p. 57.
2.   Philip Kotler, *Marketing Management*, 11th ed. (Upper Saddle River, NJ: Prentice Hall, 2003), p. 287.
3.   Canterbury Financial Group: "Ideal Client Profile," accessed September 8, 2006, from www.canterburygroup.com/idealclient/index.html.
4.   Wood, p. 64.
5.   Headstart, "Region III Demographics," accessed September 20, 2006, from www.headstartinfo. org/English_lang_learners_tkit/Immigration/regionIII/pennsylvania.htm#table2.
6.   American Marketing Association: "Dictionary of Marketing Terms," accessed September 25, 2006, from www.marketingpower.com/mg-dictionary-view1822.php.
7.   Kotler, p. 308.
8.   Ibid.
9.   Peter Montoya and Tim Vandehey, *The Brand Called You: Personal Marketing for Financial Advisors* (Costa Mesa, CA: Millennium Advertising, 1999), p. 30.
10.   Harry Beckwith, *Selling the Invisible* (New York, NY: Warner Books, Inc. 1997), pp. 113–114 (an adaptation of questions Beckwith recommends).
11.   Edmunds.com, "Manufacturers Histories: Honda," accessed September 28, 2006, from www.edmunds.com/reviews/histories/articles/65329/article.html.
12.   Andrew H. Thompson with Lee Rosler, *The Feldman Method* (Chicago, IL: Dearborn Financial, 1989), p. 30.
13.   Ibid, p. 32.
14.   Pamela Yellen, "Web Site Exclusive: Secrets of Successful Prospecting," *The Advisor Today*, August 2001, accessed October 5, 2006, from www.advisortoday.com/archives/2001_august_web.html.
15.   Thompson and Rosler, p. 50.
16.   Ibid, p. 31.

3

Selecting the Prospect

Learning Objectives

An understanding of the material in this chapter should enable the student to

---

3-1. Describe the three types of prospects and the prospecting sources for each type.

3-2. Explain the basic concepts of the 10 prospecting methods covered in this chapter.

3-3. Describe three important aspects of a prospecting system.

3-4. Explain three important record-keeping tasks related to prospecting.

Chapter Outline

INTRODUCTION

Chapter 2 outlined the advantages of target marketing and discussed the process for identifying and selecting target markets from within your natural markets. Completing the steps explained in that chapter will result in defining the group or groups of people with whom you feel working will be both profitable and enjoyable. This chapter describes the next piece of the prospecting process: selecting prospecting sources and effective prospecting methods in order to identify names and contact information for specific prospects from your target markets.

prospecting sources
prospecting methods

Often, the terms prospecting sources and prospecting methods are used interchangeably. Although there is often overlap between the two, technically they are different. The best way to understand the difference is to think of the following relationship between prospects, prospecting sources, and prospecting methods. Prospects are people you are looking for, *prospecting sources* are where you will find prospects, and *prospecting methods* are the different ways to access the prospecting sources for specific names and contact information.

Understanding prospecting sources will help you select appropriate methods to identify and select individual prospects.

Perhaps a crude analogy will help. Think of prospects as oil. Oil can be found below the ocean floor, in the desert, underneath mountains, and so on. These are the sources for oil. The methods for accessing the oil will vary due to the characteristics of the source. The same holds true with prospects. The nature of the source affects the methods to obtain prospects from it. Thus, understanding prospecting sources will help you select appropriate methods to identify and select individual prospects.

The following discussion describes prospecting sources, explains prospecting methods, and reviews the elements for organizing your prospecting efforts. Let us start by looking at prospecting sources.

PROSPECTING SOURCES

There are three types of prospects:

- people who know you favorably
- people recommended by those who know you favorably (referred leads)
- people who do not know you at all

We will discuss the prospecting sources that correlate with each of these three types.

People Who Know You Favorably

The best source for prospects comes typically from the category of people who know you favorably, basically prospects from your natural markets. By identifying and selecting target markets from your natural markets, you will have already pinpointed a few prospects in this category. Compile a list of the people who know you favorably. Then look at your target market and/or ideal client profile and circle the names of anyone who substantially meets the profile (obviously, no one will meet all of the criteria; use your judgment). The following are a few suggestions of sources from this category.

> People who have bought from you in the past already know you and have expressed confidence in your abilities. They will probably grant you an interview and listen to you.

Buyers and Current Clients

Even if you have been in the financial services business only a short time, comb your client database for buyers and clients who are members of your target markets. Cross-selling anyone who has bought from you in the past is a natural step. These people know you and have already expressed confidence in your abilities. They will probably grant you an interview and listen to you.

Example: When you approach current prospects from among your clients, try saying something like:

"John, I believe you have been pleased with the service I have provided you and your family. I would like the opportunity to explain another service I can provide—(business insurance, retirement, estate, or other) planning."

"Mary, I need your help. As you know, my company and I have worked hard to meet your personal financial needs in a cost-effective manner. We also have an excellent reputation for helping businesspeople solve the insurance problems they face. I'd like an opportunity to be of service to you in this area."

Friends and Family Members

Many times advisors overlook this source of prospects. For some advisors, the closer they are to someone, the harder it is to approach that person for fear of jeopardizing the relationship. To help negotiate this pitfall, remember the value of what you sell. If you sell life insurance, could you face a friend's family if that friend dies without life insurance? Also remember to keep the discussion of your business separate from the course of everyday life. Perhaps make it a rule not to bring up business at family and social gatherings. Instead, call your family and friends like you would any other prospect.

Businesses You Patronize

It is a rare day when you or a family member does not patronize a business. Is there any reason why you should concede those business owners who are members of your target markets to some other advisor? Who has a better right than a customer to ask for an interview? Why shouldn't a business owner devote some attention to a good customer who wants to discuss something important with him or her?

The businesses that you and your family patronize are a good reservoir of personal and business prospects. These are natural markets that many advisors ignore to some degree. Certainly, no advisor should expect to make a life insurance sale, for example, in exchange for buying a new television set or having a suit cleaned. But there is no reason why you can't ask the business owner for an interview and for referrals. Others will ask. Why shouldn't it be you—a customer of the prospect? If you are a regular customer, the business owner may be pleased to have a chance to reciprocate. Most business owners prefer doing business with their loyal customers rather than with strangers. Can you imagine their frustration (and yours) if—shortly after purchasing life insurance from a stranger—they learn that you handle business insurance (or some other area of planning)?

> **There is no reason why you can't ask the owner of a business you patronize for an interview and referrals. Others will ask. Why shouldn't it be you?**

Community or Organization Contacts

Think about the places in your community that you have contact with people in a favorable manner. This could be the local chamber of commerce, local government, church, synagogue, mosque, charitable or volunteer organizations, PTA, and so forth. If you have good standing within the community or organization, the people with whom you have contact who are members of your target market are good prospects. Chapter 4 discusses how to create awareness and build prestige in community and organizational contexts.

People Recommended by Those Who Know You Favorably

The second category of prospecting sources is people recommended by those who know you favorably. We will look at two sources for these types of prospects.

Nominators

nominator

Essentially, take anyone from the previous category of people who know you favorably and ask him or her to be a *nominator*, someone who provides referrals by giving names to you and/or by giving your name to others.

In addition, advisors whose practice does not compete with yours are excellent nominators. This is especially true when the advisor works with the same target market but sells different products. However, it also works when the advisor sells the same products but has a noncompeting niche.

Example:	Zip Financial Services works only with prospects who have at least $500,000 in assets. You work with prospects who have between $100,000 and $500,000 in assets. People with less than $500,000 in assets often contact Zip for help. Instead of turning them away, Zip could refer these prospects to you.

Orphans Assigned by Your Company

orphan

An *orphan* is an existing account that you did not personally produce and for which there is no active advisor assigned. Orphans are created when advisors in your company retire, die, are reassigned, or leave the company. The clients and buyers they personally produced and serviced are then reassigned to other advisors in the company.

Every now and then, you will be given a list of orphans. Take advantage of them. Although orphaned clients and buyers are total strangers to you, they have an affiliation and relationship with your company. Your company's reputation acts in the same way a nominator's would, providing a favorable basis for setting an appointment to meet with you.

People Who Do Not Know You at All

You cannot afford to ignore people you do not know as prospects. Without this type of prospect, you could potentially run out of names.

Obviously, your objective is to work mainly with the first two types of prospects: people who know you favorably and those whom they refer. But you cannot afford to ignore people who do not know you at all. Without this type of prospect, you could potentially run out of names.

If you are in a position in which you can wait to contact prospects who are total strangers, a good approach is to find ways to move them into one of the first two categories. For example, look for ways to meet them and get acquainted. Or find a mutual friend or business contact who would be willing to introduce you. In either case, create a presence in the community and organizations in which they live and participate. (This will be discussed more in the next chapter.)

There are many prospecting sources in this category, including

- directories and membership lists
- local papers and publications
- public records
- business information sources
- direct marketing lists

Directories and Membership Lists

Directories and membership lists give you names and contact information on groups of people with something in common. The lists could be for a church, synagogue, mosque, social organization, a business association, civic organization, and so on.

You should exercise great caution in how you utilize names from directories and membership lists (or any of the other sources described in this section). Using the directories or lists to contact people indiscriminately could potentially harm your reputation. The names and contact information are not collected to give salespeople a prospect list. Other less intrusive and subtle uses of these lists will be more effective in the long run.

Local Papers and Publications

Local papers and publications often have information about local residents who are celebrating births and weddings and about businesses that are publicizing promotions and accolades. Depending on your target market, you may find this information beneficial. For example, if you are targeting small and mid-size businesses, you will often see announcements regarding these firms in hometown papers. You can learn a little about the business, who owns it, what it does, and so forth, giving you clues about how you may be able to gain access to the business owner on a favorable basis.

Public Records

Again, depending on your target market, public records may provide another source for leads. Through a court house or local records department you can obtain lists of marriage licenses, voter registrations, real estate records and deeds, divorce records, new corporation registrations, and so forth.

Business Information Sources

For advisors whose target markets include businesses and business owners, business information sources are another possible source of prospects. For example, Dun and Bradstreet sells a service that enables you to craft your own list by industry, location, and credit risk. If you are targeting tax-exempt organizations, you can use the online version of IRS Publication 78 to search for organizations in your area.

Direct Marketing Lists

Lists are collections of names and contact information (phone numbers, addresses, and sometimes e-mail addresses). They can be generated by your company or a third-party vendor. Reputable lists are based on various characteristics (income, assets, age, and so on) garnered from completed warranty cards, contests, and so forth.

Prospecting Sources for Strangers

Directories and Membership Lists
- White pages (for people)
- Yellow Pages (for businesses)
- City business directories
- State industrial directories
- State medical directories
- State law directories
- Insurance (HMO, POS, PPO) network directories
- Business directories (Polk)
- Trade association or professional directories
- Chamber of Commerce membership lists
- Membership lists for local civic, sales, or social organizations

Public Records
- Marriage licenses
- Voter registrations
- Real estate records

Local Papers and Publications
- Birth and wedding notices
- Notices in newspapers and legal publications giving new business firms, changes of ownership, additions to plants, and so forth
- Trade association or professional journals
- Legal notices

Business Information Sources
- Dun and Bradstreet
- Hoovers
- Standard and Poor's Register of Corporations
- Moody's Investor Services
- Publication 78 (online list of IRS-approved charities)
- Thomas Register and Thomas Regional (www.thomasnet.com)

Direct Marketing Lists
- Company-provided lists
- Third-party vendors

When purchasing a list from a commercial vendor, the better you can describe your target market, the better the list will be.

Be cautious, however, when purchasing lists from commercial vendors. The lists can be expensive if you are not careful. The better you can describe your target market (use any applicable segmentation variables), the better the list will be. Also, be sure that the list has been "scrubbed." This means that any do-not-call names and undeliverable addresses have been eliminated. Additional information you might want to find out is as follows:

- When was the list last updated?
- What is the undeliverable rate?
- What is the source of the leads?
- Does the list have current phone numbers?
- Do you have any exclusive rights?
- Does the vendor have a list of repeat customers? (This indicates higher quality.)
- Does the vendor use a 5- or 9-digit zip code? (5 digits are often undeliverable)
- Does the vendor delete duplicate entries? (This is called "merge purge.")
- Does the vendor delete incomplete names?

The bottom line is that you need to plan, and you need to ask a lot of questions before you buy any given list.

PROSPECTING METHODS

> **Identifying potential sources of prospects from your target markets does not guarantee a steady stream of qualified prospects. You must gain access to those prospects.**

Identifying potential sources of prospects from your target markets does not guarantee a steady stream of qualified prospects. With each prospecting source, you must determine an effective way to gain access to the prospects in a manner that, ideally, produces a constant flow. The prospecting methods discussed in this section are as follows:

- contact database searches
- servicing contacts
- networking
- referrals
- centers of influence
- Tips Clubs
- personal observation
- cold canvassing
- direct mail
- seminars

Before reviewing these prospecting methods, it is important to give you some insight into how this topic is approached. This will help you select appropriate prospecting methods to use in your practice.

First, the type of prospecting source will influence the type of method you use to obtain a constant stream of names. Some methods are more appropriate for one type of source. However, because selecting an appropriate method will depend on the characteristics of the specific source, what follows is a discussion of prospecting methods without a direct reference to the source or sources for which it is considered "most suitable."

Prospecting requires some creativity. By not pigeonholing a method to one type of source, you can evaluate its merit without any bias. Frankly, it probably will be clear to which source a particular method will apply.

Second, the personal nature of the business means that personal methods for prospecting are emphasized. You will find that the methods toward the beginning of this discussion are geared more to prospects recommended by people who know you favorably. The methods toward the end are more likely to be used with prospects who do not know you at all.

Finally, some prospecting methods are also effective for creating awareness and building prestige, as well as functioning as a preapproach and even an approach for an appointment. Seminars fall into this category. Although this prospecting method is mentioned in this chapter of the text, it is discussed more fully in the next chapter because of the tremendous opportunity that seminars provide to create awareness and build prestige.

Contact Database Searches

The first prospecting method any advisor who has a contact database filled with clients should start with is to run contact database searches. Using criteria from your target market and/or ideal client profile, filter the contacts in your database to find prospects among your current clients (and contacts). Keep in mind that the more criteria you use, the more restrictive your search will be and the fewer names that will result.

Servicing Contacts

Service can be initiated by the client (a change to a policy, contribution amount, mutual fund account, and so on), or it can be initiated by the advisor (an annual review, for example). In either case, your contact with the client can be the basis for identifying new needs in the client's financial situation. (This concept of prospecting by servicing is the topic of chapter 7.)

Networking

networking

Another prospecting method that you can use to tap a prospecting source is networking. *Networking* is the process of continual communication and sharing of ideas and prospect names with others whose work does not compete with yours, but whose clients might also be eligible to become your clients. Conversely, your clients may need their products too. It's a two-way street. Networking involves the following:

- sharing knowledge, resources, and contacts
- receiving advice and assistance from people you know

- giving advice and assistance to people who know you
- leveraging your time to increase productivity
- seeking out and building long-term, prosperous relationships

On the other hand, networking is not any one of the following:

- selling something
- seeking monetary compensation, donations, or funding
- a business transaction
- manipulating others

Traditional networking regards people as contacts, describing a process in which you call people you know to find out whom they know. The classic rationale is illustrated in figure 3-1 and explained as follows:

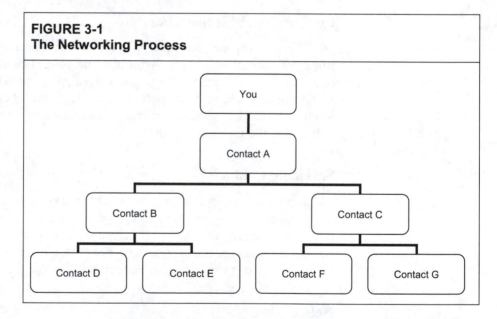

FIGURE 3-1
The Networking Process

- You know Contact A.
- Contact A knows other contacts and gives you the names and numbers of Contacts B and C. You get permission to call them, using Contact A's name as a reference.
- You call Contacts B and C, talk to them, indicating that Contact A gave you their names and that because you were in the process of seeking advice and counsel, A thought that B and C would be good people to speak with for advice or possibly any additional leads.
- You meet with Contacts B and C, and by the end of those meetings, you learn what might be happening in each of their respective worlds

and the names of four more individuals. You call them (as well as the other leads from Contact A), and the process continues. It becomes an endless chain.

- From one individual, you received six potential prospects!

Principles of Networking

On its face, networking is simple, but to gain the maximum advantages from it requires more than just collecting names and phone numbers. In fact, to reap the full benefit from networking, it is critical to approach it like the planting of a vegetable garden. These principles apply to both:

- Take a long-term view.
- Sow good seed.
- Cultivate and nurture.
- Focus on quality.
- Be patient.

On its face, networking is simple, but to gain maximum advantages from it requires more than just collecting names and phone numbers.

Take a Long-term View. Planting a garden today does not provide food today. A garden is planted today to meet future needs. In the same way, advisors must see networking from a long-term perspective. They should not expect their networking activities to produce prospects immediately.

Sow Good Seed. The gardener always reaps what he or she has sown. If the gardener plants carrots, he or she will reap carrots—not corn. Thus, if advisors want to reap goodwill, they must sow it. This means seeing networking as a means to help others by being a "gratuitous information resource provider,"[1] or what authors Bob Littell and Donna Fisher have termed "NetWeaving" in their book *Power NetWeaving: 10 Secrets to Successful Relationship Marketing.*

Littell and Fisher go on to explain, "NetWeaving is all about putting other people together in win/win relationships that will solve problems, satisfy needs, or result in new or expanded business opportunities."[2] In other words, match the person who needs a plumber with the outstanding plumber you know. Or do some pro bono work for someone.

Cultivate and Nurture. When gardeners plant seeds, they have to nurture them. That involves, among other tasks, fertilizing, weeding, and watering, which entails an expenditure of time and energy. Likewise, effective networking requires a commitment of resources (time, energy, money, and so on) to cultivate and nurture relationships. Building relationships requires more than passing and collecting business cards. It isn't the occasional phone call, the holiday greeting card, or remembering names of individuals you met at last week's soccer game. It is getting to know people and becoming their friend. Fortunately, if you choose the right kind of people, the task is rather enjoyable!

Focus on Quality. It follows from the last point that the gardener should limit the size of the garden based on the time and energy that he or she can commit to tending it. Plants that are healthy will produce more and better fruit than unhealthy ones. In other words, the gardener needs to focus on quality, too.

Like the gardener, advisors need to allocate their resources wisely to ensure the quality of their networking relationships. Remember, you cannot possibly build deep, lasting relationships with everyone whose name you can cram into your address book. One good friend who will help you is often worth more than dozens of names of people whom you barely know and who have little in common. Focus on quality.

> **One good friend who will help you is often worth more than dozens of names of people whom you barely know and who have little in common.**

Be Patient. This point bears repeating: Be patient. There is nothing humanly possible that gardeners can do to accelerate the growing process. They can water, weed, and fertilize the ground, but when the fruit is ready is a function of time and nature. Gardeners must be patient. Similarly, advisors must be patient and expect no immediate returns. Instead, advisors need to believe in the law of the harvest that a person will reap what he or she sows, letting nature take its course.

Getting Started with the Networking Process

The networking process involves making contacts and building relationships. But before we discuss more about building relationships, let's think about the dynamics involved in building contacts: the most visible and elementary part of networking

Start with People You Know. Although you may be tired of this question, you must ask it continually: Whom do you know who is a member of a target market, works with members of a target market, or knows people who can help you? Through your friends and family you probably already have a strong network in place. Some of the more typical entries on your list:

- immediate family members (parents, siblings, grandparents)
- extended family members (uncles, aunts, cousins)
- previous teachers (both high school and college)
- former classmates from high school and college
- bosses from past jobs
- your landlord
- your lawyer and accountant
- salespeople, beauticians, and barbers
- a member of the clergy

- members of your religious congregation
- your banker
- others with whom you volunteer or serve your community
- fellow members in your professional association

Extend Your Contact Base. According to Harvey Mackay, author of *Sharkproof,* to extend your contact base you should collect five contacts a day. It is difficult, but it is possible. As with prospecting itself, you can deal with total strangers or individuals recommended by people who know you favorably.

> **If you are calling a complete stranger who is busy, you had better have a compelling means of linking yourself to that individual.**

If you are calling a complete stranger who is busy, you had better have a compelling means of linking yourself to that individual. You might both belong to the same professional organization. But your commitment to that organization may be much stronger. Other members may not be as involved; they may not feel the same ownership that you do; they may belong only because their employer makes it a requirement. You do not know. Therefore, do not be surprised if you call a fellow member, and the response is, "Yes, we are both members. So what?" His or her lack of commitment to the organization makes it easy to dismiss your call.

Obviously, your best connection with a stranger is through someone else you both know on a favorable basis. Contact a stranger using the name of one of his or her good friends, and he or she is more likely to receive your call and meet with you, based on the fondness and respect for your mutual friend.

But there is an important fact about networking: *Not everyone you meet will feel comfortable putting you in touch with everyone he or she knows.* Let's face it. Suppose you were to give a complete stranger one of your friend's names, telling him or her to use you as a reference. What are you really doing? You are putting your reputation—and your relationship—on the line!

What happens if your friend does business with this stranger, who then turns out to be unreliable and untrustworthy? How does that make you look? That can be a difficult situation for you. Have you ever met people to whom you would not *dare* refer to trusting colleagues? We all have.

Some people are more trusting than others. But you have to use discretion when meeting a new person and introducing him or her to your friends. After all, the two of you just met. The contact may know a little about you, and, ideally, knows the person who gave you his or her name (that relationship can have a strong bearing on your meeting, as you will see). Still, there is something missing . . . a deeper familiarity . . . an element of trust . . . a relationship.

> **Building relationships is more powerful than networking. Here's why: Relationships are ongoing and more sincere.**

Build Relationships. Building relationships is the much deeper side to networking, and it is a little more challenging. It is also better and more powerful. Here's why: Relationships are ongoing and more sincere.

You usually have some kind of continuing contact or involvement with a person with whom you have a relationship. It does not mean you see or talk to the person every day or necessarily every week. Your association, however, does reach beyond a first-time meeting.

When you meet someone, find a way to stay in touch. Do not wait until you want something. You need to create trust, which takes time. Trust is built on continuing interaction and involvement. As people begin to know and understand you, they learn what is important to you. They evaluate your words and actions, and over time, they gain an understanding of the depth of your character to the point that they will help you instinctively.

Yes, relationships do take time to develop. Do not be discouraged by this. The good news is that there is a shortcut. If you do not have many deep relationships in place, you can start building them by leveraging the relationships that other people have. These people are what we call centers of influence (COIs). We will discuss COIs a little later.

Referrals

Referrals are the first of three prospecting methods used to generate prospects from a nominator. Successful advisors have found referrals to be the most efficient and effective means to generate a perpetual supply of prospective clients. For this reason, a large portion of this chapter is devoted to examining the referral process.

Definition

referral

A *referral* is a person to whom you are introduced in one fashion or another by someone who knows and values your work. Think of referrals as borrowed trust. Obviously, trust must be earned. That's why a relational, client-focused approach is critical in creating a steady stream of prospects.

> **Until word of mouth can fill your appointment book, you will need to solicit referrals.**

In their highest form, referrals are unsolicited. The client enthusiastically tells people about you in a compelling way, in much the same way that he or she tells friends about a must-see movie. As a result, the client's friends call *you*. Although this is idealistic, it does happen, and the more it happens, the easier selling becomes. But until word of mouth can fill your appointment book, you will need to solicit referrals.

Referral Reluctance

Unfortunately, asking for referrals is difficult for some advisors. There are many reasons for this; the most common is the fear of rejection. It is helpful to keep in mind that the whole referral process is based on trust. If you have acted in the client's best interest, without pressure, helping him or her to understand and plan for the future, go out on a limb. If the client bought your product or seemed appreciative of the process even if he or she did not buy, there is a good chance the client trusts you. The worst the client

> One thing is for sure:
> The answer is always
> no if you do not ask.

can say is no, and chances are he or she will not. But one thing is for sure: The answer is always no if you do not ask.

Principles of Asking for Referrals

The referral process, like any other process, is most effective when planned in advance. That means knowing what you will say and when you will say it. Because every person has a unique style and every situation is different, instead of giving you a prescribed set of steps, here are several principles to consider as you plan your referral process.

Be Referable. Referrals are not an entitlement. You must earn them. Successful advisors know how to make themselves referable. Providing excellent postsale service is critical to this end because many of your referrers will be current clients. Consider the following:

- Stay in touch with clients and other potential nominators. It can take time to gain someone's trust to the point where he or she is comfortable giving you referrals. Continue to bring your advice, wisdom, and other resources to the relationship.
- Provide newsletters, birthday calls, service calls, annual reviews, and friendly notes to enhance business friendships.
- Do not expect opportunities without building value for others. If you do not add value, you are no longer necessary.
- Give referrals and you will get referrals. Present yourself as a resource center (see previous discussion on networking).
- Be a teacher; find a way to help people. Determine what is important to them.

Pave the Way. Asking for referrals should not come as a surprise to your prospect or client. Pave the way as early as possible in your relationship by telling prospects you would like to help others they know tackle the same types of issues. You can do this when you start the fact-finding process.

Example: "Jane, as we go through this process and you are satisfied with the work I do for you, you might think of some others like yourself who might benefit from my services. I would love to be able to help them out as well. So if you think of anyone, please let me know."

Paving the way accomplishes several things. First, it sets the expectation that you will earn the prospect's trust by providing a service he or she values.

Second, it positions your request in terms of the prospect's point of view. Does the prospect know someone who really should sit down and think about his or her financial future? Whom does the prospect love or care about?

Third, it plants a seed that will encourage the prospect to think of names. It also creates an awareness that will make asking for referrals later an expected and natural result of the prospect's satisfaction.

Recognize Opportunities. There are many times during the normal course of your activities as an advisor when you have an opportunity to ask for referrals, including the following:

- after the sales interview is complete. You have demonstrated your service and value. You have established rapport and trust with the client.
- at policy delivery. You have satisfied your client with your professionalism.
- when you provide a service to your client
- when you conduct an anniversary review

Remembering to Ask

Write a note to remind yourself to ask for referrals. Some advisors will put the initials A.F.R. (Ask for Referrals) on a yellow post-it note and attach it to the first page of the blank application or other paperwork concerning a client.

Some say you should constantly request referrals. If you receive a warm response and good referrals, continue this approach. Be careful, however, about creating tension for the client. High pressure can lead clients or prospects to suspect you will do the same to the referral, which might have a negative impact on their relationship with the people they refer. Usually, they will not give you referrals if they feel this way.

Regardless of your philosophy, look for signs that the client values your work. The most obvious time to ask for referrals is when the client agrees to buy the product. However, look for other moments in the selling/planning process that point to the client's perception of you as a trusted advisor, regardless of whether or not he or she purchases a product. Remember to ask for referrals after you hear the client say things like:

- "I should have done this a long time ago."
- "(Ben) should know about this."
- "I didn't know that!"
- "Thanks for helping me out. I really appreciate that you took the time."

Getting Referrals (by Bill Cates)	
STEP 1	**DISCUSS THE VALUE RECOGNIZED**
Advisor	*Martha, at this point I'd like to take a few minutes to review our process thus far. Have you found this process to be valuable, and if so, in what way?*
Martha	Absolutely. First, just taking the time to really think through my situation has been helpful. I'm learning so much. Plus, I now feel we're on the road to some great strategies for my retirement and the education of my children. I guess peace of mind has been a byproduct of all this.
STEP 2	**TREAT THE REQUEST WITH IMPORTANCE**
Advisor	*I'm glad to know you're seeing the value in our work. With that in mind, I have an important question to ask you.*
Martha	Okay.
STEP 3	**GET PERMISSION TO BRAINSTORM**
Advisor	*I'm wondering if we can brainstorm for a few minutes about people you care about who should know about the important process we deliver. Could we do that for a few minutes?*
Martha	Sure, I guess so.
STEP 4	**SUGGEST NAMES AND CATEGORIES**
Advisor	*Great. We're just brainstorming here. Let's begin with something I call the "Whom do you know" questions. Quite often, I am able to really help folks who fall into this category. So let me ask you a few questions to trigger your thinking.*
Martha	Okay, shoot.
Advisor	*(Then the advisor proceeds with various questions describing the type of prospect for whom he or she is looking.)*
	Whom do you know . . . ?
TEACHING POINT	
	There are many ways to implement this four-step process to ask for referrals. The above script is meant to trigger your imagination and help you come up with your own script.

Bill Cates is the author of *Unlimited Referrals* and the president of Referral Coach International (www.ReferralCoach.com). Copyright © 2001 by Bill Cates. Reprinted with permission.

If you fail to plan, you plan to fail.	***Know What You Will Say.*** If you fail to plan, you plan to fail, so think about what you want to say and put it in writing. This will eliminate the fear of not knowing how to ask. Here are some things to keep in mind as you develop your script:

- Reflect the value the client has received from the process. (A good idea is to probe for this information by asking, "What about this process have you found helpful?")
- Emphasize that you consider the client's stated value of the process extremely important.
- Ask if there are people in the client's life who may need and value the same service.
- If the client stumbles, present some alternatives with a simple "Do you know someone who . . . ?" Fill in the blank with a description of what you are looking for in a prospective customer. Use a description of a prospect from your target market or ideal client profile.

After we have covered the rest of the principles, we will review some of the different approaches advisors use to ask for referrals.

Prepare to Handle Concerns. Not everyone will provide referrals right away. People will usually have some concerns. If you meet resistance, try to find out what the concern is without pressuring the client. Perhaps it is a misunderstanding that you can explain to his or her satisfaction.

Here is one approach:

- Acknowledge the concern.
- Ask why the client feels the way he or she does.
- Answer the concern and ask for the referrals again.

Example:	Mel	"I'm not comfortable with that."
	Advisor	*"Mel, I can appreciate that. Do you mind if I ask why you aren't comfortable?"*
	Mel	"I don't know. I guess I'm afraid of offending my friends."
	Advisor	*"Others have expressed the same concern. That's natural. Is there a particular part of the process that concerns you?"*
	Mel	"No, I guess not."
	Advisor	*"Mel, the way I conducted business with you is the way I conduct all my business. I will treat anyone you refer to me the same way. Okay?"*
	Mel	"Okay."
	Advisor	*"Great, now whom do you know . . . ?"*

Handling Resistance (by Bill Cates)	
STEP 1	**AGREE AND VALIDATE THE CLIENT'S POSITION**
	Because your client's perception is his or her reality, the client will resist your attempts to change that view if you're not careful. After you validate the client's position with some supporting statement, don't erase that goodwill by using the word "but," which negates everything you said just before it.
STEP 2	**ASK PERMISSION TO EXPLORE**
	You want to fully understand the nature of the client's resistance and what the deeper truth might be. You need to be low-key here. I've found that when I ask permission to explore—have a quick conversation about it—my clients are not threatened and don't get defensive.
STEP 3	**REFRAME THE CLIENT'S VIEW OF THINGS WITH PERMISSION**
	Reframe the client's view of things gently, with his or her permission. If, in the exploration step, you realize that the client may be willing to consider other ways to look at things, you want to gently help him or her do that. If you can accomplish this by educating the client further or through other quick thinking on your part, move to Step 4. On the other hand, if the resistance is deep-rooted, you probably want to move on to Step 5.
STEP 4	**GAIN AGREEMENT AND MOVE ON WITH THE CONVERSATION**
	In this step, you basically ask if the client can see and adopt the new "frame" or perspective you've presented. If the client can, then continue to explore whom he or she knows who may also benefit from knowing you.
STEP 5	**IF RESISTANCE IS DEEP, BACK OFF**
	Because obtaining referrals involves relationships, you don't want to weaken your relationships. If a prospect is adamant against giving you referrals, you're better off retreating and pursuing this another day. I've known many advisors who have asked for referrals, encountered some resistance, backed off, and then received many referrals from that same client later.

Bill Cates is the author of *Unlimited Referrals* and the president of Referral Coach International (www.ReferralCoach.com). Copyright © 2001 by Bill Cates. Reprinted with permission.

If the client persists in
his or her reluctance,
do not pressure. Not
everyone will feel
comfortable giving
referrals.

If the client persists in his or her reluctance, do not pressure. Not everyone will feel comfortable giving referrals. Exit gracefully. You could run the risk of damaging your long-term relationship. Remember, the client could still represent repeat sales. Also, "no" now does not mean "no" forever. Perhaps as you get to know each other better, the client will change his or her mind and send you referrals.

Example:	Mel	"I'm not comfortable with that."
	Advisor	*"Mel, I can appreciate that. Do you mind if I ask why you aren't comfortable?"*
	Mel	"I don't know, I guess I'm afraid of offending my friends."
	Advisor	*"Others have expressed the same concern. That's natural. Is there a particular part of the process that concerns you?*
	Mel	"No. It's just that I'm not comfortable giving out names."
	Advisor	*"Mel, I can understand that. If you do run into someone who would value this process and you feel comfortable, tell him or her to give me a call. Fair enough?"*
	Mel	"Fair enough."

Remember, you want
to obtain more than
just names from your
referrers. You also
want information
about each prospect.

Qualify the Names. No matter what method you use to obtain referrals, remember that you want to obtain more than just names from your referrers. After a referrer has given you a name, ask for information about each prospect such as the following:

- contact information (home address, e-mail address, and phone number)
- interests and hobbies
- occupation and employer
- personality
- lifestyle
- family situation (married, children, and so on)

Obtain as much information as you can. Use your target market or ideal client profile to guide the process.

Example:	Ariel sells long-term care insurance. Her target market profile includes the characteristic "parents with adult children." Ariel tells her referrers that she wants

to work with people aged 55 or older. When a referrer gives her a name, she asks if the new prospect has adult children, how many, and where the adult children live.

Obtain a Personal Introduction. What you really want from referrers is a personal introduction if you can get one. Thus, when you are obtaining referrals, know what you will say to ask for such an introduction. Depending on the situation or relationship, the referrer could arrange for a face-to-face meeting. For instance, the referrer could have lunch or play a round of golf with you and the prospect. If this is not an option, consider asking the referrer to introduce you by phone or to write a brief note of endorsement. In this way, you can obtain the strongest "borrowed trust," and the relationship can begin on the best possible footing.

Example:	In some situations (if the referred prospect is on a do-not-call list), Mark will ask referrers if they would be comfortable calling the prospect and introducing Mark right then and there.
	On other occasions (if the referred prospect is not on a do-not-call list), he will ask the referrer to jot a quick note on an index card that Mark will later mail with an introductory letter followed by a phone call. The notes are very short: "Joe, Mark just helped me protect my family's financial future. Thought you would benefit from meeting him."

Assuming the referrer agrees to a face-to-face introduction, when you meet the referred prospect, try not to bring up business initially unless the prospect does. Instead, use this time to qualify the prospect as he or she gets to know (and qualify) you. Ask the prospect about his or her business or personal interests with genuine curiosity. The goal of the personal introduction is to earn the right and pave the way to a more formal meeting later on.

Keep in mind that even though you have met the prospect, if he or she is on a do not call list, you can contact the prospect via the telephone only with his or her express written permission (verbal permission is not enough under current legislation). (See chapter 5 for more information about the Do Not Call Law.) Thus, you will need a strategy for asking for an appointment. See figure 3-2 for some options and possible advantages and disadvantages of each option. Whatever your strategy, it is a good idea to gain the referrer's approval beforehand. He or she should not be surprised by anything you do or say.

FIGURE 3-2
Appointment-Setting Options for Introductions

Option	Advantages	Disadvantages
Ask for the appointment at the end of the intro-duction—get an e-mail address and/or *business* phone number to confirm the appointment.	• You do not have to obtain permission to call. • You can confirm by e-mail or calling the prospect at his or her place of business.	• It could make the prospect or the referrer uncom-fortable.
Give the prospect your business card and ask the prospect to call you.	• There is no pressure on the prospect. • It is comfortable for the referrer.	• You cannot call the prospect. • The prospect may never call you.
Ask the prospect for his or her e-mail address and e-mail him or her for permission to call or set the appointment via e-mail.	• There is low pressure on the prospect. • There is the possibility of obtaining permis-sion to call. • There is some means to contact the pros-pect legally.	• Many people do not check their e-mail, or they may delete your e-mail by accident.
Ask the prospect for permission to call and have him or her put it in writing.	• You have permission to call the prospect to set an appointment.	• It could be a bit awk-ward. • It could make the prospect or referrer uncomfortable.

> **Make the referral a priority. Then follow up with a note to let the referrer know the result.**

Follow Up with the Referrer. Make the referral a priority. Then follow up with a note to let the referrer know the result. Some advisors have used token thank-you gifts or perhaps a lunch for clients who have referred business to them. There are some people in the financial services industry who feel that gifts are not appropriate because they cheapen the value of the work they do. Ultimately, you will need to make your own determination of how you say thank you. The important thing is to remember to do something.

Use the Right Approach to Ask for Referrals

Many times advisors fail to obtain referrals not because the referrer does not want to help the advisor, but because the advisor's request is generic—something like, "Do you know anyone who needs my services?" Such a request requires the referrer to create a mental list of criteria—something he or she does not feel comfortable doing. What follows are some proven ways to structure your request to help the referrer bridge the gap by describing specific types of prospects with whom you want to work.

The List Approach. Make a list of people from your target market whom the referrer might know. For example, if you think target market members live in the same neighborhood as a referrer, do a crisscross directory search (see www.imcpl.org/resources/guides/business/crisscross.html). Given enough pre-selected names, the referrer will probably know at least one of the people about whom you ask. When your client recognizes a name, the client often will say, for example, "Oh yes, I know John, but I don't think he needs retirement planning."

You will find it helpful to reassure your referrer that you will never imply to the new prospect that the referrer believed that the prospect should purchase something. You simply would like to meet the prospect on a favorable basis to determine if you might be of service.

You might add, "Client, when I next see Prospect, do you mind if I mention that I handled (or serviced) your account?" When your client says this would be acceptable, you have a useful referral.

Whom Do You Know? You are asking for an introduction to people who fit your target market or ideal client profile. Think of any relevant triggers that are observable. Phrase your questions to focus the referrer on prospects who meet these criteria. Taking this approach can help to eliminate prospects who are not qualified to buy and to assist your referrer by providing characteristics you seek in referrals. For example, you can ask, "Whom do you know who just

- got married
- moved into a new home
- had a baby
- sent his or her first child off to grade school
- remodeled a home
- sold a home
- changed jobs
- started or expanded a business
- began saving for a child's college education
- had his or her first grandchild
- started to plan for retirement
- retired or is retiring soon

The following are other questions you can ask:

- Who in the neighborhood is getting ready to purchase, or has recently purchased, a new home?
- Who in your company has recently been promoted?

- Whom do you know who could become a leader in your profession in the next 10 years?
- Who is the most successful person you know?
- Of the people you know, who is most likely to succeed?

Then be patient while the potential referrer thinks. Use gentle prodding if you must, but remember to wait. When you ask for referrals in a way that allows the referrer to imagine someone whose characteristics match the criteria in your questions, he or she will likely respond. Have pen and paper ready to record names.

Associations. Another way to seek valuable referrals is to ask clients or friends about their involvement with local professional or industry associations. Business owners and other successful people are often on boards of and active in local organizations.

When you learn that your contact is a member of an organization, you might ask, "Mary, do you know the president or vice president of (your association) personally?" If you can obtain a referral to an association officer and ultimately secure this respected individual as a client, your prestige with other association members will be high. This will significantly enhance your ability to meet with additional group members.

3–2–1

What is the biggest reason people don't get referrals? Right—they don't ask for them! You have to be prepared. You shouldn't find yourself in the car after an interview saying, "Oh gee, I forgot to ask for a referral."

Here is a sure-fire method to get referrals. Take out a piece of stationery marked with the client's name and the numbers 1, 2, and 3. Say something like:

"Well, Bill, you're a successful person, and I've really enjoyed doing business with you over the years.

"You know, all my clients now are personal referrals. If you'll recall, the way I met you was on a very favorable basis.

"You remember that Ralph Referrer suggested I get in touch with you. If it hadn't been for Ralph, you wouldn't have the fine financial program you now have.

"Just as Ralph got us together, I would like to know if there are two or three people you do business with who are the same caliber of person and whom you think I might be able to help."

Bill replies, "Okay, there's Tom Jones."

Now, don't put Tom up in the number 1 spot. Put him in the number 3 spot. Mentally, it really works. Put the first name you receive at number 3, and you'll get the other two names. If you start at number 1, you may get one, possibly two names. You'll probably never get three names.

Also, when the client is giving you names, don't ask anything about Tom Jones. You want to get the other two names first. Then you have time to hear a little bit about Tom. He's a good guy; he plays golf—all the information that you typically want.

When your source gives you a name, the ice is broken. He or she will generally give you lots of additional information if you ask for it. Do ask for it.

Sometimes your contact will be unable to refer you to any of the key officers of an organization. Try gently for referrals to other group members. Many successful advisors develop "nests" within particular industry or professional groups in this manner.

When your source gives you a name, the ice is broken. He or she will then generally give you lots of additional information if you ask for it. Do ask for it. Find out everything the referrer knows about the referral. Again, consult the characteristics of your target market or ideal client profile.

Centers of Influence

center of influence

The second of three prospecting methods for generating prospects through a nominator is the center of influence. By definition, a *center of influence* (COI) is an influential person who knows you favorably and agrees to introduce or refer you to others. Clients may become effective COIs for you, just as COIs may become clients, but this is not necessary to the relationship you need to establish.

In general, you will find that COIs are

- active in a community or sphere of influence
- sought out for advice by the community or within their sphere
- good communicators
- givers, not takers

COIs generally know you well enough and have enough confidence in you to refer you to prospects. An important aspect of a COI is that he or she has a personal interest in helping you succeed. Again, COIs may or may not be clients. They may be friends of the family or people with whom you have worked in the past, either in business or in a community activity.

Identify a Center of Influence

Go through your lists of clients and identify the ones who meet the criteria for being a center of influence as described previously. Then, record what you know: age, profession, social circles, associations, hobbies, interests, college, degrees, and so on. Compare these to your target market profile(s). Choose the COIs who are best positioned to help you find prospects in your target market(s).

Cultivate the Relationship

You typically already have a good relationship with a potential COI and he or she already perceives you as a competent financial advisor. If this is the case, you are ready to approach him or her. But you may find that you do not

---

Mentoring: A Powerful Role for a Center of Influence

A mentor is usually a center of influence whom you have "adopted" as an informal teacher and advisor. Most of today's successful advisors attribute some of their success to at least one mentor who guided and, more important, introduced them to other prospects. (Mentoring is discussed in detail in chapter 8.)

By seeking out a COI's advice, you are already fulfilling one of your mentor's very important needs—the need to be needed. Regardless of what career stage people are in, they all like to be appreciated, to feel needed, and to make a positive contribution to others. People get a natural sense of satisfaction when they know they have made a difference. Think about a time when you "saved the day" for someone else. It was a very satisfying feeling.

When your COI refers or introduces you to other prospects, you are carrying the "weight" of your COI with you. For example, let's suppose you know an influential corporate executive. Because you are well acquainted, you ask the executive for advice as you build your client base. You ask to be introduced to people who may also benefit from your products and services.

The executive gives you the names of five other people. When you call each person, it is the executive's influence (and relationship) that wins you an audience with the prospect—and it will remain so until the prospect is impressed with and accepts you for who you are and what you offer (this is the case, regardless of whether or not the prospect becomes a client).

When seeking centers of influence as mentors, you will want to look for individuals who

- inspire genuine respect and for whom you feel a level of trust
- display character traits and values similar to your own
- can connect you with other significant people and meaningful opportunities

---

know the potential COI very well and/or that he or she may not be comfortable with your skill as a financial advisor. In that case, you will need to spend more time cultivating the relationship. Again, think long-term. Plant seeds today so you can eat tomorrow.

Find Ways to Help. The relationship will develop properly if you think first of the center of influence's interests and what you can do for him or her. By doing so, you will find your COI willing to promote your cause in return.

> **Your relationship with a COI is a two-way street. There is something in it for both parties, not just something for you.**

The relationship is a two-way street. It is a cooperative effort in which there is something in it for both parties, not just something for you. There should be some direct benefit for the center of influence. Remember also that he or she should be pleased with the relationship with you, and that you have earned the right to his or her help.

Prove Your Competence. Ideally, the assistance you give to the center of influence will be related directly to the work that you do. If not, you may have to wait for an opportunity. In the meantime, be a resource for relevant information. Identify the types of financial ideas and concepts that you understand and that the COI would find interesting. Every now and then (do not overdo it) send him or her an article (especially one you have written). Bide your time until you feel you have gained the trust you need to approach the COI for possible referrals.

Meanwhile, find ways to borrow his or her influence indirectly by working with the COI on matters that pertain to your target market. For example, if you are a long-term care specialist cultivating a relationship with an elder-care attorney, offer to put together a presentation on long-term care insurance for a seminar he or she is hosting.

If business-related opportunities are difficult at first, then help the COI in a non-business-related project that involves a common interest or passion you share. Cosponsor a benefit for a charity you both support. Volunteer together at a local soup kitchen. Find some way to demonstrate your character and competence.

Approach the Center of Influence

Approach the center of influence for help when you feel that he or she respects and trusts you. This will invariably involve setting up a meeting to talk to the COI, during which you will outline your goals and objectives and formally ask for assistance. One strategy is to approach the COI by focusing on how what you can do will benefit him or her.

Example: "Pat, you have a great reputation in the community, and I would feel comfortable referring clients to you. I would like to get together with you to brainstorm ways we can help each other build successful practices. Would breakfast sometime next week work for you, or would a lunch be better? My treat."

Or, if you are working with a community leader, focus on how what you can do will benefit the community.

Example: "Kim, you really command the respect of the seniors in this community. One of the critical questions many seniors are asking is, 'How do I pay for long-term care?' Unfortunately, a lot of people are making decisions about long-term care based on inaccurate information. I would like to discuss this issue with you, along with ways you could help the seniors in this community make well-informed decisions. I am not going to try to sell you anything;

> I only want to show you how I can help you help the community. Would next Wednesday be a good day to meet, or would another day work better for you?"

In addition, when you are ready to ask the COI for the names of prospects, prepare a list ahead of time of as many people you can think of whom you believe the COI knows or is likely to know. Then review the list with the COI and have him or her identify and qualify the prospects he or she knows. Similar to the discussion of referrals, the goal is to gain personal introductions to prospects. Plan how you will ask for them.

Tips Clubs

Tips Club

A third way to obtain leads from a nominator is the Tips Club. A *Tips Club* is composed of local salespersons who represent such diverse industries as insurance, real estate, banking, office products/furniture, and so on. The club is formed solely for the regular exchange of information on prospects. Each member of the Tips Club shares his or her own expertise, business connections, and social contacts with the group. The resulting exchange of ideas and leads can multiply each member's referred leads.

Example:	One of your clients is in the market for a new car. You give that lead to the car salesman in the Tips Club. ------------------ You know a client who is thinking about buying a new home or who wants to explore the possibility of refinancing her home mortgage. You give this lead to the realtor and/or the banker in the Tips Club.

Usually, a Tips Club has only one member per industry. Each member must come to the meeting with at least one prospect for someone. If a member misses three meetings in a row or attends three in a row without a prospect, he or she is no longer a member of the club.

You may want to start your own Tips Club or ask other successful salespeople in your community if they belong to a club you might join.

Personal Observation

Make the most of your opportunity wherever you go, and you may stumble onto something by pure luck.

There is a saying, "Wherever you are, you are there." You may be thinking sarcastically, "That's *real insightful*." But the point is to make the most of your opportunity wherever you go, and you may stumble onto

personal observation

something by pure luck. The key to tapping these hidden opportunities is *personal observation*: watching, listening, and engaging.

Think about your target market and/or ideal client profile. What are some observable ways you could conclude that someone will likely fit it? Those are the characteristics you look for or determine by what people say. Create some low-key questions that can help you find out valuable information without being intrusive. Many such questions are part of everyday conversations:

- How many children do you have? (if a child is with the person)
- Are you from the area?
- Where do you work?
- What do you do for a living?

> **Personal observation revolves around mastering the art of listening and the art of small talk.**

As you can see, personal observation revolves around mastering the art of listening and the art of small talk. The key to impromptu conversation is to begin with something you may have in common or something interesting about the person. For example, "Looks like we both like the Yankees," or "That's a real sharp outfit. Where did you buy it?" Then, start to converse about the innocuous and listen carefully. As the old adage says, "Luck is when preparedness meets opportunity."

Example 1:	Kenny's train is running late. A woman standing nearby is obviously waiting for the same train. Kenny engages the woman in some small talk that starts with a comment about the train. In talking, Kenny finds out the woman's 28-year-old son is getting married this weekend. Kenny's target market is young married couples. He gives the woman his business card.
Example 2:	Denny, an extrovert, is a pretty good golfer. He goes golfing at least once a week at the public golf course. He calls ahead to the person who schedules tee times and asks to be placed with a threesome. Between holes, Denny engages the other golfers in small talk. At the end of the game, he has exchanged business cards with all of them.

Cold Canvassing

cold canvassing

Cold canvassing refers to the old-fashioned knocking-on-doors method of prospecting. It is challenging, time consuming, and rarely as productive as getting referred leads from clients or centers of influence. There are situations,

There are situations when going out and knocking on doors can generate some new and exciting opportunities that might not otherwise have occurred.

however, when going out and knocking on doors can generate some new and exciting opportunities that might not otherwise have occurred, especially for advisors who work in the business market. In addition, cold canvassing is both a prospecting method and a way for an advisor to create awareness about his or her practice.

Example: Helen wants to target business owners. She drives to the lobby of a local business building and makes a list of the businesses on the wall directory.

Helen goes back to her office and researches the companies, using Dun and Bradstreet's online resources. She prepares a 5-minute introduction and puts together an introductory kit, which she will give to each business owner.

Many advisors make successful use of canvassing as a part of their prospecting repertoire even in selling to individuals and families, especially in the wake of the national Do-Not-Call Registry. This works especially well if a target market is well represented in a particular neighborhood. The objective of cold canvassing is simply to identify possible prospects and to let people know who you are.

Example: Harvey sells property and casualty insurance along with other financial products. Recently, he sold a homeowners policy in a neighborhood that had swing sets in almost every backyard. The buyer, Ellen, was reluctant to provide Harvey with any referrals, but she did give him permission to refer to her as a client. So Harvey walked around the neighborhood introducing himself to the neighbors. He told people that he insures Ellen down the street and if there is damage to her house when she is away, to give him a call. He gave each neighbor a refrigerator magnet with his contact information, which included his cell phone number.

Direct Mail

direct mail

Another prospecting method is *direct mail*, which involves sending letters with reply cards that prospects can return if they are interested in an appointment or more information. Sometimes, the letter will offer a small gift, such as a road atlas or a free booklet, to prospects who respond to the direct mail letter and agree to a free consultation with the advisor. If prospects are not on a do-not-call list, advisors may follow up with a phone call to set an appointment.

Before devising a direct mail plan of your own, investigate what programs are available through your company. If a program is available, the company will prepare and mail letters to prospects whom you select—in many cases saving you time and money.

Remember that it takes multiple exposures to information before people respond to a letter or an idea. Sometimes, the reason a letter is effective is that it arrives around the same time that another more compelling trigger occurs, making the information acutely relevant. Because advisors do not usually know when such events occur, they can implement a technique often referred to as wave mailing, which involves sending several pieces of mail over a period of time to the same prospect.

Example :	You receive a little pamphlet in the mail from the water company warning you that ruptured water lines are your responsibility as the homeowner. The pamphlet tells you that you could pay thousands of dollars to fix the problem. The good news is that the water company offers insurance for this very problem that costs about $100 per year. Are you interested?
	But what if your neighbor has a rupture in her water line and spends $3,000 to have it repaired? Shortly afterwards, you receive the same pamphlet from the water company in the mail. Now are you interested?

Some advisors use a compressed time frame, mailing three or four letters to the same prospects over a 2- or 3-week period. Others extend the amount of time to a 3-month period. Regardless of the period you select, at the end of the period, you call for an appointment.

The most important thing to remember about direct mail is that you must persevere. Inconsistency will not work. You must give it an ample chance to succeed. When you use it regularly, you will lay the groundwork for future appointments.

Seminars

seminar

A *seminar* is a prospecting method in which you, alone or as a part of a team of professionals, conduct an educational and motivational meeting for a group of people who are interested in the topic presented. Seminar selling permits advisors to meet a large number of potential clients in one location and at one time.

This technique is a hybrid: part prospecting method and part preapproach or awareness strategy. A more detailed treatment appears in the next chapter.

ORGANIZING YOUR PROSPECTING EFFORTS

> **Even with the best sources and methods, prospecting still comes down to execution and discipline.**

Selecting effective prospecting sources and methods is essential to a successful practice. But even with the best sources and methods, prospecting still comes down to execution and discipline. Simply put, you must consistently invest the right amount of time and execute the right prospecting methods with the right sources.

For most successful advisors, the key to achieving consistent and adequate levels of activity is the organization resulting from an efficient prospecting system. These advisors also keep accurate records to ensure that they are maintaining adequate levels of activity as well as to evaluate their prospecting methods and sources. Let us take a look at each of these topics, beginning with prospecting systems.

---

25 Points a Day

It is a constant challenge to stay focused daily on the activities that are most productive and will contribute toward the achievement of your business goals.

A technique that many advisors use to quantify their daily activity goals is a simple point system that assigns values to the prospecting and sales activities they need to do every day. One version of this system is called "25 Points a Day." The objective is to accumulate at least 25 points each day (125 points per week) by performing some combination of the activities below. (Total points are the number achieved for each activity multiplied by points per activity.)

Activity	Number Achieved	x	Points per Activity	Total Points
Obtain referred lead		x	3	
Make a phone contact		x	1	
Secure a new appointment		x	2	
Make a new face-to-face contact		x	1	
Complete a fact finder		x	2	
Complete a closing interview		x	3	
Complete a sale		x	5	
			TOTAL POINTS FOR DAY	

Regardless of all the other responsibilities you have as a professional advisor, if you make accumulating 25 points per day your first priority, it will become a positive work habit that gives you a genuine sense of accomplishment.

Prospecting Systems

Consistency is essential to ensure adequate levels of prospecting activities. Thus, you must develop a system that guarantees a steady stream of new prospects and prompts you to contact them. It can be manual or electronic. However, the system should enable you to manage contact information, allow quick access to that information, and include a way to identify the prospects you should call first.

> Consistency is essential to ensure adequate levels of prospecting activities.

Manual or Electronic System

There are many prospecting systems. One well-known method is the One Card System© by Al Granum. Granum's system includes ways to monitor all prospecting activities, log and sort names and reasons to call, and even track progress and effectiveness.

For advisors who would prefer a computer-based system, there are software programs that have integrated methods like the One Card System©. Some utilize the portability of personal digital assistants (PDAs) and enable you to carry information about your prospects in the palm of your hand. Many companies offer their own client administration systems to organize and track prospecting and client service activities, or they contract with providers to make these systems available to you. These methods will be discussed more in chapter 8.

Managing Contact Information

Whether you use a manual or electronic system, you should track the same basic contact elements. At a minimum, you will want to develop file records on a computer or on individual 3″ x 5″ index cards. Track information such as name, contact information, date of birth, family status, occupation, possible needs, and so on.

If you use a card system, you should indicate at the top of each card the date you want to call that person, and file your cards in that order so you will always be able to retrieve the cards at the moment you should make the call. If you have access to a computer, the same records—and more—can be set up with a variety of key fields noted for numerous sorting and retrieving possibilities. It is helpful to keep records of contact dates and summaries of conversations so that you can quickly refresh your memory and the prospect's regarding the progression of your relationship.

See chapter 8 for a detailed discussion of contact management systems.

Quick Access to Information

Whichever system you use, it should allow quick and accurate access to prospect information. Do not use long pieces of paper that can become

Organization is the key to prospecting continuity and success.

unwieldy or little notes that can get lost. Every prospect deserves an individual card or an individual record in a contact management database.

If you are not currently using a prospecting system, find out what your company recommends. Whatever you do, do not try to prospect without a system. Organization is the key to your prospecting continuity and success.

Identifying Priority Prospects

A good prospecting system will include some way to identify which prospects you should contact first. This means that the system should allow you to evaluate and prioritize prospects. It also should allow you to group prospects by priority and call date.

The information you track using your prospecting system should include the most relevant characteristics for both identifying target market members (and ideal clients) and indicating their need for your products and services. If possible, the system should also track characteristics that might indicate the level of importance that prospects place on meeting specific financial needs. Using this information, you can evaluate and group your prospects by the priority level (low, medium, high) you assign them.

One way to do this is to create a spreadsheet using as column headings the most important characteristics of a qualified prospect (see figure 3-3). (Two columns are intentionally left blank for you to identify important characteristics of your own qualified prospects.) It is especially important to think about the favorable basis upon which you will contact the prospect. Most likely, this will be a referrer, your own personal contact with the prospect, or a common interest

---

FIGURE 3-3
Prospect Prioritizer

A/B/C	Name	Source	Target Market	Specific Needs	HH Income			Favorable Basis	Call Date	Appt. Date

you and the prospect have (membership in an organization, college alma mater, and so on). If you do not have a favorable basis or you have one that is weak, consider finding a way to create a favorable basis before you approach the prospect. This could be through finding someone to refer you, networking, prestige-building activities (to be discussed in the next chapter), or other means.

Finally, assess whether there is a date that would be an opportune time to contact the prospect (see the following example). Update the prospect's record accordingly. For the rest of the prospects, the priority level and your goal for the number of calls or face-to-face contacts will determine the call date.

Example: Maggie has the name of prospect given to her by an influential attorney. Although the prospect is not in her target market, Maggie values the relationship she has with this attorney. That prospect is an "A" prospect and will be one of the first prospects she calls.

Maggie ascertains that people are most interested in Medicare supplement policies about 6 months before they reach age 65. Thus, the call dates for all Medicare supplement prospects are 6 months before their 65th birthday.

For existing clients, Maggie uses 3 weeks before their annual review date as the call date.

Record Keeping

> **Using only your intuition to make prospecting decisions is like being lost in a forest and relying solely on instincts, ignoring the map and compass you brought with you.**

Do you know your best prospecting sources? Do you know which prospecting methods work best for you? Do you know if you are on track to meet your goals? How do you know? Is it based on a gut feeling or actual numbers? Although your intuition is an indispensable part of decision making, you should never ignore hard data when such information is readily collected and analyzed. That would be like being lost in a forest and relying solely on your instincts, ignoring the map and compass that you brought with you.

Clearly, you can make better decisions if you have data to analyze. Thus, it is important to track your activities and results. Then you can make knowledgeable decisions regarding how you invest your time, energy, and money. You will have a basis for planning your daily, weekly, and monthly prospecting and other marketing activities, as well as for diagnosing problem areas that you need to shore up. Accurate record keeping also provides the basis for long-term planning, which enables you to set annual and other longer-range goals (topics touched on in chapters 6 and 8).

Tracking Prospecting Sources

One of your daily tasks should be to summarize information about your prospects, specifically prospecting sources (and if not too onerous, prospecting methods). Of course, if you have a contact management database, you will be able to do this quite easily. The reports feature of the database program will allow you to create a summary at the click of a button. You may need to add customizable fields to record the date to call, prospecting source, prospecting method, and target market.

If you do not have a computer program (even a simple spreadsheet would work—as mentioned earlier, www.openoffice.org currently offers a free office suite package, including a database and a spreadsheet program), the task can be done manually. Use a chart like the one in figure 3-4. Delineate between your target market(s) and your general market.

Tracking Use of Time

Before discussing the prospecting and sales numbers you must track to set goals and evaluate progress, it is important to mention, albeit briefly, the necessity of planning and tracking the use of time.

Do you have a plan for how much time you will spend prospecting and approaching prospects? Do you know how well you are meeting that goal? If not, make it a priority to plan the usage of your time and to track how you are actually spending it. It would be very valuable to know not only how much time you spend on prospecting, but also how much time you spend on various methods and sources. In addition, budgeting and tracking your time can ensure that you spend time to network and build the relationships necessary for harvesting future prospects. Furthermore, you will need to base decisions regarding target markets, prospecting methods, and prospecting sources on regular analysis (weekly and monthly) of your time.

The actual practical steps for managing your time will be explained more fully in chapter 6.

Tracking Prospecting and Sales Numbers

A third set of numbers to monitor is your prospecting and sales numbers. If you keep and review these numbers daily and weekly, it will help you stay on track to meet your goals. Prospecting and sales numbers you will want to track weekly include

> If you monitor your prospecting and sales numbers daily and weekly, it will help you stay on track to meet your goals.

- phone/face-to-face contacts made
- appointments seen
- qualified prospects obtained
- fact-finding interviews conducted

FIGURE 3-4
Tracking Weekly Activity

Weekly Prospecting Activities

Phone/face-to-face contacts		Appointments seen		Qualified prospects obtained	
Fact-finding interviews conducted		Closing interviews conducted		Sales completed	
Referred leads obtained		First-year commissions generated			

Source of Prospecting Leads

	Mon	Tues	Wed	Thurs	Fri	Sat	Sun	Total
Client								
Target								
General								
Orphans								
Target								
General								
Referrals								
Target								
General								
COIs								
Target								
General								
Lists								
Target								
General								
Network								
Target								
General								
Unsolicit.								
Target								
General								
Other								
Target								
General								

- closing interviews conducted (Note: If the closing interview and fact finding take place during the same interview, indicate as both a fact finder completed and a closing interview conducted.)
- sales completed (Note: If you sell two products in one interview, count that as two sales. Keep track of the type of product sold and its purpose.)
- referred leads obtained
- first-year commissions generated

prospecting and sales effectiveness ratios

From these numbers, you will be able to calculate your basic *prospecting and sales effectiveness ratios*. Here are some of the more common ratios:

- *total contacts to appointments*. Divide the number of phone and face-to-face contacts you made by the number of appointments. This tells you the total number of contacts it takes to generate one appointment.
- *appointments to qualified prospects*. Divide the number of appointments by the number of qualified prospects. This determines the number of new initial appointments you need to obtain one qualified prospect.
- *qualified prospects to fact-finding-interviews*. Divide the number of qualified prospects by the number of fact finders, yielding the number of qualified prospects you need to complete one fact finder.
- *fact-finding interviews to closing interviews*. Divide the number of fact-finding interviews by the number of closing interviews. This produces the number of fact-finding interviews you must have to generate one closing interview.
- *closing interviews to sales*. Divide the number of closing interviews by the number of total sales made. This number represents the number of closing interviews needed to sell one financial product.
- *first-year commissions (FYCs) to sales*. Divide the amount of your FYCs by the number of sales. This tells you the average amount of FYCs that one sale generates. If you divide the total FYCs you wish to earn in a year by your average FYC per sale, the result is the number of sales you will need to achieve that FYC goal.

For all but the last ratio, the lower the ratio, the more efficient and effective you are. You can track your prospecting and sales ratios using the information recorded as shown in figure 3-4. You can do this on paper (manually), on a spreadsheet, or through a report option available in many contact management database programs. In addition to tracking these numbers for the current week, you will want to track the same results on a monthly and annual basis. In time, you will begin to see the strong and weak points of

---

Activity = Money

By tracking your prospecting and sales activity numbers, you can see what each activity is worth and make strategic decisions. For example, suppose your total first-year commissions are $50,000. If you divide $50,000 by 2,000, which is the number of phone contacts you made to produce that business, the result is $25. That amount is the average value of each phone contact you made during that period.

As one advisor said, "Knowing that every phone contact is worth $25.00 gives me the motivation to pick up the phone."

It doesn't matter what the prospect says. Just make the call. The sales ratios will do the work for you.

---

your prospecting efforts, and you can make the necessary adjustments to strengthen your game.

(The use of these prospecting and sales effectiveness ratios will be discussed further in chapter 6 as they relate to goal setting and in chapter 8 as they relate to business planning.)

CONCLUSION

Think about where you can find members from your target market; this will lead you to good sources for prospects. Once you have identified prospecting sources, you will need to apply effective prospecting methods to generate a steady flow of qualified leads. This chapter covered 10 different methods for doing so. The right methods will depend on the nature of the source, your own strengths and weakness, and your target market's personal preferences. It is important not to fixate on which methods work best for other people in other situations. Be open-minded and use a little creativity.

Obviously, all things being equal, your best sources are those that produce people who know you favorably. Unfortunately, you will most likely run out of names if you prospect only for this type of prospects. Thus, you will need to turn to strangers. There are two effective strategies for working with strangers. Either find a way for them to get to know you favorably, or identify someone they know who already knows you favorably and will refer them to you. Therefore, the majority of this chapter was devoted to networking, referrals, and centers of influence.

Of course, identifying the best sources and best methods is most effective when you have a good prospecting system to organize your prospecting efforts. It is important to have a method that enables you to manage contact information, provides quick access to that information, and identifies priority prospects to call.

It is also essential that you keep score. This is the only way you can stay on track and make adjustments to achieve your goals. Track your prospecting sources weekly so you know which sources are providing leads. Plan and record the usage of your time to ensure that prospecting activities are a part of your daily schedule. Finally, tabulate your prospecting and sales numbers, from which you can derive your prospecting and sales effectiveness ratios. These ratios are critical for annual planning.

CHAPTER THREE REVIEW

Key Terms and Concepts are explained in the Glossary. Answers to the Review Questions and Self-Test Questions are found in the back of the textbook in the Answers to Questions section.

Key Terms and Concepts

prospecting sources	Tips Club
prospecting methods	personal observation
nominator	cold canvassing
orphan	direct mail
networking	seminar
referral	prospecting and sales effectiveness
center of influence	ratios

Review Questions

3-1. Describe the relationship between prospects, prospecting sources, and prospecting methods.

3-2. Define the three types of prospects, and identify two sources for each type.

3-3. Explain two principles for selecting prospecting methods.

3-4. Identify three steps for getting started on building a network.

3-5. Discuss the eight principles for asking for referrals.

3-6. Explain one of the approaches for asking for referrals.

3-7. Explain how to cultivate a relationship with a center of influence.

3-8. Identify three important aspects of a prospecting system.

3-9. Discuss three record-keeping tasks and explain their importance.

Self-Test Questions

Instructions: Read chapter 3 first, then answer the following questions to test your knowledge. There are 10 questions; circle the correct answer, then check your answers with the answer key in the back of the book.

3-1.   Which of the following sources will most likely generate prospects who know the advisor favorably?

   (A)   the membership list of an organization for which the advisor is the president
   (B)   the wall directory at a local office building
   (C)   a list of engineering firms the advisor purchased from Dun and Bradstreet
   (D)   local school board members, one of whom the advisor met once

3-2.   Which of the following is one of the purposes of networking?

   (A)   to sell something
   (B)   to share knowledge, resources, and contacts
   (C)   to manipulate others
   (D)   to seek monetary compensation, donations, or funding

3-3.   Della asks her client, Stella, for referrals. Stella says that she does not know anyone who needs Della's help. Which of the following is the best way Della can respond?

   (A)   "That's fine. If you do think of someone, give him or her my business card."
   (B)   "Are you sure you can't think of anyone?"
   (C)   "Stella, the more time I spend looking for new prospects, the less time I have to help you. Can't you think of anyone?"
   (D)   "You don't need to know if people need my help. Just tell me, whom do you know who . . . ?"

3-4.   Which of the following is a prospecting and sales number that advisors will want to track?

   (A)   average prospect income
   (B)   interview length
   (C)   referred leads obtained
   (D)   prospect family status

3-5. Which of the following is (are) a major type of prospect?

    I.   People who know the advisor favorably
   II.   People who contact the advisor for an appointment

    (A)   I only
    (B)   II only
    (C)   Both I and II
    (D)   Neither I nor II

3-6. Which of the following is (are) excellent as a method for obtaining prospects from a would-be nominator?

    I.   Referral
   II.   Center of influence

    (A)   I only
    (B)   II only
    (C)   Both I and II
    (D)   Neither I nor II

3-7. To utilize the prospecting method of personal observation well, the advisor must learn to master which of the following?

    I.   Public speaking
   II.   Cold canvassing

    (A)   I only
    (B)   II only
    (C)   Both I and II
    (D)   Neither I nor II

3-8. All the following are examples of people who know the advisor favorably EXCEPT

    (A)   the owner of the dry cleaning business that the advisor patronizes
    (B)   an orphan client assigned to the advisor last week
    (C)   the advisor's best friend
    (D)   someone who has already bought a product or service from the advisor

3-9.   All the following advisors are completing a task that will improve their referability EXCEPT

(A)   Bill, who teaches a class for his company's new agents
(B)   Jill, who refers her client to an estate attorney she knows
(C)   Phil, who sends out a newsletter to his best clients
(D)   Will, who helps a client buy a car

3-10.   All the following are examples of minimal prospect information advisors should track EXCEPT

(A)   address
(B)   date of birth
(C)   natural market
(D)   family status

NOTES

1.   Robert Littell and Donna Fisher, *Power NetWeaving: 10 Secrets to Successful Relationship Marketing* (Cincinnati, OH: The National Underwriter Company, 2001), p. 4.
2.   Ibid., p. 5.

Creating Awareness

Learning Objectives

An understanding of the material in this chapter should enable the student to

4-1. Define and describe various ways an advisor can increase his or her social mobility as a means for creating prestige.

4-2. Describe four other prestige-building activities that promote an advisor's personal brand.

4-3. Explain how to plan, execute, and follow up a seminar.

4-4. Expound on the principles for selecting, writing, and utilizing preapproach letters.

Chapter Outline

INTRODUCTION

Ideally, advisors will work with prospects who know them favorably or are referred to the advisors by those who do. More than likely, however, advisors will have to identify and approach prospects who are complete strangers. As discussed in chapter 3, when working with prospects who are total strangers, an advisor has three options:

- Cold call the prospects.
- Find someone who will recommend the advisor to the prospects.
- Create a favorable awareness of the advisor's personal brand and the need for the advisor's products and services.

> **Cold calling complete strangers is a recipe for frustration for most advisors.**

Cold calling complete strangers is a recipe for frustration for most advisors. Not only can the rejection associated with it demoralize the advisor, but the growing number of consumers adding their names to the national Do-Not-Call Registry also jeopardizes the advisor's long-term viability if he or she relies predominantly on cold calling. Unfortunately, even leaning heavily on referrals may not adequately generate the number of prospects an advisor needs to meet his or her financial goals. The fact is that there are some prospects for whom the advisor will not find nominators to recommend him or her.

Consequently, the solution is for the advisor to plan and implement strategies to create awareness of his or her personal brand through prestige-building activities. In addition, the advisor must educate and motivate prospects regarding the financial and emotional needs the advisor's products can satisfy. The advisor can achieve education and motivation through preapproach activities, which the advisor can use with all prospects, not just for complete strangers.

This chapter will discuss both prestige-building and preapproach techniques. Although discussed as two distinct marketing activities, there is overlap between them. The techniques you choose and how you carry them out should be directly related to the positioning statement and value propositions referred to in chapter 2, as well as the prospecting sources and methods identified in chapter 3.

PRESTIGE BUILDING

prestige building

Prestige building is your public relations campaign to position your personal brand favorably in your target markets. The limit to prestige-building activities is your imagination. This section will examine several techniques: social mobility, personal brochures, Internet websites, newsletters, and advertising. Use them as a starting point and innovate based on your strengths and personal characteristics.

Social Mobility

The most effective way to promote your personal brand to your target market is by increasing your social mobility. From a socioeconomic standpoint, *social mobility* is "the ability of individuals or groups to move within a social hierarchy with changes in income, education, occupation, etc."[1] For our purposes, we will utilize the concept of movement from this definition. Thus, social mobility in this textbook refers to a person's movement within and impact on a society or community. The result of social mobility is a reputation, good or bad, within the community. This is why it is the most appropriate tool for creating awareness of your personal brand.

There are a variety of ways to create social mobility. All of them allow prospects to see you active in the community, be it the business community, the schools, or other areas. What follows are some of the more common ways to increase social mobility. Ultimately, you must select methods that are consistent with your personal brand and most effective for reaching your target markets.

Community Involvement

The first method to think about is *community involvement*, specifically in social, civic, or charitable organizations and causes in the community that are important to both you and your target markets. Obviously, you cannot do it all. Make wise and careful choices. Let us look first at some guiding principles for community involvement, then at the various levels in which you can participate.

Guiding Principles. When deciding where you should involve yourself in the community, consider the following principles. First and foremost, align yourself with organizations and causes that you would support, regardless of your target market's interests. Although you are trying to create a perception, you want it to be based on reality. You do not want people to think you have a deep concern for animals if you really do not. Such disingenuous positioning will be discovered and your integrity will be questioned.

Second, conduct some due diligence with the organizations and their leaders. Look for matching values and consider any conflicting ones. You certainly want to support organizations and leaders you believe in and trust.

Third, evaluate your capacity for involvement. How much time have you budgeted for creating social mobility? Knowing your time limitations will help you decide which community interests you should pursue and the level of your involvement.

Fourth, aim for visibility and not shameless self-promotion. People have negative impressions of those who like to tell everyone how good they are

social mobility

Social mobility is the most appropriate tool for creating awareness of your personal brand.

community involvement

Align yourself with organizations and causes that you would support, regardless of your target market's interests.

and what good they are doing. As the Jewish proverb says, "Let another man praise you and not your own mouth; a stranger, and not your own lips."

Sponsoring and Giving. The easiest level of involvement in terms of time and energy is sponsorship and giving. Find community causes for which you and your target market share a passion, and consider supporting that cause or organization financially. If your target market is parents with children and you love baseball, sponsor a little league baseball team. Donate money to the league for new equipment. There are endless opportunities for giving. Again, do not worry about getting credit. Word gets out.

> **Do not worry about getting credit for giving. Word gets out.**

Volunteering. As a volunteer, your commitment level is minimal (although you should follow through on any commitments you make). Yet volunteering is an effective way to build relationships with people. Your involvement with an organization bonds you to other members or volunteers with that same organization. By working together to reach a common goal, you build relationships spontaneously and naturally. Volunteering is a way for you to do some common good, as well as to get to know the heart and soul of an organization before you go to the next level: joining.

Joining. If you are new to the financial services industry, be careful about your time. Talk to other advisors who have been in the business for a few years to see what a realistic expectation is for the amount of time you will have for joining an organization. You do not want the organization's activities to draw you away from your business activities.

In addition, join only if you are serious about supporting the charter and goals of the organization. If your heart is not really into something, it will show in the quality of your work and in how you relate to people. That will reflect badly on you and do much more harm than good.

Joining (or volunteering) does not mean you have to associate with members every weekend, or even see them on a weekly basis. If you have the opportunity to interact with them regularly, which membership (and volunteering) allows, you will become better acquainted. People will discover your talents, your values, and your character, which you cannot usually accomplish through a single 30-minute meeting.

Leading. Members in organizations have opportunities to lead by holding an office or chairing various initiatives the organization sponsors. This level of involvement offers the greatest visibility and thus requires a great deal of time and attention. The benefits, however, can be far reaching. First, it is an opportunity to showcase your talents and abilities without uttering one word of self-promotion. Second, the leaders of these organizations are usually the more successful and influential people in your community. And, depending on the purpose of the group, they have learned that success involves contributing to the good of the community. In other words, they are most likely good centers of influence.

Writing

Would you like to see your name and picture in the newspaper without having to purchase an advertisement? If you have some writing ability and your target market reads a certain publication, then write an article and have it approved by your compliance department. Print publications of all sizes are seeking quality content that serves the interests of their readers. Their need is your opportunity to get your name and company in front of a large number of prospects in a highly cost-effective manner.

When you write an article or a book, you are regarded as an authority. This gives you the credibility you need to build relationships with prospects who are total strangers. It requires only an investment of your time (admittedly, a substantial commitment in some cases) to write and research publishing opportunities.

When you publish an article in a community newspaper, a trade magazine, or an industry newsletter, you can leverage its marketing value beyond the time that the publication is in circulation. If you collect or purchase extra copies, you can send them to prospects, providing them with valuable information, with the added bonus that your name is in the byline. Thus, you are positioning yourself as an expert. When the time comes for the prospect to call someone, you have increased the odds in your favor!

If you want to get published, consider these points:

- Research publications that your prospects and clients read.
- Explore these publications. Contact the editors, and inquire about their needs. Remember, they are not interested in promoting you (even if you may be an advertiser). They are interested only in serving their readers—help them achieve that objective, and they will help you achieve yours.
- When writing, don't pitch your product or service. Editors will discern this and not run your piece. Your article should offer useful information. If you can tie your advice to current events or market trends, it is even better. Place yourself in the prospect's position—what does the prospect care about? What does the prospect want? Address these issues.
- If you're feeling stuck, tell a story. Has one of your clients or associates experienced a life event that would relate to others?
- Another approach to writing is to present solutions or a message in numbered points, such as "The Ten Secrets of Estate Taxation" or

> **When you write an article or a book, you are regarded as an authority. This gives you the credibility you need to build relationships with prospects who are total strangers.**

"Five Ways to Know if You Are Adequately Insured." Such a method makes writing easier for you, and the learning easier for the reader.

- Leverage the article for all it is worth. Although you may not be able to reprint the article word for word (check with the publication regarding its guidelines), you can quote from it or perhaps link to it from your website.

- Make sure credit is given where it is due. This not only includes other experts or publications you may reference, but also yourself. At the conclusion of every published article there should be two or three sentences about you—who you are and what you do—and at least an e-mail address and preferably also a phone number. Note that you are available for any questions or to speak publicly.

- Get approval from your compliance department.

Speaking to Groups

speaking to groups   ✓

Speaking to groups entails giving presentations to organizations or businesses as a way to build your prestige, as opposed to generating immediate appointments and sales. Whether you work in a big city or a small town, there are many officers of social, civic, and fraternal organizations in your community who need a speaker for next week's or next month's meeting. Sometimes, a human resources representative may be looking for someone to teach a financial topic as part of a low-key educational opportunity offered to employees. Both of these situations are ways for you to increase awareness of who you are and what you do.

> Do not dismiss the opportunity for public speaking without at least obtaining some training and trying it out.

Cutting Your Teeth. For most people, public speaking is one of the major phobias of life. Perhaps you feel that way, too. Although you should choose social mobility activities that are in line with your personal strengths, do not dismiss this opportunity without at least obtaining some training and trying it out. Winston Churchill was purported to have had a lisp and found public speaking difficult. Today, he is regarded as an orator of great renown. His secret? The same as for other great men and women: practice and perseverance.[2]

If your target market is accessible through speaking engagements, strongly consider taking a class on public speaking and joining a speech club, such as Toastmasters International (www.toastmasters.org), to practice giving speeches. Start with speeches on simple material and proceed to ones that deal with your financial expertise and knowledge. Then try other nonthreatening venues in front of people who know and love you.

Identifying Material. Even if you do not feel ready to make speeches, begin building files or writing outlines. Think about the topic areas that would interest your target market. Then keep an ear out for opportunities to speak informally to groups whose members are people from your target markets on topics about which you feel confident.

Knowing Your Information. The best part of speaking is that it is interactive. You have a live exchange, a dialogue, with your audience. This gives you the opportunity to build rapport and have a spontaneous give and take with people who could become your clients. Because, however, some aspects of speaking may not always be rehearsed, it is important for you to *know your information,* and know it well. The depth of your knowledge and understanding will reveal themselves in how well you field questions and reach a shared understanding with your audience.

Educating and Motivating. The goal of speaking engagements is first and foremost to establish your reputation as an expert. Thus, focus on educating and motivating your audience. Outline the problems clearly, but create a sense of curiosity and a need to seek solutions.

Promoting Your Practice. Keep promotional information to a minimum. It is acceptable to have a "word from our sponsor"; just keep it short and put it at the end. If people want more information, they can contact you.

Having said that, members of the audience should take with them—at a minimum—your business card and/or personal brochure (discussed later). Ideally, they should also take something else, such as a Q & A brochure, a handout with your presentation slides on them, or some other handout that will remind them of your important message points. Obviously, your contact information should appear on everything you distribute.

Other Media Opportunities

There are other media opportunities, including local radio and television. There are many financial advisors who host their own 1-hour radio or television show on local public access channels. Others find their way on to local radio talk shows and local television news, and into newspaper articles as financial experts. Let local media know you are available, and inform them of the topics on which you can provide expert opinion. If they call you, the exposure is free and will help establish you as an expert in your field.

Personal Brochure

A second important method for creating an awareness of your personal brand is a personal brochure (for some captive agents, this may not be allowed). You may have seen these brochures from other successful professionals or advisors, or from the service providers who create them. The *personal brochure* is typically a one-page (usually front and back) document

personal brochure

that introduces the advisor. Treat it as the prospect's first impression of you. Therefore, it should impress, inform, and create interest.

Personal Brand

Review your positioning statement and value proposition. The theme of your personal brochure should promote your unique position in your target market's mental list of advisors who do the type of work you do. Pin down the most compelling reason that members of your target market should work with you.[3] Often, the essence of this unique position is boiled down to a slogan of sorts. For example, an advisor, a former GM human resources employee, whose target market is employees from General Motors, uses the slogan, "Steering GM employees to financial freedom."[4]

Once you have identified the main theme of your personal brochure, everything else should be consistent with it, including the information and design.

> Pin down the most compelling reason that members of your target market should work with you.

Standard Information

A personal brochure should include the following standard business information:

- company name
- company address
- company phone number
- e-mail address
- web address
- products and services offered

Personalized Information

The amount of personalization runs the gamut. There are some advisors who subscribe more to a resumé approach, listing their professional accomplishments and credentials with very little, if any, personal information. Other advisors utilize more personal information, communicating both their credentials and their personality. How much personalization you use will depend on your target market. It is important that you know what information members in your market want to know about their financial advisor.

Some common personal information that many advisors share in their brochures, includes the following:

- photographs of the advisor and/or family
- pictures of the advisor taking part in his or her hobbies and interests

- list of causes and organizations the advisor supports
- description of values and convictions relevant to the target market

The theory behind including such personalized information is that it attracts prospects with whom the advisor wants to work and repels those whom the advisor would like to avoid.[5] The theory rests on the fact that individuals generally would rather conduct business with people who are most like them.

Credentials, Experience, and Accomplishments

> **Design your brochure from the perspective of what members in your target market want to know.**

Remember to design your brochure from the perspective of what members in your target market want to know. Include accomplishments and credentials that *they* feel are important. If you went to a cardiac specialist, how would you feel if her resumé indicated that she was the highest paid doctor on the East Coast? You might care. But you would definitely care to know that she was recognized as one of the top 10 cardiac surgeons in the country.

Example:	You qualified for Million Dollar Round Table (MDRT) honors. Do you include this in your personal brochure? Ask, "Is it important to members of my target markets? Will it attract them?" If you are working with high-net-worth markets, this information will probably be important (successful people want to work with successful people). But it might repel target markets in the middle class because the connotation might be that you made a lot of money at their expense.

Most important, present your experience and credentials in a way that emphasizes the benefits to the prospect. Prospects may not care that you have earned your LUTCF or ChFC. That is, unless you can tell them how it benefits them. Okay, it is impressive that you have been working with business owners for 10 years, but what does that mean to the prospect?

Compare the statements in the following examples.

Example 1:	In 2003, Julie earned her LUTCF designation from The American College.
	The American College's LUTCF designation that Julie earned in 2003 included three courses related to retirement and long-term care planning. These courses have equipped Julie to help her clients clarify

their retirement goals, spot potential problems, and provide effective ways to achieve their dreams and protect their families' financial futures.

Example 2: Sue has been working with small business owners for 10 years.

Sue's 10 years of working with small business owners means she has the experience to identify and address financial problems that often are overlooked.

> **Instead of meaningless accolades, communicate your service and expertise in terms of the results you have achieved for clients.**

Avoid the clichés regarding your outstanding service and expertise. Everyone is "top rated," provides "exceptional service," has "integrity," and is "uniquely qualified." Reading personal brochures that many advisors use could make you wonder if they all live in Lake Wobegon, Garrison Keillor's fictional Minnesota town, "where the women are strong, the men are good looking and all of the children are above average." Instead of meaningless accolades, communicate your service and expertise in terms of the results you have achieved for your clients.

Example 1: Kelly has helped many of her clients save a minimum of $100 per month for their children's college education. The benefit? Their children graduate with a lower student loan balance and, consequently, a better starting financial position in adult life.

Example 2: In 2 years, Mark has helped more than 100 moms and dads implement life insurance plans to ensure that their children's basic needs are met—no matter what!

Layout and Printing

A promotional brochure is the first impression many prospects will have of you. Unless you have a graphic design bent, you should strongly consider delegating the layout and design to a professional. Yes, it will cost you something. But it will be worth it. Spend a little money here and put your best foot forward.

Some companies offer their advisors a template from which to create a compliance-approved sales promotion piece. In addition, there are vendors who will design and print brochures. If you go with the vendor option, choose one wisely. A good idea is to look at other sales professionals' brochures, select a few that you find impressive, and obtain the names of the companies that created them.

In either case, verify the brochure's appeal with selected members of your target market before you go to print.

Compliance

As with other materials, your brochure must meet your company's compliance standards. Make sure you engage your compliance department early in the process. Keep in mind some of the obvious, but important, do's and don'ts:

- Do make clear any company affiliations.
- Do list any financial products you sell.
- Do identify the business you are in (insurance, mutual funds, and so on).
- Don't claim designations you have not earned.
- Don't exaggerate your accomplishments.
- Don't refer to yourself or your services in a way that would mislead the public.

Uses for Your Personal Brochure

The personal brochure is a tool that you can use in many different situations so long as those situations are congruent with the purpose of the brochure: to introduce you to the prospect. Here is a quick list of ways to use the personal brochure:

- Send it along with a cover letter as a preapproach.
- Use it as one of your handouts at a seminar or public speaking engagement.
- Give a few copies to centers of influence to pass on to potential prospects.
- Gain permission to place it in strategic locations frequented by your target market (chamber of commerce, a doctor's office, a book store, senior center, and so forth).
- Pass copies out at a trade show or an event you sponsor.

Internet Website

Does the ubiquitous nature of the World Wide Web, the proliferation of websites, and high-speed Internet access mean that you should join the fray?

The World Wide Web with its graphical and text interface as we know it was born in the early 1990s. It is absolutely amazing to see the evolution of this technology, and the best is yet to come. The World Wide Web has revolutionized communication, giving people of all walks of life access to information (and misinformation) at the click of a button.

But does the ubiquitous nature of the World Wide Web, the proliferation of websites, and high-speed Internet access mean that you should join the fray? If so, what should you attempt to accomplish by your website? How should you go about doing it?

In discussing these three questions, we acknowledge two things. First, not all advisors are able to implement what follows. For you, the information discussed here will help you understand what other advisors can offer. There are some avenues to explore with your company that may enable you to establish some kind of presence on the World Wide Web using a standard website created and maintained by your company. Second, for those of you who have the flexibility to create a website, remember to consult with your compliance department.

Assessing the Need for a Website

According to an April 2006 report by The Pew Internet & American Life Project, 73 percent of all adults in the United States use the Internet (the Internet means, among other things, surfing the World Wide Web and/or corresponding via e-mail) and 42 percent of all households access the Internet through broadband connections (DSL, cable, and so on).[6] In addition, consider the following statistics:

- Eighty-seven percent of the people who are shopping online for auto insurance and are willing to switch carriers are going online for rate quotes; 58 percent of them are willing to purchase online.[7]
- Six percent of all sales made by agents of one large auto insurer were from referrals generated from its website.[8]
- Although only 2 percent of all life insurance premiums are due to policies sold through direct marketing channels (including the Internet), they represent 20 percent of all policies sold.[9]
- Thirty-one percent of all online consumers have researched life insurance on the web. For annuities and investments, it is 18 and 38 percent, respectively.[10]

Furthermore, as the purchase of cell phones and other wireless devices that offer Internet access grows, more and more people will turn to the World Wide Web for research and on-the-spot information to answer a question like "Does my auto insurance agent sell life insurance, too?" Thus, the short answer to the question, "Do I need a website?" is "Yes, unless your target market *never* uses the Internet."

For many captive advisors, there is no reason *not* to have a website. Costs, if any, are minimal because it is a template-generated website offered by your company. For independent advisors, the costs are minimal simply to have at least a personal brochure-type website.

The real question, and one that will have a different answer for each advisor, is "What kind of website do I need?" Ultimately, the only way to know is

> **The real question, and one that will have a different answer for each advisor, is "What kind of website do I need?"**

to ask members of your target markets. But, in general, if your target markets are composed of younger adults, your website will probably need to be more involved. Consider that a LIMRA study found that the percentage of Generation X (those born from 1965 through 1976) who shop online as reported in a LIMRA study in 2002 was 90 percent. It was 87 percent for Generation Y (only those born from 1977 through 1984 were considered for this study).[11]

Setting Objectives for a Website

As you evaluate what content you need to include on your website (provided you are afforded this opportunity), think through its purpose. Specifically, consider the following five possible objectives:

- to create an emotional connection
- to establish credibility
- to sell
- to educate and motivate
- to service

Create an Emotional Connection. This is the most basic and universal objective for a website. It is the same as the main objective for a brochure. You want to help prospects feel comfortable with you before you contact them for an appointment or *vice versa*. In a nutshell, you want to introduce yourself and what you do. Most advisors achieve this on their home page and in an "About Us" area of the website.

Basically, the "About Us" area should include the same information as your personal brochure. Some advisors include a replica of their personal brochure online. Here is a brief list of information to provide on the About Us page:

- your name
- an overview of what you do (communicate your value proposition from the prospect's perspective)
- your story (telling where you came from and why you are doing what you are doing)
- credentials and relevant experience
- contact information
- professional organizations to which you belong
- any employees or support staff (you might consider using first names only for privacy concerns)

This page is also a good place to include a personal photograph of you and/or your family and of any employees and support staff. (You should consult with an expert regarding the need to obtain a signed waiver from each employee or support staff before you post their pictures online.) It's also an appropriate place to put the mission statement for your practice, if you have one.

Establish Credibility. Your website should also aim to establish your credibility. The inclusion of your credentials and experience on the "About Us" page is a good example of information that will boost your credibility. Frame your experience and credibility from the perspective of their value to the prospect (just like in your personal brochure). In addition, you may want to do the following:

- Create case histories (using fictitious names, of course) developing the problems clients faced. Emphasize the potential ramifications of the problems. Assess them honestly (do not exaggerate). Provide a general description of the solution.
- Post testimonials of clients. Avoid fluff: A testimonial could read, for example, "Working with John has been a pleasurable experience." But what did you do for these clients? How has your work improved their financial position? An emotional appeal is good, but you need to substantiate it with facts. You could combine testimonials with case histories.
- List the carriers and affiliates with whom you work. Link to their websites.

Sell. By now you probably have an inkling that selling over the Internet has not yet made the big splash many experts predicted. There are many reasons for that, including a concern for privacy, the complexity of the products (especially insurance), and so forth. (Keep your ears open—this could change in the future.)

Educate and Motivate. So if people do not go online to buy, what do they do? Online consumers who visit a financial services website are coming mainly to obtain information. As reported earlier, although for all ages, not more than 50 percent of online consumers research information for insurance, annuities, and investments, for those in Generations X or Y, it is 69 and 67 percent respectively.[12] Thus, the pages on your website devoted to education are critical. Here are a few suggestions of what to include:

- *articles about financial topics* about which your target market wants information. Your company may have some articles, or many website developers who work only with financial advisors will provide content for you. If you can write and have time to do so, this is the best option because you will know what specific topics to address.

- *financial calculators, worksheets, and spreadsheets* that will help members of your target market make financial decisions
- *links to other websites* (Bankrate.com, Kiplinger.com, Money.com, MorningStar.com, MotleyFool.com, SavingforCollege.com, and so forth). You should obtain permission to link to a home page (the page that comes up when you put in www.sitename.com). Permission is required for links to any pages other than the site's home page.

Again, make sure all materials on your website are approved by your compliance department.

In addition, consider other issues that affect your target market. Presumably, you have much in common with members of your target market. Therefore, you do not have to do a lot of extra work to discover and understand them. For example, if many people in your target market own or will purchase a home in the near future, information and links about home ownership and buying a home might be an added value. Also, if your target market is within a fixed locale, you might include community information and local merchants.

weblog (blog)

Some advisors have gone so far as to create their own *weblog*, or *blog* for short. A blog is an online journal of sorts that allows for interaction between you and your web audience. Advisors who blog typically will post an article from a financial publication or online news source and give their opinion. Readers can then ask questions and make comments (if the advisor selects this option). Postings need to be frequent; weekly is a good goal. This strategy is more appropriate if you are dealing with a target market that is regional or national and one that is interested in blogs.

Service. Because this textbook is focused on prospecting, we will hit this topic briefly. Think about the types of service transactions that can be requested online. An e-mail form can be created to request a transaction. In addition, you can instruct the client about the information needed to process a change. This will expedite the process and increase your efficiency.

Creating a Website

Now that you have some background information, let us talk about what you need to do to create a website.

First, talk to members in your target market. Find out the following:

- Where (books, websites, magazines, radio, television, and so on) do they go to obtain information about (your products and services)?
- What websites do they visit to gain information about (your products and services)?
- What information would they like to see?
- Have they ever used online financial calculators? Which ones? How did they like them?

Think about how much time it will take to update your website. A site that doesn't change may not be effective for some target markets.

Second, identify the different areas you want to include on your website. Think about how much time it will take to update your site. A site that doesn't change may not be effective for some target markets. Here are a few areas you might want to include:

- home page
- index
- site map
- About Us page
- company/carriers
- glossary of financial terms
- search engine (to search the site)
- articles and information
- stock quote
- financial calculators
- event calendar (for seminars and other events)
- driving directions
- referrals to other advisors

Third, look for a professional website developer. If you do not know one, consider a company that specializes in developing websites for financial advisors. Some of these companies provide articles and financial calculators for additional cost. Unless you are planning to offer portfolio tracking, insurance quoting, and so on, the cost for developing the website will be fairly reasonable. You will not need to spend thousands of dollars.

Fourth, do not use third-party information without securing permission. As already mentioned, it is a good practice to obtain permission simply to link to other websites. In addition, observe the copyright laws. Give attribution when someone else's material is used. As always, run all of your content through your compliance department.

Track whether or not prospects and clients have been to your website, and obtain feedback for suggested improvements.

Fifth, when the website is finished, put the web address on everything that goes out of your office. Publicize it on your newsletters, business cards, advertisements, personal brochure, and so forth.

Finally, track whether or not prospects and clients have been to your website, and obtain feedback for suggested improvements.

Newsletters

Newsletters are another way to build prestige and create interest in your products. Although traditionally geared to relationship building and marketing to current clients, newsletters can be used with prospects as well.

One cost-effective method is to send your newsletters to selected prospects who have indicated an interest in them. If you are writing some of the articles, this will be an opportunity to demonstrate your expertise.

The same holds true if you send out your newsletters via e-mail and/or post them to your website. The added bonus here is that typically, an electronic newsletter is less expensive to produce than a printed one. If you do send your newsletter electronically, put a link on your website that allows people to sign up for it online.

Finally, you can send a prospect a newsletter that addresses a topic or need in which he or she has expressed an interest or concern. You could send this newsletter with a note, "It was nice to meet you. I thought you might be interested in this." Even if you have not written any of the articles, this method demonstrates that you listen to people. Furthermore, the relevant article(s) may make the prospect more aware of a need and more open to meeting with you.

Advertising

advertising

Advertising is the use of persuasive messages communicated through the mass media. The ultimate goal of advertising is to create new clients. Unfortunately, the effect advertising has on client creation may be difficult to measure. Remember that prospects often need multiple triggers before they are moved from awareness of their need to seeking relevant information. In situations in which multiple triggers were required, which trigger was unnecessary?

In addition, prospects want to work with advisors with whom they have some level of comfort and trust. All things being equal, they will want to work with an advisor they have at least heard of rather than a total stranger. Advertising promotes in a prospect a level of familiarity with the advisor. It may predispose the prospect to a favorable response when the advisor does finally approach him or her.

Advertising promotes a level of familiarity with the advisor. It may predispose prospects to a favorable response when the advisor approaches them.

Yet having said all that, the case for advertising is not universally accepted by successful advisors. Judging from the arguments on both sides, this is definitely one awareness strategy whose effectiveness depends largely on the target market and its interaction with various advertising media.

Determine How Your Target Market Identifies Potential Advisors

Determine how members in your target market find advisors like you. Do they use the phone book, or do they almost exclusively use referrals? The type of products you sell could influence this. For example, homeowners and auto insurance are products for which prospects often will shop for the lowest price, using online rate quotes and/or a telephone directory. Advertising in a telephone directory (online or book) would make sense for these types of products. The bottom line is that if your target market turns to a particular source for financial advisors' contact information, that is a good place to advertise.

Consider Any Places Your Target Market Frequents

Consider advertising in any "places" a high concentration of your target market frequents. By places, we mean physical locations as well as radio shows, radio stations, websites, and so on. For instance, one very successful life insurance agent began advertising on a radio show geared toward the traditional senior market. Eventually, she gained an endorsement from the show's host, and her annuity sales skyrocketed.

Select Advertisements Relevant to Your Target Market

Select advertisements that are appropriate to your target market. If you are able to create your own marketing messages, consider retaining the services of an advertising firm for assistance. As always, check with your compliance department for approval and guidance.

Track Your Advertising Efforts

Track prospect interaction with your advertising efforts. Ask prospects (regardless of the source) if they have seen any of your advertisements (in addition to whether they have been to your website). Some advisors ask for this information using a form much like the one you complete when you go to a doctor's office for the first time (see figure 4-1 for some sample questions).

FIGURE 4-1
Sample Questions from a Prospect Information Form

How did you hear about us? Please check <u>all</u> that apply.

❑ Yellow Pages

❑ Heard your ad on radio station
_____

❑ Saw your ad in _____

❑ Phone directory on the Internet

❑ Received your letter or postcard

❑ _____ told me about you

❑ Your company's website agent locator

❑ Saw your billboard at _____

❑ Link from website _____

❑ Other_____

PREAPPROACHES

preapproach

Although there is overlap between prestige building and preapproaches, we delineate between the two. Whereas prestige building deals mainly with creating an awareness of who you are and what you do, preapproaches focus more on heightening a prospect's interest in your products and services. Thus, a *preapproach* is any method used to stimulate the prospect's interest and precondition him or her to agree to meet with you. This section will provide information on seminars and preapproach letters. The bulk of our attention will focus on seminars.

Seminars

A seminar is a prospecting method in which you, alone or as a part of a team of professional advisors, conduct an educational and motivational meeting for a group of people. Seminars are distinct from speaking to groups in that a seminar's objective typically is to produce appointments. Thus, there is a predetermined method for asking for appointments after the seminar has concluded. Sometimes, however, seminars are designed with prestige-building objectives in mind.

Advantages of Seminars

> **Seminars give advisors a way to identify prospects, build prestige, preapproach prospects, qualify them, and approach them for an interview —all at one time.**

Seminars are the marketing version of a Swiss army knife. This all-in-one marketing method gives advisors a way to identify prospects, build prestige, preapproach prospects, qualify them, and approach them for an interview—all at one time. Not surprisingly, seminars offer several key advantages, many of which are related to their ability to accomplish multiple marketing tasks.

Increased Efficiency. Due to the multitasking nature of seminars, they are a very efficient marketing tool. For instance, one of the main objectives of an initial interview is to qualify the prospect in terms of his or her need and motivation to meet that need. Typically, such a task necessitates a presentation of sorts, which takes time. Assume that this portion of the initial interview requires 30 minutes to complete on average. This means that a 2-hour seminar with four people (or four couples) takes the same amount of time to achieve the same objectives as four initial interviews—not including travel time to and from the appointments, if applicable. Seminars with more participants increase the efficiencies of this approach, reducing the amount of time needed to conduct the initial interviews.

Effective Prestige Builder and Preapproach. Assuming that you and any other presenters give educational and motivational presentations, seminars build your credibility as an expert. What other method allows you to demonstrate your knowledge of the financial problems prospects face and the solutions you can provide? Moreover, not only can seminar attendees evaluate your expertise, they can also observe your personality and other intangibles that build trust. What better way for prospects to judge your personal brand than to interact with you?

In addition, think about how much more effective even a 1-hour seminar is in helping prospects identify financial needs than a postcard or letter. You can raise more than one probable need, which increases the odds that you will find the trigger to make the prospect want more information and eventually buy your products. Try that with a postcard!

Maximize Public Speaking Skills. If you are someone who enjoys presenting and teaching, seminars are a way to leverage these abilities. Some advisors are natural-born public speakers, making this form of marketing extremely enjoyable and productive.

If you have an aversion to public speaking, you can still take advantage of seminars by arranging for other advisors, agents, company specialists, and experts to speak. A great format is to share the presentation with them, giving them the bulk of the speaking time. Be sure, however, to take an active part of the presentation so that new prospects can gain a sense of your professionalism and expertise.

Segue Naturally into an Appointment. The natural conclusion of a seminar is either to ask prospects for appointments or to set the expectation that you will be doing so. In either case, you have a basis for asking.

Planning a Seminar

> **If a seminar is not well planned, you risk losing credibility with the people who attend.**

Seminars do not just happen. They require a great deal of thought and preparation to attain the degree of professionalism crucial to their success. If a seminar is not well planned, you risk losing credibility with the people who attend.

It is helpful to have a written game plan—a step-by-step description of what you need to do and when you need to do it. A checklist, such as the one provided in figure 4-2, lets you see at a glance where you are and what remains to be done.

Define Your Objective. The first step in the planning process is to set your goal or objective. Before you do anything else, ask yourself what the seminar will accomplish. Your goal should be specific, attainable, and measurable. For example, you may set a goal of making 10 appointments with seminar attendees. Or your goal may be to provide an informational seminar for 15 of your best clients. The goal you set is important because it will affect other decisions regarding the seminar details.

Set a Budget. It is especially helpful to establish a budget and work to stay within this constraint. Food is generally not served, but if you wish to provide refreshments, keep them simple and inexpensive. If you are using a hotel or restaurant, you may need to resist the pressures to serve lavish meals. Remember you are selling your knowledge and expertise—not giving away food.

FIGURE 4-2
Seminar Checklist

Arranging for a Meeting Room

- Has the meeting been confirmed?
- Who's your contact?
- Have you met this person?
- Have you seen the facility?
- Have you done a walk-through to see what it will be like when prospects arrive and go to the meeting room?

Guest List

- Do you know who will attend?
- Will anyone from your company be there?
- How many prospects will your budget cover?

Seminar Promotion

- Have you arranged a schedule for printing promotional material and material to be distributed during the seminar? Does the schedule include time to proofread printed material to avoid potentially embarrassing errors? Does it allow for delays at the print shop?
- Have letters or formal invitations been sent at least 4 weeks prior to the seminar?
- Have the invitees been called during the week just prior to the seminar to confirm their attendance?
- Have announcements been sent to local newspapers, radio stations, and other media outlets?

Getting There

- Do prospects know the date and precise time of the seminar?
- Do they have directions to the meeting rooms?
- Will they have to use a special entrance?
- Have you asked the hotel or office complex for signs with directions?
- Will you have a check-in desk?
- How will early arrivals be handled?
- What are the provisions for parking?
- Will your prospects need a means of identification such as name badges?
- Who will look after late arrivals?
- Will you need a message board?

Meeting Facilities

- How will the room be arranged?
- Do you want chairs only or chairs and tables?
- Will you need reserved seat signs?
- Are you supplying paper, pens, and pencils?
- Will you need a microphone? If so, do you have an extra one in case one cannot be located at the meeting facility?
- Will you need a lectern?
- Will the speakers need help with any equipment during the presentation?
- Is the lighting adequate?
- Will there be an overhead projector?
- Will there be a screen?
- Will you have spare projector bulbs?
- Will you need a flip chart, writing pad, or magic markers?
- When the overhead projector is on, do the main room lights need to be turned off? If so, who will do this?
- Have you supplied water and glasses for the speakers?
- Who will make the introductions?
- Will there be a coffee break?

Seminar Content/Speakers

- What topics will be covered?
- Who are the best speakers to address each area?
- How long should each person speak?
- What is the most appropriate format for the presentation (for example, lecture, round table discussion)?
- How much compensation (if any) should the speakers be offered?

Problems

- Whom do you contact if there is a problem?
- Will the house audio/visual technician be available?
- What do you do if someone becomes ill?
- Where are the fire escapes located?
- Are there any security measures required?
- How will people make outgoing telephone calls?
- Who will handle the incoming calls?
- How will you deal with cell phones ringing in the middle of the presentation?
- Do you have a reliable assistant to help you run the meeting?
- Have you thought of everything that could possibly go wrong?

Invite the Right Audience. Seminars involve target marketing. The people you invite should have a common interest or need. This allows you to focus on the specific needs of the audience. For example, if you sell long-term care insurance, you would have more success targeting either retirees or preretirees, but not both at the same time. With retirees, the context of the seminar could be health care needs, whereas with preretirees, a context of planning and saving for retirement would be relevant.

Once you have identified a target audience, you will need to determine how many people to invite. Begin by setting a goal for the number of attendees you wish to have. This will depend on the purpose of your seminar. Keep in mind that it takes roughly 15 to 20 attendees to justify the expense of holding a seminar, depending on the selection of facilities and other frills. You will learn from experience how many people to invite to have the desired number of attendees. One rule of thumb is to invite 10 people for each desired attendee. The ratio of invitees to attendees will improve as your seminars become better known.

Example 1:	The purpose of Manny's seminar is to build his reputation in the community as a financial planner. His objective is for as many people as possible to see him share the podium with a group of respected local experts on retirement. His goal is to have 50 attendees.
Example 2:	Helen's seminar series is geared to prospects she does not know at all. Her guidelines are to keep the attendees to a manageable number so that she can maximize contact and ensure she can meet with each interested prospect within 3 weeks after the completion of her seminar series. She feels that a goal of 15 to 20 prospects is in order.
Example 3:	Ichiro decides to hold a seminar for CPAs and other noncompeting financial professionals to educate them on the need for long-term care insurance. Ichiro believes that a smaller number of attendees will be most effective. Thus, his goal is 10 attendees.

The next task in planning the seminar is to create a list of names. Your list should come from the prospecting sources you have previously identified. One of the more popular sources is list directories from third-party vendors. Review the information in chapter 3 on prospecting sources.

Choose the Right Topic. The content of your seminar should be a blend of technical information and motivational material. How much of each

If you give too much information, you may bore your prospects or they may have no need to make an appointment with you.

depends on the needs of the prospects you invite. The program should be technically accurate and informative. If you give too much information, however, you may bore your prospects or they may have no need to make an appointment with you. In addition to being educational, the program should motivate attendees to meet with you for more information.

After you decide on a topic, your next step is to figure out whether material is available or if you must develop it yourself. Many companies have seminar material for their advisors and require advisors to use it. If your company allows it, you can purchase seminar presentation material. If you decide to develop and write your own seminar material, set aside enough time to research your topic thoroughly, have adequate resources to produce it, and allocate time for your compliance department to review it.

On the other hand, you may decide to use the services of a company expert or another advisor such as a broker, CPA, or attorney. (Remember to follow your company's compliance procedures for situations in which you bring in an outside speaker.) You may even decide to help organize and cosponsor seminars in which other advisors are the seminar leaders. This may be a good choice until you have enough experience to conduct one yourself. If you pair up with another advisor, you may need to agree on some form of compensation such as a split in the commission. If you both invite prospects, you could offer to pay for the cost of the seminar, especially if the other advisor is doing the majority of the speaking.

Consider a Team Approach. One of the most effective ways to develop and present a seminar is to use a team approach. For example, for an estate planning seminar, the team could consist of a life insurance agent, an estate planning attorney, and/or an accountant versed in estate planning issues. In effect, prospects are offered a complete package of services. So in our example, the life insurance agent could open with an overview of estate planning issues. The attorney could then outline the law and legal documents necessary to implement an effective plan. Next, the accountant could review investment strategies, costs, and financial planning issues. Finally, the agent could explain how various products solve needs and dovetail with both legal and financial planning.

Choose a Date and Time. This may seem mundane and unimportant, but choosing a date and time is critical. The optimal day and time will depend on the needs of your audience. Avoid holidays and competing events. You may plan a terrific seminar but if it competes with the NBA finals, the county fair, or the latest popular reality TV show, your attendance may be adversely affected. Also, consider the time element. It would be foolish to hold a seminar for working prospects during working hours. Similarly, it may be unwise to hold a seminar for seniors late in the evening.

Select a Convenient Location. The site you select should be convenient for the members of your targeted group. Parking may be a prime consideration in urban and suburban areas. Also, the location should be neutral. It is generally recommended that you do not use your office. The fear here is that invitees may see your office as a high-pressure environment and therefore fail to attend. An outside space creates an environment of neutrality and objectivity.

Furthermore, the accommodations should match the size of your group. A small group in a large room gives the appearance that a lot of people decided not to come. On the other hand, too small a room may result in people leaving your seminar because they feel overcrowded.

The traditional approach to seminar selling has been to list one date at a hotel location. You may wish to consider two variations to the traditional approach. First, rather than scheduling your seminars in a hotel, consider using a college campus or local library, which may be more consistent with a low-key consultative approach; it may also cost less. Second, you may find that you obtain better results by providing prospects with a choice of both date and location for your seminar. For example, an invitation might list the following choices for a retirement planning seminar:

Tuesday, January 10, at 8:00 p.m.
at the Manhasset Library Community Room

Wednesday, January 11, at 8:00 p.m.
at the Nassau College Boswick Hall

Thursday, January 12, at 8:00 p.m.
at the Great Neck High School, Room 202

Here, the invitation offers three dates and three locations within a 15-mile radius. By offering a variety of choices, you are less likely to lose an interested prospect because of family scheduling conflicts or a reluctance to travel to a specific location.

Announce the Seminar. Once you select a target audience, it is time to consider the invitation. The invitation should clearly inform the prospect that the seminar will be educational in nature. It should provide the topic, date, time, and length of each seminar session as well as any fees to be paid. The seminar title should be clear and relate to the perceived needs of the targeted audience. If your seminar will include workbooks or data forms, the invitation should highlight the benefits of these materials to the prospect. You may wish to plant a seed for follow-up by informing the prospect that each attendee will be entitled to a personalized complimentary analysis of his or her situation. Also, if an expert will be featured, note this on the seminar invitation together with the expert's credentials and accomplishments.

In addition to the invitation itself, your letter should contain a response mechanism for more information—a telephone number, e-mail address, or a stamped, self-addressed postcard, as well as a number to call to reserve a seminar seat. Some financial advisors send a second (reminder) mailing. For best results, follow up replies with calls confirming attendance a day or two prior to the presentation.

Finally, it is important to monitor both the mailing of the invitations and the response rate. Careful monitoring will allow time for you to make adjustments, if necessary. The mailing itself should usually begin at least 4 weeks prior to the seminar date. When you are sending large numbers of invitations, seed the mailing with your name and those of a few trusted friends or colleagues. This gives you a mechanism to determine whether or not invitations have reached their intended destinations.

Check the Facilities. Checking and rechecking the facilities you have chosen will help your meeting run smoothly. Try to visit the facility while another meeting is in progress. This will give you the opportunity to evaluate the lighting, the sound system, and the visibility of any screens you will use with a projector. You can assess how well everyone in the room can see the speaker and judge whether the ambiance of the room reflects the feeling you wish to convey to your audience. Consider what audiovisual equipment or visual aids, such as an easel or whiteboard, you will need before you begin calling facilities.

Feedback Mechanism. Plan to use a feedback mechanism. Some seminar presenters ask attendees to sign in, giving their names, addresses, and phone numbers. Others design an evaluation form that asks the attendees for this information, as well as for feedback on the quality and usefulness of the presentation. See figure 4-3 for a sample postseminar questionnaire.

Addressing Miscellaneous Details. It is important to pay attention to details; they send your prospect a message about yourself. Remember to bring name tags, pens, paper, and handouts. Do not skimp on quality. For example, handouts of the highest quality that include your name and address convey the message that you are professional and that you are willing to put your name on the work you do. Having paper and pens available sends the message that what you say is important enough to write down.

> **It is important to pay attention to details; they send your prospect a message about yourself.**

Preparing the Presentation

An integral goal of seminar marketing is to enhance your image in your marketplace. Therefore, it is important to take all steps possible to heighten the professionalism of your presentation. Use a written script and rehearse it in preparation for the seminar. When conducting the actual seminar, however, it looks more professional if you speak from an outline or note cards, not a script.

FIGURE 4-3
Postseminar Questionnaire (Long-Term Care Seminar)

We would greatly appreciate your help by answering the following questions. Your responses will guide us in improving future seminars and determining if our organization may be of assistance to you in the future.

1. Which speaker did you find the most helpful?

   _____

2. Which subject(s) covered during the seminar was (were) of most interest to you?

   _____

   _____

3. Briefly list any subject areas that were not covered in the seminar that you think should have been covered.

   _____

   _____

   _____

4. You may have some questions that you did not have an opportunity to raise during the question-and-answer session. Please briefly list these questions below. We will make every effort to answer these questions for you or to find experts who can provide responses.

   _____

   _____

   _____

For the following statements, circle the response that best describes how you feel.	Strongly Disagree		Agree		Strongly Agree
I fear depending on family or friends for my care as I get older.	1	2	3	4	5
If I ever need long-term care, I want the freedom to decide what type of care I will receive and where I will receive it.	1	2	3	4	5
I am afraid I may not qualify for long-term care insurance if I wait too long.	1	2	3	4	5
I want to protect my assets from the potentially high costs of long-term care.	1	2	3	4	5
I am concerned about taking care of my parents or another relative should they need long-term care.	1	2	3	4	5

Name: _____   **Address:** _____

Phone: _____   _____

Fax: _____   **E-mail:** _____

A written script is particularly important if there are several presenters. It provides a framework for the presentation, familiarizes the presenters with the material, and helps them develop answers to potential questions that may arise from the audience.

If you are working as part of a team, be sure that each speaker is aware of the overall scope of the program to achieve a cohesive presentation and avoid repetition. Have each team member submit a draft of his or her material a month before the seminar. This allows time for necessary revision if the speaker's material does not fit the seminar content.

Presenting the Seminar

If properly presented, a seminar helps to sell you as a competent professional. The following list gives you a few pointers in making an effective seminar presentation:

- Begin and end on time.
- Speak to your audience.
- Get to the point and stay focused. Your audience wants to know what you can do for them. They do not want to hear war stories.
- Keep your goal in mind during your presentation. This will help you stay on track.
- Speak from an outline or note cards; do not read or memorize your speech.
- Be conversational, friendly, and enthusiastic.
- Use visual aids, if appropriate, but keep them simple.
- Avoid offensive jokes, stories, comments, or language.
- Move around as you speak. Do not remain in one spot the whole time.
- Relax, smile, and enjoy yourself. People like to work with professionals who take satisfaction in their work.
- Ask for feedback during your presentation in the form of questions and after your presentation in the form of a critique.

Conducting an Effective Follow-up Campaign

Seminars present you as a knowledgeable professional to a group of potential prospects. Most effective seminars are low-key and avoid overt attempts to sell products or specific services. Any one-on-one selling typically occurs after the seminar. For this reason, the follow-up phase of the seminar takes on real significance. Many advisors end their presentation by telling their audience that the advisor will contact each attendee to answer any questions that might result from the seminar. Others bring their appointment book to the seminar and schedule appointments right then and there.

Study Feedback from the Seminar. As mentioned earlier, many advisors utilize a postseminar questionnaire to analyze prospects' interest. The mere fact that an attendee takes the time to complete the questionnaire is often a sign of interest.

In addition, some advisors rely on observation and target those people who have asked questions during the seminar presentation. These advisors mix with the audience during breaks and identify prospects who seem to be most interested in the presentation. They also ask each of the seminar speakers to identify attendees who asked the speakers questions after the seminar. (Remember that nametags for all attendees will help you and your speakers in this process.)

Approach the Prospect for an Appointment. Next, contact seminar attendees by telephone or in person to schedule appointments to discuss individual needs. Ideally, this follow-up should occur within 1 or 2 days of the seminar. Below are some sample scripts. Principles for writing your own telephone approach scripts are addressed in chapter 5.

Example 1:   John, at the seminar you recently attended, we had a question-and-answer period. Because the time allocated for this purpose was fairly brief, I am calling to schedule a meeting with you so we can discuss your specific questions or concerns. Would the evening or daytime be more convenient for you?"

Example 2:   Ann, several of the speakers from the seminar you attended this week indicated that you had some excellent comments regarding some of the areas discussed. I'd like to spend some time with you covering those areas. Would the a.m. or p.m. hours better suit your schedule?

Many advisors recommend that you contact no-shows as well (people who indicated they were coming to the seminar but did not show up). Even when these advisors have been unable to schedule an appointment, they keep no-shows on their mailing lists and send them newsletters or bulletins on ideas that might be of interest. Then they invite these prospects to the next seminar and generally get a positive response.

Making Seminars Cost Effective

Finally, for seminars to be a cost-effective marketing tool, you must repeat programs regularly. Only with repeat seminars will you be able to justify the heavy commitment of time that is required to develop a viable seminar process and an effective seminar program. If your seminars acquire

> If your seminars acquire the reputation for being informative and valuable, they will be well attended by qualified prospects.

the reputation for being informative and valuable, they will be well attended by qualified prospects.

Preapproach Letters

preapproach letter

A *preapproach letter* is a letter or postcard mailed (or e-mailed) to a prospect with the goal of introducing the advisor and arousing the prospect's interest in meeting with the advisor. Properly used, preapproach letters can be a powerful aid to your approach.

Advantages of Using Preapproach Letters

One of the advantages of the preapproach letter is that it creates a sense of familiarity. The very fact that you can start by saying, "I wrote to you last week, mentioning that I would call on you" gives you a basis for contacting the prospect. In addition to preparing prospects to receive your call, effective use of preapproach letters can result in other advantages to your prospecting efforts, as follows:

- They make prospects more receptive because they create an awareness of possible financial needs and problems that prospects should address.
- They help develop good work habits. By sending a definite number of letters on a regular basis, you commit yourself to calling on prospects every week. This gives you a measure of outside discipline—discipline that requires a regular schedule of prompt follow-up calls. When you use direct mail regularly, you let the law of averages work in your favor.
- They relieve some of the pressure on the first call. The prospect has seen your name at least once.
- They improve the ratio of interviews to calls. All things being equal, a previous contact with the prospect, however slight it was, means you have a better chance of obtaining a fact-finding interview than if you made a completely cold call.

Consider the impact of the following preapproach letters.

Example 1:   ***Preapproach Letter for Buy-Sell Agreement***

Dear (Prospect),

As you know, a sale made in distress is a costly proposition. When an entire business is liquidated in a forced sale, especially on a piecemeal basis, the losses are often 50 percent or more.

Yet this is precisely what does happen, in many cases, when a business owner dies. You may have already given some thought to this problem and its possible effects on your family.

I would like to show you some of the insurance ideas that other business owners are using to solve this problem. I plan to call on you in a few days to see if we can arrange an appointment.

Sincerely,

Charles Stonewall, CLU, FSS

Example 2:   ***Preapproach Letter for Life Insurance***

Dear (Prospect),

The events of Hurricane Katrina have demonstrated that a lack of planning can make a difficult situation worse. The same holds true in the case of the death of a parent. Financial difficulties will only intensify the emotional pain that the surviving loved ones feel.

However, proper planning now will mean an easier adjustment period and greater financial security for those left behind.

I help families like yours design an emergency plan using life insurance and other planning techniques. Several families I have worked with will attest: Planning made all the difference for them.

If you are interested in meeting with me, please return the postage-paid reply card and I will contact you promptly. I look forward to meeting with you soon.

Sincerely,

Mercedes Hightower, CLU, FSS

Choosing Effective Letters

Keep in mind the following guidelines when you are choosing from among your company's inventory of standardized mailings:

- Select letters that reflect your target market's needs.
- Generally, the shorter the better.
- Postcards often are more effective than letters because there is no envelope to open.

If you feel your company's standard letters are not adequate, obtain company approval to draft your own. Your three main objectives in writing a preapproach letter are to

- trigger your prospect's interest by highlighting briefly a problem that needs to be solved
- communicate the relevant parts of your value proposition related to meeting that need
- prepare the prospect for your call or request for his or her response (if the prospect is on a do-not-call list)

| Letter writing is an art. Thus, expect an effective letter to take a fair amount of time to compose. |

Letter writing is an art. Thus, expect an effective letter to take a fair amount of time to compose. Before you write your letter, review the information on positioning your products in chapter 2. Here are a few additional guidelines to consider:

- Aim to write something that grabs the prospect's attention.
- If possible and necessary, establish a basis for your contact by referring to how you have heard of the prospect.
- Describe the most probable and acute financial need that prospects face and you can address.
- Link that financial need to an appropriate emotional need.
- Do not overstate the need; that is manipulative.
- In communicating your value proposition, keep it to one or at most two sentences. Avoid platitudes and clichés.
- Confirm the credibility of any statistics you use, and use them responsibly and appropriately (do not overuse them).
- Pay strict attention to wording, grammar, spelling, and punctuation.
- Ensure that the letter conveys an image of professionalism by using quality stationery and typeface.
- Ask a current client from your target market to review the letter's message and appearance.

Figure 4-4 is a guide to help you draft your own preapproach letters.

Preapproach Letter Logistics

Think about the logistics related to mailing preapproach letters. Sometimes the little things can make a big difference. Evaluate and try some of the following:

| No letter is good enough to do a selling job by itself. Follow-through is crucial to the success of any mail program. |

- No letter is good enough to do a selling job by itself. An efficient and effective system of follow-through is crucial to the success of any mail program. Do not send letters to more prospects than you can follow up with the following week.

FIGURE 4-4
Writing a Preapproach Letter

Template for Drafting Letter	
Basis for contact (if applicable)	Met the prospect at a fund raiser for leukemia
Financial need/ problem	To pay for children's college tuition and other expenses
Corresponding emotional need	A desire for children's financial security and success
Problem definition	College tuition and expenses are costly now. What will they be in 10 or 15 years? How much debt will your children be in to pay for college?
Value proposition	I can help you find money today to save for tomorrow's education, minimizing debt and giving your children the opportunity to get started in their adult life on a more solid financial footing.
Closing	I will be calling in the next week to see if I can help you.

Letter to Prospect

Dear (Prospect),

It was a pleasure to meet you last week at the Chester County Find the Cure fund raiser.

You mentioned you had three young children. You probably have considered the importance of their college education. Indeed, over a lifetime, the difference between a bachelor's degree and a high school diploma is estimated to be more than a $1,000,000 (Source: U.S. Census Bureau statistics).

But consider that 1 year of college costs an average of over $12,000 for a public college (over $30,000 for a private one) in 2006.* What will the cost be in 10 or 15 years? How much debt will burden your children after they graduate?

I help families like yours find money today to save for tomorrow's education costs. By planning today, you can minimize your children's student loan debt, enabling them to get started in their adult life on a more solid and secure financial footing.

I will be calling in the next week to see how I might be able to help you.

Sincerely,

Winston Dunn, ChFC, FSS

* Source: College Board's Trends in College Pricing 2006.

- Consider addressing letters by hand. Some advisors have found that handwritten addresses increase the probability that prospects will open the letters.
- Affix postage with an individual stamp rather than a postage meter. Some advisors highly recommend the use of commemorative stamps.
- Consider including an attention-getter in the envelope. One advisor includes a dollar bill to pay the prospect for reading the letter. Another advisor includes a Band-Aid with a health insurance preapproach letter.
- E-mail is another way to send a preapproach letter. If you are dealing with a technically savvy target market, e-mail may be more effective than regular mail.

Third-Party Influence

A variation of the preapproach letter is the third-party influence piece. Many advisors have found that marketing material from their own companies or commercial services can be valuable in creating awareness. Various publishing houses have available such items as a series of tax letters, legal bulletins, and other mailing pieces that are of interest and value to business owners and executives. Obviously, it would help to establish prestige if you are the one who is sending prospects such publications. You could also enclose articles of this sort with a preapproach letter.

Prospects who receive third-party material from an advisor will probably recognize that the advisor has an appreciation of financial matters. They may remember the advisor's name, especially if they receive material from the advisor on a regular basis (every month or every quarter, for example). Thus, they are more likely to be willing to listen to what the advisor has to say.

CONCLUSION

> Obviously, if advisors had to depend only on cold calling, they would starve.

Face the facts. Most advisors have to contact strangers to succeed in this business. Obviously, if advisors had to depend only on cold calling, they would starve. Instead, by planning and executing a prestige-building strategy, advisors can publicize their personal brands to their target markets. The key, as has been one of the refrains of this book, is knowing their target markets and taking stock of their own personal attributes and abilities.

Probably the most important aspect of prestige building is increasing social mobility. Advisors should allot time to activities that give them personal exposure to their target markets. Advisors can do this by finding an activity, cause, organization, and so on that they and a high proportion of their target markets have in common in which they can get involved, join, or even lead. In addition, advisors should make use of other prestige-building techniques such as a personal brochure, Internet website, newsletters, and advertisements.

As important as personal branding is, it typically cannot stand on its own. Prospects in a target market can think the world of the advisor but not see the need and/or feel motivated to purchase the advisor's products. Thus, prospects will not be inclined to grant the advisor an appointment. As a result, advisors need to use preapproach methods that will precondition prospects to respond favorably when the advisors approach prospects for an appointment. The most powerful, albeit time- and energy-intensive, preapproach is the seminar. It does a little of everything: build prestige, preapproach, and qualify prospects. Nevertheless, a well-written preapproach letter can also be effective in creating a sense of need.

As with other marketing activities, make sure you track their effectiveness. Ask people how they heard about you. Ask customers and clients what triggered their awareness of the financial needs they addressed through you. Cull the best activities and spend your time and energy on them.

CHAPTER FOUR REVIEW

Key Terms and Concepts are explained in the Glossary. Answers to the Review Questions and Self-Test Questions are found in the back of the textbook in the Answers to Questions section.

Key Terms and Concepts

prestige building	weblog (blog)
social mobility	advertising
community involvement	preapproach
speaking to groups	preapproach letter
personal brochure	

Review Questions

4-1. Describe the four levels of participation related to community involvement.

4-2. Discuss five important aspects for the advisor to consider if he or she is looking to use writing to create social mobility.

4-3. Explain how to get started in speaking to groups.

4-4. Marcie has completed her Chartered Advisor in Philanthropy, a designation that equips her to advise clients and prospective donors about long-term planning for financial assets. How should she describe this on her personal brochure?

4-5. List the five objectives for a website, and briefly describe how an advisor could approach achieving them.

4-6. Explain the four advantages related to seminars.

4-7. Describe five important tasks to complete in planning a seminar.

4-8. Discuss how to write an effective preapproach letter.

Self-Test Questions

Instructions: Read chapter 4 first, then answer the following questions to test your knowledge. There are 10 questions; circle the correct answer, then check your answers with the answer key in the back of the book.

4-1. Which of the following statements regarding writing to increase social mobility is correct?

  (A) The advisor's goal is to publish; he or she should not worry if the target market does not read the publication.
  (B) Writing is a great opportunity to pitch the advisor's products and services.
  (C) The advisor should purchase extra copies of an article to send to prospects as a preapproach.
  (D) Fortunately, the advisor does not have to worry about his or her company's compliance department.

4-2. Which of the following is a method for the advisor to create social mobility?

  (A) joining the local Rotary club
  (B) sending a preapproach letter
  (C) selling a policy to a relative
  (D) training new agents from his or her company

4-3. Which of the following statements regarding advertising is correct?

  (A) The advisor should ask prospects with whom he or she meets how they heard about the advisor.
  (B) Advertising is most effective if it is geared to the general public.
  (C) It is very easy to measure the effect advertising has on client creation.
  (D) The case for advertising is universally accepted by successful advisors.

4-4.   Which of the following statements regarding personal brochures is (are) correct?

     I.   A personal brochure should include standard business information.
    II. Advisors should relate their experience and credentials in terms of the benefits to prospects.

    (A) I only
    (B) II only
    (C) Both I and II
    (D) Neither I nor II

4-5.   Viable locations at which to host a long-term care seminar aimed at seniors include which of the following?

     I.   The advisor's office, despite the limited parking
    II. A conference room at the local senior center

    (A) I only
    (B) II only
    (C) Both I and II
    (D) Neither I nor II

4-6.   If advisors write their own preapproach letters, they should consider doing which of the following?

     I.   Use a lot of statistics to make their point.
    II. Avoid appealing to the emotions; just give the facts.

    (A) I only
    (B) II only
    (C) Both I and II
    (D) Neither I nor II

4-7.   All the following are guiding principles that an advisor should follow in determining where to get involved in the community EXCEPT:

    (A) Make sure his or her values match those of the organization.
    (B) Evaluate his or her capacity for involvement.
    (C) Select organizations that allow members to promote themselves with few restrictions.
    (D) Choose organizations that the advisor would support regardless of his or her target market.

4-8.   All the following are possible objectives that an advisor might try to achieve through a website EXCEPT

   (A)   creating an emotional connection
   (B)   selling
   (C)   servicing
   (D)   asking for referrals

4-9.   All the following are important for the advisor to remember when giving a presentation during a seminar EXCEPT:

   (A)   Remain in one place to avoid distracting.
   (B)   Begin and end on time.
   (C)   Do not read the presentation to the audience.
   (D)   Be conversational, friendly, and enthusiastic.

4-10.   All the following are characteristics of effective preapproach letters EXCEPT:

   (A)   They generally are short.
   (B)   They are handwritten to make them personal.
   (C)   They reflect the target market's needs.
   (D)   They often are postcards rather than letters.

NOTES

1.   social mobility. Dictionary.com. Dictionary.com Unabridged (v 1.0.1). Based on the *Random House Unabridged Dictionary*, © Random House, Inc. 2006, accessed October 24, 2006 from http://dictionary.reference.com/browse/social mobility ().

2.   "The Young Politician," The Churchill Society, accessed October 24, 2006, from http://www. churchill- society-london.org.uk/19011913.html.

3.   Peter Montoya and Tim Vandehey, *The Brand Called You: Personal Marketing for Financial Advisors* (Costa Mesa, CA: Millennium Advertising, 1999), p. 93.

4.   Timothy G. Herbert, accessed October 24, 2006 from http://www.timothyherbert. com/new/ timothyherbert/

5.   Montoya and Vandehey, p. 35.

6.   Mary Madden, "Internet Penetration and Impact April 2006," Pew Internet & American Life Project, accessed October 27, 2006, from http://www.pewinternet.org/pdfs/PIP_Internet_ Impact.pdf.

7.   Lori Chordas, "Interest Grows in Online Auto Insurance Quotes," *Best Review* (Oldwick, NJ: 2005), vol. 106, iss. 3, p. 106.

8.   Lavonne Kuykendall, "Buying Auto Insurance Online; Consumers Do Research on Web, but Few Close Sale Without Agent," *Wall Street Journal*, Eastern ed. (New York, NY: February 25, 2006), p. B-4.

9.   James O. Mitchel, "LIMRA's MarketFacts Quarterly," LIMRA Intl., (Hartford, CT: 2006), vol. 25, iss. 3, pp. 13–14.

10.   Mary M. Art, "Consumer Internet Use for Insurance and Investments in the United States Phase II: The Practices of Online Consumers in the Buying Process," LIMRA Intl. (Windsor, CT: 2003), p. 62.

11.   Mary M. Art, "Generations X and Y Online: The Wired Generations," LIMRA Intl. (Windsor, CT: 2002), p. 10.

12.   Ibid., p. 24.

5

Approaching Prospects

Learning Objectives

An understanding of the material in this chapter should enable the student to

---

5-1. Identify the objective and explain the importance of face-to-face approaches to prospects.

5-2. Explain how the national Do-Not-Call Law applies to advisors' telephone prospecting activities.

5-3. Identify the four principles of telephoning.

5-4. Describe the three main elements of preparation to telephone prospects.

5-5. Identify the seven components of an effective telephone script.

5-6. Explain techniques used to handle prospects' questions while on the telephone.

5-7. Describe techniques for handling a prospect's objections on the telephone.

5-8. Identify the important aspects of a system to track results during phone sessions and to quantify telephoning effectiveness ratios.

5-9. Describe the procedure to confirm and reschedule appointments on the telephone.

5-10. Explain the use of prequalifying questions on the telephone.

Chapter Outline

INTRODUCTION

> **No matter what method you use to contact prospects, you are trying to achieve one primary goal: to get an appointment.**

After identifying prospects, your next step in the selling/planning process is to contact them to arrange an appointment. No matter what method you choose to contact prospects—face-to-face or telephoning—you are trying to achieve one primary goal: to get an appointment. This will involve awakening your prospect's interest in you and your services.

Typically, advisors most often use the telephone to approach prospects for appointments. This chapter will focus mainly on methods to develop efficient telephoning work habits and demonstrate confidence when using the telephone. It will also review some basic techniques of effective telephone approaches and methods to prequalify prospects during the appointment-setting call.

First, however, we will discuss face-to-face approaches that can be used in the personal and business markets. Because advisors use this approach method much less frequently than the telephone approach, our discussion of face-to-face approaches will be more limited than our discussion of telephone approaches later in this chapter.

FACE-TO-FACE APPROACHES

> You must be able to describe who you are, what you do, and why someone should meet with you to discuss the financial products, concepts, and services you have to offer.

There are many face-to-face opportunities for you to obtain appointments with prospects. You must be aware of where these opportunities present themselves and be prepared to capitalize on them when they do.

In both the personal and business markets, there are proactive and reactive approach techniques that you need to develop and have available to prospect effectively. Whether the situation arises by chance or you actively seek it, you must be able fluently to describe who you are, what you do, and why someone you encounter should meet with you in an appointment setting to discuss the financial products, concepts, and services you have to offer.

Face-to-face approaches can take place on a proactive basis in personal observation, networking, or canvassing situations, or on a reactive basis in social, travel, or recreational settings. In the former, you should prepare an approach that targets a specific idea or concept that you think will interest a prospect in a certain personal or business market. In the latter, you should position yourself in a manner that stimulates the prospect's curiosity enough to want to speak further with you about financial issues at some time in the future. In either case, your objective is to arrange a formal appointment with the prospect.

This section will examine two types of approach talks for prospecting in the personal market, as well as an idea-based canvassing approach to small business owners within the business market.

Prospecting in the Personal Market

The Elevator Speech

As an advisor, you should always be prepared for planned or unplanned opportunities to describe with passion, precision, and persuasiveness to others who you are and what you do. Often, these chance encounters can lead to future clients and referrals if handled professionally. Consequently, you need to know the right words to say to make somebody's acquaintance so you can create just the right impression.

elevator speech

To create a lasting first impression that showcases your professionalism and allows you to position yourself, compose a brief talk that advertises the results you help your clients achieve. Known also as an *elevator speech*, this talk is designed to create interest and give the prospect a quick synopsis of what you do and how you can help people. An elevator speech is a short (15 to 30 seconds, 150 words or fewer) sound bite that succinctly introduces you and helps you to connect with someone you have just met. Use your elevator speech whenever you want to introduce yourself to a new contact—for example, in such impromptu prospecting settings as when you are in an

---

Tips for Writing an Elevator Speech

- *Define your audience.* Will you deliver your elevator speech at a chance encounter with a stranger, an organization meeting, or some other setting?
- *Define your objective.* For example, the objective may be to establish name recognition and permission to contact this person at a later date.
- *Determine content or subject matter.* What descriptive words will you use that leave the right impression and create the right image of you? Determine the key points you want to make with each target audience so that your contacts are intrigued.
- *Emphasize the benefits, services, and/or features* that you provide to your clients.
- *Create an opening sentence that will grab the listener's attention.* The best openers leave the listener wanting more information.
- *Practice your speech* in front of the mirror and with friends. Rehearse it until you can say it naturally and confidently. You do not want to sound canned.

---

elevator, shopping at the supermarket, waiting in line at an ATM, or purchasing your morning latte.

To be effective, an elevator speech should be well rehearsed so that it sounds spontaneous and natural but is also flexible enough that you can adjust it to meet the situation. It spotlights your uniqueness, it focuses on the benefits you provide, and it should be delivered effortlessly.

The presentation begins with a brief description of what you do for a living and for whom you work. Then you should mention some key differentiators that give unique value to your clients. Generally, you should close the talk with a "bridge" that offers a link to future communication with the contact. Your objective is to get permission to talk to him or her again, either on the phone or face to face in an appointment setting.

Example: "Hi, I'm Bob Smith, an advisor with (name of your company). I work with business owners and other professionals who are often too busy to focus on their financial affairs. Most of them have already worked with a number of advisors at different times and under different circumstances during various stages of their financial evolution. Consequently, they lack a coherent and unified approach to managing their financial affairs. Besides offering innovative products and concepts to help you achieve your financial and business goals, I can help to coordinate the efforts of all the professional advisors with whom you currently work.

"Perhaps we can arrange a convenient time to talk, so that if you're ever in the market for what I do, maybe you'll think of me. Do you have a business card?"

(Offer the prospect your card.)

10-Second Commercial

Socializing provides an excellent way to create awareness and interest in what you do. One prospecting opportunity is to customize your response to the question, "What do you do for a living?" Have a short response that is relevant and interesting to the person to whom you are talking.

A response like "I sell long-term care insurance" tells the person how you make money; it also tells the person specifically what you can do for him or her. However, it is neither interesting nor attention grabbing.

Contrast such a response with the *10-second commercial*, a variation of the elevator speech, which creates both interest and attention.

Example: Suppose you are talking to a 55-year old married man. You might say something like this:

"Do you know what conservationists do?"

(Answer: "They preserve the environment.")

"Well, I do something similar with retirement nest eggs. I help couples like you and your wife put plans into place that preserve their retirement savings from the costs of long-term care. Do the two of you have a plan like that?"

Here are some tips for creating a 10-second commercial:

- Ask a "positioning" question that is relevant to a prospect's need for your services. The question positions your response as a solution to a problem or answer to a question.
- Follow up with your 10-second commercial, stating your value in terms of the results you achieve for your clients.
- Be creative and make it interesting, but follow your company's compliance guidelines. Exercise caution and do not misrepresent the work you do or products you sell.
- End with a question that measures the prospect's interest.
- Have a business card to give the prospect if you feel it is appropriate.

The key is to customize your 10-second commercial to the prospect and refer to a need that he or she might have. By doing so, you are creating not only awareness of your products and services but also interest—and interest will get you an appointment.

Prospecting in the Business Market

The business market approach has the same purpose as the personal approaches—namely, to obtain an appointment. In addition, the types of approaches that work to obtain personal market appointments should also work to obtain business market appointments. One type—the idea-based approach—entails presenting one or more specific ideas that may be of distinct value to your prospect.

Idea-Based Approach

When you approach prospects, sometimes you present a particular idea in the belief that this idea will help solve a specific problem your prospect faces. More frequently, your presentation is intended to stimulate your prospect's curiosity. The following is an example of the idea-based approach for use with small business owners.

Example:	"Good morning, I'm Janet Smith with (name of company). The reason for my visit is to introduce myself and tell you that my work involves helping employers provide comprehensive benefit packages for their employees without feeling that they are overpaying. What I would like to do is meet with you at a convenient time to discuss some interesting ideas I have been sharing with other business owners. These ideas might help your business save money and at the same time reward your employees with a benefit plan that they like. Is later this week or next week more convenient?"

Summary

The objective of all face-to-face approaches is to obtain an appointment with the prospect on a favorable basis. The same is true of approaches made on the telephone, which are discussed in the next section.

THE POWER OF THE TELEPHONE

Telephoning a prospect for an appointment is typically the most efficient way to fill your appointment calendar. It is inexpensive, it enables you to

> **The only purpose of the telephone approach is to get an appointment for a face-to-face meeting with your prospect.**

make many contacts in a short time, and it is interactive, allowing two-way communications with your prospects.

The only purpose of the telephone approach is to get an appointment for a face-to-face meeting with your prospect, which is why the approach is so efficient. Knowing this will help you keep your calls focused, simple, and effective.

This is your initial approach with many of your future clients, so it is important that you are prepared and professional and that you set the proper tone for a productive future relationship.

A steady flow of appointments with prospects makes all the difference between a practice in which you expend your energy in a constant struggle to establish yourself versus one in which you can concentrate on serving your clients in a satisfying financial services career. This difference is often not numbers alone; it is consistency from day to day, week to week. Because prospecting is an ongoing daily activity, it follows that phoning those prospects is, too.

Telephoning Compliance

Do Not Call (DNC) Regulations

The Federal Communications Commission (FCC), the Federal Trade Commission (FTC), and the majority of states have adopted rules regarding unsolicited telemarketing calls without the consumer's prior consent, or sales calls to persons with whom the caller does not have an established business relationship. The FCC rule makes use of the *Do Not Call Registry* established by the FTC and sets forth the ways that telemarketers must check the list to ensure compliance. The term "telemarketer" refers to any person or entity—including a financial advisor—making a telephone solicitation.

Do Not Call Registry

Do Not Call Law

The regulations under the *Do Not Call Law* are designed to protect consumers from telemarketing abuses. Generally, a telemarketer's call to a person on the national or state DNC list is subject to a fine of $11,000 or more per violation.

There are many restrictions placed on telemarketing. Some of the important ones are as follows:

- Sales calls to persons who have placed their residential or mobile phone numbers on federal or state DNC lists are prohibited.
- Calls cannot be made before 8 a.m. or after 9 p.m.
- Stiff penalties are placed on violators.
- Sellers must maintain an in-house DNC list of existing customers who do not want to receive sales calls.
- Sales callers must, at the beginning of every sales call, identify themselves, the company they represent, and the purpose of the call.
- Telemarketers may not intentionally block consumers' use of caller identification.

For many, these restrictions are seen as a major obstacle to their prospecting efforts. For others, however, the restrictions are viewed as an opportunity to explore avenues of marketing that play within the exceptions to the Do Not Call Law.

These exceptions apply when there is

- *an established business relationship.* A business relationship exists when there is a product or service in place, and it continues for 18 months after that product or service is no longer in effect or active. Several states have stricter requirements. If a consumer contacts an advisor, whether by phone, mail, or in person, to inquire about a product or service, an existing business relationship exists for 3 months after that inquiry. (Referrals do not satisfy the established business relationship exceptions and are not a basis to call someone on the list.)
- *prior express permission.* Advisors may make calls to a person on the DNC lists if they have a signed, written agreement from the consumer in which he or she agrees to be contacted by telephone. Written permission if it is received prior to the call is valid indefinitely unless revoked by the consumer.
- *a business-to-business relationship.* The DNC regulations do not apply to business-to-business calls.
- *a personal relationship.* Calls may be made to people with whom an advisor has a personal relationship, including family members, friends, and acquaintances.

A Professional Phone Skills Trainer's Viewpoint*

Many advisors have expressed frustration because of the Do Not Call Law. They have called me, lamenting, "What should I do?" since the passing of this new legislation.

Here are some ideas for making calls to "easier" leads:

- Seminars are a terrific forum for getting interested people to sit and listen to you talk.
- You can also canvass a business area and introduce yourself to local business owners. Make sure you have a good, short face-to-face introductory presentation that will lay the groundwork for the appointment-setting phone call that follows.
- Try calling the people in your natural market such as friends, family, and acquaintances whom you have been too afraid to call. Those folks will accept your phone call, won't they?

I am sure that overall, this Do Not Call Law will help the truly professional financial advisors to rise to the challenge.

* Gail B. Goodman, president, ConsulTel, Inc./www.phoneteacher.com

Because of the DNC regulations, you need to capitalize more effectively than ever on the opportunity to make telephone appointments with prospects who are not on the national DNC Registry. For prospects who are on the registry, you must think creatively and utilize every opportunity within the law to make appointments with them.

In addition to the federal Do Not Call Law, most states have rules that must be observed regarding solicitations using the telephone. It is your responsibility to know how they apply to you in your everyday business activities.

Also be aware that telemarketing by charitable organizations is exempt from the Do Not Call Law. Furthermore, people can be solicited at their place of employment because business phone numbers are also exempt from the law.

> **All advisors are responsible for knowing and following the compliance rules of the companies with which they do business.**

Last, all financial advisors are responsible for knowing and following the compliance rules of the company or companies with which they do business. Accordingly, any advisor who uses the telephone scripts or techniques for handling prospects' questions and problems presented in this text must get approval from his or her company's compliance department and/or manager prior to using them.

A Word on Cold Calling

The world of cold calling seems headed for extinction. Perhaps in the business market cold calling will always have a foothold because business owners sometimes appreciate a new idea from an unknown source, especially if they perceive that it can help improve their bottom line. But with the Do Not Call Law, the personal market has moved away from accepting calls from strangers. Most individuals buy financial products from people they know or people to whom they are referred, so these are the best prospects to cultivate for calling.

Financial advisors are more successful in developing a thriving practice when they pursue the leads that are most likely to result in an initial appointment. Therefore, you should give the "easier" leads precedence over the "harder" leads in your everyday prospecting activities.

Four Principles of Phoning

The four principles below set the foundation for successful phone calling.

Principle #1: When You Are on the Phone, Do Not Sell

This principle is the most critical of the four because financial advisors are salespeople, and many sales companies tend to focus more on sales training than on phone training.

Two Baskets of Skills

Phone Skills   Sales Skills

Imagine that you have two separate baskets of skills. The basket on the right contains your sales skills. It holds a multitude of tools that your company's training department has available for you. Your phone skills are in the left basket. Unless you've been to the same number of phone classes as you have to sales classes, it is likely that you don't have as many tools in that basket as is in your sales tool basket. One objective of this chapter is to fill your phone skills basket with the tools necessary to make you more effective on the phone so that when you are phoning, you use a phone skill, not a sales skill.

Unfortunately, because many financial advisors are weak on the phone, they tend to use their sales tools when they are calling because it makes them feel more comfortable. These selling methods, however, won't help you make enough appointments to satisfy your prospecting objectives. You need to focus on phoning language and phoning words, and to recognize the environment you are in when you are on the phone.

The phoning environment is very different from the sales environment. When you visit a prospect in a face-to-face appointment, it is usually a relaxed situation because he or she has agreed to give you a certain amount of time, presumably knows the topic you will be discussing, and is ready, willing, and able to have that extensive conversation. When you are making a phone call, however, you are interrupting somebody's day at the precise time that you decide to make the call. In general, therefore, setting an appointment is more difficult than selling.

Principle #2: People Are Irritated by Telephone Solicitations

At the beginning of every phoning session, remember these three words: tension, irritability, and defensiveness. When the phone rings, the individual you are calling experiences an increase in all three. The first question the individual asks himself or herself is, "Who are you?" The second question is, "What do you want?" Later in this chapter, when scripting is discussed, you will realize the importance of answering these two questions without delay.

To be successful in phoning, you must use techniques to reduce tension, irritability, and defensiveness as quickly as possible. You do not want your prospect to hang up on you. By navigating your way through the prospect's potential negative reactions, you can get him or her to listen to what you have to say. You can then ask the prospect for an appointment, and explain why meeting with you will be beneficial.

Principle #3: Not All Leads Are Created Equal

It is no secret that cold calling prospects is much harder than calling referrals or someone you know. This is true whether you are calling

> **Easy leads are more willing to take your call and give you an appointment. They are not necessarily more wiling to buy a financial product from you.**

individual households that are not on the DNC list or businesses. However, a secret to success on the phone involves cultivating easy leads.

Easy leads are defined as people who will more readily take your call and are more willing to give you an appointment. They are not necessarily individuals who are more willing to buy a financial product from you.

The easier leads are people whom you know, people who were referred to you, and people whom you have met before (at a trade show, networking event, or seminar, for example). Although these categories of people are easier to call and to make appointments with, keep in mind that you still have to prepare yourself with the right words in order to be effective.

Principle #4: There Is No Such Thing as "Call Reluctance"

> **Call avoidance arises from either a lack of training or a lack of scripting.**

The reason some advisors are hesitant to get on the phone is that they don't have an effective script—one that feels right to them and contains words they are comfortable saying. If those elements are missing, financial advisors do not want to make the prospecting phone call. This is not call reluctance. It is call avoidance—arising from either a lack of proper training or a lack of good scripting.

If anyone who has a classic sales personality (which usually means that the person is verbally adept) thinks that he or she will sound awkward or inarticulate on the phone, the individual is not going to place the call. It is in conflict with the person's self-image, which entails being a smooth and fluent talker. People shy away from activities or behavior that clashes with their self-image.

Another factor at work when someone lacks confidence or feels that he or she will sound silly on the phone is that it increases the individual's telephone anxiety level. When people are anxious, their verbal skills shut down. This makes them less able to find the words that will enable them to be effective on the phone, which, in turn, only makes them more anxious. The cycle, however, can be broken.

The Four Principles of Phoning

1—When you're on the phone, don't sell. Use your phone skills.

2—People are irritated by telephone solicitations, so when the phone rings, their tension, irritability, and defensiveness naturally increase.

3—Not all leads are created equal. Focus on working the easy leads, which are people more willing to meet with you for an appointment.

4—There is no such thing as call reluctance. There are only advisors who have been asked to get on the phone without proper coaching, training, and practice.

To eliminate this crippling anxiety and negative behavior on the phone, advisors must be trained properly and given words that will make their phoning skills more effective. This improves their phoning confidence and reduces their phone anxiety.

Thus, there is no such thing as call reluctance. There are simply many intelligent, outgoing, articulate advisors who are anxious about making phone calls and therefore need training, scripts, and practice. Advisors who have the confidence in their phoning skills and know that they sound professional on the phone are more willing to make calls.

PLANNING AND PREPARATION

This discussion of telephoning, then, doesn't start with picking up the phone. It starts much earlier—with planning and preparation.

Planning

Developing the self-discipline to make prospecting telephone calls each and every week is a necessary skill for your success.

Scheduling Your Phone Time

The first planning step in phoning is to schedule your phone sessions on your calendar every week, several weeks in advance, at the beginning of each new month or year, before you fill up your calendar with appointments, meetings, classes, and other activities. Most advisors, however, schedule their phoning time *after* they fill their calendar with appointments and other meeting commitments. This approach is backwards.

There are many different theories about the best times to phone prospects for appointments. Some successful advisors like evenings; some like early in the week; others swear by Saturday mornings or Sunday evenings. By keeping track of the results of your phoning sessions, you will determine what works best for you.

> **After you schedule a phone session, it is every bit as real as a face-to-face appointment. You cannot cancel it. You may only reschedule it.**

After you schedule a phone session, it is every bit as real as a face-to-face appointment. You cannot cancel it. You may only reschedule it and, then, not very often. These telephoning appointments should be for either a fixed number of calls or for a fixed time. Once you reach your goal, you are finished with that session.

The challenge with proactive phone scheduling is to control your response to other events that may occur in the natural course of your business week. These can undermine the disciplined acquisition of appointments with new prospects, which is the key to your success in a financial services career. It is easy to find yourself at the end of a busy work week in which you may have made several sales but didn't find time to get on the telephone to schedule appointments for the next week. If, however, you have scheduled your phoning time in advance, it becomes habitual to keep these appointments with yourself. In fact, you will probably feel guilty if you don't keep them.

Most advisors tend to book phoning sessions in 1-hour blocks of time. It is more effective, however, to book 90-minute time slots, which will give you time to prepare for phoning and to assemble your script, leads, and counting sheet. (See appendix A, section 5, for a sample counting sheet.) Also, it takes time and a few calls to get into the rhythm of phoning. Therefore, 90-minute phoning segments are much more effective than 1-hour segments.

Preparation

Another important part of being proficient on the phone is preparation. Having lots of people to call is only one aspect of this preparation.

There are three key elements in being prepared for phoning. You must be

- mentally prepared
- physically prepared
- verbally prepared

In addition, you must be disciplined. The next section of this chapter examines the three elements above and the concept of discipline as it relates to telephone preparation.

Being Mentally Prepared

Your attitude is going to affect how well you can set appointments on the phone because the way you feel is the way you sound. You must come across as being in a good mood and enthusiastic about the call.

If being comfortable on the phone does not come naturally to you and you must act as if you enjoy it, then act.

If being comfortable on the phone does not come naturally to you and you must act as if you enjoy it, then act. After all, actors in a play must get into their roles when they are on stage despite what may be going on in their personal lives. So even if you have to fake being in a good mood, that's fine as long as it works for you in projecting a positive attitude.

If you're lucky enough to be able to change your mood at will, that is all the better. Some people can instantaneously adapt their emotions to the situation or setting in which they find themselves. If you are capable of this behavioral technique, you need to apply it when phoning.

Another way to mentally prepare is to talk yourself into feeling the necessary enthusiasm. Tell yourself that you're in a good mood, that the people you call will be receptive, and the phone call will go well. Anything you tell yourself is true!

If you're traveling to the office to place your calls, make your car environment conducive to helping you get into the right mood. Listen or sing along to your favorite CD. If you like soothing music or the recorded sounds of nature, have those ready as well. Use whatever works best for you. Another helpful method is to enjoy some "happy food." (Two examples are chocolate and coffee.) Do whatever it takes to get away from negative thoughts or feelings and get to a happy—or at least neutral—place.

Being Physically Prepared

The second preparation step is to arrive at your telephoning appointment ready to go with all your tools assembled for use. You have to manage and plan your physical phoning environment. Make sure your office and desk have the necessary equipment handy to make calls. You need to assemble the following tools to be ready to get on the phone:

- leads with correct phone numbers
- counting sheets
- appointment book or palm pilot
- pen or pencil
- clean desk (create a sufficient space)
- clear, working phone

Obviously, your leads must have the correct phone numbers of the prospects you will be calling. It is a good practice, too, to have more names than you expect to call. You should approach a 90-minute dialing session with a minimum of 75 names. You will also need a counting sheet to keep track of your activities on the phone. You must have an appointment book or palm pilot, or both. And, of course, you need to have a writing instrument such as a pen or pencil.

You will require a clear space on your desk that equals the width of your shoulders and the length of your arms. Remove all papers from this space such as unrelated documents, piles of correspondence, or post-it notes that may distract you. You also need good telephone reception and a phone that is free of static, call dropping, and other problems.

Although you may want a cup of coffee or other beverage at your desk to sip between calls, don't eat, chew gum, or drink during the call itself. These sounds are amplified over the phone and convey an image of carelessness and lack of professionalism.

Once you pick up the phone and begin your calls, don't hang up until you have written down an appointment date. Hanging up breaks your momentum and concentration.

Once you pick up the phone and begin your calls, don't hang up the receiver until you have written down an appointment date. Hanging up breaks your momentum and concentration. Keep the phone in your hand and keep going.

If you use a cell phone, be careful of when and where you call. When phoning prospects from your car, it may be wise to pull off the road so you can be fully engaged in your phone conversation. (Some states even mandate that callers cannot use a hand-held phone while driving.) Make sure your cell phone service is good. Don't make a cell phone call from a noisy area such as a shopping mall or airport. How you sound over the phone plays a crucial role in creating a good first impression with a new prospect.

Being Verbally Prepared

> You will never get a second chance to make a first impression.

The third critical preparation element is knowing what you are going to say. *You will never get a second chance to make a first impression.* This call is your introduction to your future clients. Be prepared! Actors memorize their lines until they are second nature. This leaves them free to focus on their costars and create the magical chemistry the audience enjoys in top-notch productions. You must memorize your script and ways to handle prospects' responses for exactly the same reasons. You can't sound relaxed and confident if you're forced to "wing it" with the first response that comes to mind. Knowing your script will allow you to concentrate on your delivery, using a measured pace, an upbeat tone, and a natural flow of words and ideas.

Remember, when composing any new phone scripts, make sure they are approved by your company's compliance department.

Parts of a Phone Call. There are really two major parts of a phone call. The first part is what you say to a new prospect; this is known as the script. The second part is what the prospect says back to you; this is known as the response.

Script. Because you are initiating the call, you should never get on the phone until you know what you're going to say. Be prepared with a script. Although you hope that the prospect doesn't interrupt you in the middle of your script, it sometimes does happen and you have to be equipped to handle your responses.

Response. When the prospect says something in answer to your request for an appointment, your response must be intelligent, and it must direct the conversation back to scheduling the appointment. Although some advisors are uneasy about using scripts, even more are anxious about handling prospects' responses. Think of your phone conversation as a tennis match: After you serve with your script, the prospect volleys. The pace quickens, and you must respond intelligently, leading the prospect to scheduling an appointment with you. Avoid babbling.

(Handling prospects' questions and objections will be discussed fully in a subsequent section of this chapter.)

Discipline

The last part of preparation is discipline. Discipline is critical. You must do the things that need to be done, when they need to be done.

You need to be honest with yourself about whether or not you have been willing to "put your feet to the fire" when it's time to get on the phone. Evaluate whether you have contributed to some of the challenges you have encountered in getting appointments. Rethink your level of discipline and strive to improve. Resolve to use the techniques discussed in this chapter to make your scripts and responses more effective. Then practice diligently.

Some timeless words of wisdom bear repeating here:

Successful people have the habit of doing things that unsuccessful people don't do. They don't like doing them necessarily either, but the strength of their purpose overrides their dislike of the task.

STRUCTURE OF THE TELEPHONE SCRIPT

Section 1 of appendix A contains some sample phone scripts. As you study them, you will notice that they are short, professional, and to the point. Remember, each of these scripts has only one objective: to schedule an appointment. A phone script doesn't have to be complicated. It does have to sound conversational. It is meant to be spoken, not read; therefore, the wording will be simpler than words that are supposed to be read.

It it important to distinguish between scripting and ad-libbing when on the phone. If you ad-lib, you run the risk of stumbling to find the right words, saying "um," "aaa," or "er," which can make you sound unprofessional. Also, if you don't have a phone script, you may find yourself babbling, talking in circles, going off on tangents, and not getting to the point of the phone call, which is to make an appointment. This can either bore your prospects or make them feel uncomfortable—perhaps irritable—for infringing upon their valuable time.

> **A good phone script lasts only between 30 and 40 seconds. Any longer takes up too much of the prospect's time.**

A good phone script is brief and succint, usually lasting between 30 and 40 seconds. Any longer takes up too much of the prospect's time. A 30- or 40-second script does not include any time to ask, "How are you?" The question is too wide-ranging anyway. You want to be more specific, even when calling a friend or acquaintance from your natural market. You may say, for example, "Hi, I'm calling for two reasons," and then say, "I was wondering how the new job is going." This will keep unnecessary chit-chat down to a minimum. The prospect will also remember that you said you were calling for two reasons. In the back of his or her mind, the prospect will be wondering what the other reason is.

Components of a Phone Script

A good phone script includes seven components, as follows:

- part A: your greeting
- part B: your introduction (your name)
- part C: your company affiliation
- part D: the connector (explains why you are on the phone with the prospect)
  - D1: memory jog (used when you've met the prospect before)
  - D2: referral or orphan lead (used when the prospect is someone you've been told to call, either by an individual or by your company)
  - D3: credential or specialty (used when the prospect is someone to whom you've mailed something or whom you are cold calling)
- part E: offer of information via a face-to-face appointment (the first time you mention getting together)
- part F: the benefit of the appointment to the prospect
- part G: alternative-choice close (the point at which you ask for an appointment)

Each of these seven components should appear in the same order in most of your phone calls (the only exceptions are discussed shortly). With each of them, there are a variety of options or words you can use according to the particular category of prospect you are approaching.

We will now discuss the seven components of a telephone script in greater detail.

A: Your Greeting

It is considered poor etiquette to skip a greeting. Be sure to include one. Say, "Hello," "Hi," "Good morning," "Good afternoon," or "Good evening." One caveat is not to use any slang words or phrases. Always adhere to a professional manner of speaking when using the phone.

Certain parts of the country have local greetings that may be used when appropriate. "Howdy" may sound quite friendly in some parts of the West, but it might be inappropriate in other areas of the country. Follow what's commonly accepted in your area, and ask your trainer or manager for advice in this regard.

B: Your Introduction

There are four ways to introduce yourself. The most popular is, "This is _____." The second is simply, "I'm _____." Or you might combine the two and say, "Hi, this is _____." The third way, for people you know very well is to say, "It's _____."

Avoid introducing yourself by saying, "My name is _____," which is the fourth possible introduction. It indicates to the prospect that the caller is a stranger and/or a telemarketer. This introduction merely increases the initial tension.

C: Your Company Affiliation

Clear identification of your company is a requirement of the National Association of Insurance Commissioners (NAIC). Furthermore, it is good business ethics. This component of script writing is the only one that might be movable, and we will explore situations in which you may state your company affiliation later in the script. Generally, however, to obtain appointments in your initial market, identify your company in part C. For example: "Hi, this is Jane Doe, an advisor with (name of company)."

D: Connector

The connector, which is the most important component in your phone script, should be two sentences that reiterate your relationship with the prospect (if appropriate) and explain why you are calling him or her. A phrase like "The reason for my call is . . ." belongs in this part of the script.

This step identifies your connection with the prospect. There are three different types—memory jog, referral or orphan lead, and credential or specialty. We will discuss these from the easiest to the most difficult of the three.

D1: Memory Jog. This connector is used with somebody you've met at least once prior to making the phone call. Thus, these prospects will be the easiest category of leads with whom to make appointments. Perhaps they are acquaintances from personal, social, employment, school, organization, networking, or business affiliations, or maybe you just happened to meet them in casual circumstances. Your own clients, family, and friends are also part of this group. The example below illustrates a memory jog approach.

Example: "Hi, this is Jane Doe. I stopped by your store the other day. I shook your hand and we talked about an idea for. . . . I'd like to position myself as a financial resource for you and your family."

The memory jog lets the prospect know that you are available to help him or her without your having to be specific about a particular financial product. The objective is to get an appointment at which you can complete a fact finder with the prospect.

For all prospects in the D1 category, except your clients, this is one of the occasions, mentioned above, where you will want to put Part C of the script after Part D. The sequence, therefore, would be parts A, B, D, then C. Thus, you would modify the script as shown in the following example.

Example: "Hi, this is Jane Doe. [parts A and B] I stopped by your store the other day. I shook your hand and we talked about an idea for retirement planning. [part D] You may recall that I'm an advisor with (name of your company). [part C] I'd like to position myself as a financial resource for you and your family."

D2: Referral or Orphan Lead. The one characteristic that all prospects in this category have in common is that someone else told you to contact them. With referrals, a prospect, client, friend, or center of influence suggested that you contact the individual. With orphan leads, the company for which you work asked you to contact the person to provide service to him or her.

For a referral, follow the same script sequence as with prospects in the memory jog approach: parts A, B, D, then part C. Because your main concern when calling a referral is to lower his or her initial tension, you must mention the referrer, whom the prospect will recognize, as soon as possible. (Remember, a referral, by definition, does *not* know you. If the referral knows you, he or she is a memory-jog prospect.)

The following example illustrates a referral call where a third party is the connector between you and a prospect you have never met.

Example: "Hi, this is Jane Doe. Your friend Jennifer Lopez suggested that I give you a call."

The referral call is the most popular type of connector and has proven to be the most effective. The prestige of the person who gives you the referral usually causes the prospect to want to hear the rest of your script. The fact that you share a mutual friend, acquaintance, or colleague removes much of the prospect's resistance to meeting with you.

Remember that you must get permission from the referrer to use his or her name. Also, regarding the DNC Law, remember that you cannot call a referral on the DNC Registry without receiving written permission from the referral to the referring party. Otherwise, it is considered breaking the law. If you have the referral's permission, however, you can explain why you are calling by continuing the approach above and then stating your company affiliation.

Example: "Hi, this is Jane Doe. Your friend Jennifer Lopez suggested that I give you a call . . . because recently I did some planning for Jennifer and her husband, Mick, that focused mostly on college savings ideas regarding their young children. She told me that you and your husband also have young children.

"I'm Jennifer and Mick's friend and an advisor with (company)."

With an orphan lead, the order of the script is parts A, B, C, then D. This is because the company name (not your name) is recognizable to the prospect, and the company asked you to make the call.

D3: Credential or Specialty. This approach, which establishes a connection to a person you do not know, meets the highest level of prospect resistance and has the smallest chance of success. In other words, it requires a large number of contacts to secure enough appointments to reach your activity goals.

It is most effective in a direct mail program where you send a preapproach letter that describes a product or service and follow up with a phone call. However, making calls to nonreplier prospects from a direct mail list can be frustrating because you must call only those who are not on the DNC Registry. You have a greater chance of success if you call prospects who returned a reply card that expresses their interest. In that case, you can use an approach like the following.

Example: "Hi, this is Jane Doe, an advisor with (company). I just received your response card requesting information about funding your retirement income. . . ."

The third party here is the letter you sent the prospect, which serves as the credential that justifies your calling. Your motivation for the call is to respond to his or her request for more information. It is important to state, however, that you will share specific information about the product or service only on a face-to-face basis.

If you haven't sent a preapproach letter, you are making a cold call. For this type of call, your connector is your specialty—that is, your area of expertise. Your call should sound something like this.

Example: "Hi, this is Jane Doe, with (name of company). I specialize in working with (individuals in the prospect's profession or business). I've been successful in showing people like you ideas for improving their employee benefits protection."

Your area of specialty is addressing the needs of prospects in a specific market, not selling a particular financial product. Your best hope is that the person will actually be interested in your idea at the time that you call.

> **Remember, you are offering information. The way to deliver that information is via an appointment with the prospect.**

E: Offer of Information via a Face-to-Face Appointment. Remember, you are offering information. The way to deliver that information is via an appointment with the prospect. The offer, therefore, is an invitation to get together and share some ideas with the prospect. The following is an example of this script component.

Example: ". . . Because I share this information on a face-to-face basis, I'm simply calling (this afternoon or evening) to schedule an appointment so that we can get together and explore some strategies as they might apply to your situation. . . ."

Notice that because there is no way of knowing specifically what product or service would be of interest at this time, the offer is solely to get together and explore general ideas about financial planning. It is nonthreatening and thus should not cast a negative light on the referrer.

Although the offering seems simple, it actually has three disctinct parts:

- First, it uses an "inviting" verb or verb phrase such as "get together." Similar words—"meet," "visit," "see," "sit down with," or "spend a few minutes" will do just as well.
- Second, another verb tells the prospect what you are going to do when you get together. The example above uses the word "explore." Others like "talk," "share," "provide," "discuss," and "show you" may work equally well, depending on the prospect and the circumstances. Words to avoid are "review" and "present." "Review" is too nonspecific. "Present" is a sales word with a negative connotation.
- Third, the offering tells the prospect what you are you going to show or provide to him or her. In the example above, the advisor uses the word "strategies." Other good words are "ideas," "concepts," "options," "opportunities," and "techniques" as they pertain to the financial services and products you offer. All these words allude to a face-to-face meeting.

As emphasized earlier, it is important to avoid the use of selling words on the phone. The summary below identifies some good words for use in parts D, E, and F of your phone scripts. It also identifies some words to avoid.

Composing Telephone Scripts

Good Words to Use		Bad Words to Use		
program	share	death	bought	introduce
concept	show	disability	in your area	myself
technique	talk	premium	can and will	application
idea	chat (for women)	sell	pop in	tragedy
strategy	explore	buy	drop by	review
vehicle	investment portfolio	guarantee	stop in	present
get together	financial portfolio	proven	God forbid	second
discuss	insurance portfolio	investment	contract	opinion

F: The Benefit. There has to be a compelling reason or the promise of a positive experience for a prospect to want to meet with you. Part F of the script relies on some powerful words and an expressive delivery. Use vivid language to describe the benefit to the prospect. Don't be overly specific because that crosses into selling skills. Still, you want to stimulate the prospect's curiosity. You must sound genuinely enthusiastic and convey clearly that you have something interesting and important for the prospect.

> There has to be a compelling reason or the promise of a positive experience for a prospect to want to meet with you.

Example: ". . . As a result of our meeting together, you might discover the need to improve your financial position, or at least we can confirm that you are already doing everything you can. . . ."

G: Alternative-Choice Close. In parts A through F of the script, you introduce yourself and your company, state the reason for your call, and create an interest in meeting you to obtain the benefits you offer. Now you are ready to end the call by closing an appointment with the prospect.

Give the prospect a choice of times to get together with you. Avoid being specific about the day of the week or the time of the day when you first ask for the appointment. Broad time periods appear to work best in this busy high-speed society. You might say, for example, "Would it be easier for you to meet in the morning or the afternoon?"

You also want to show your awareness of how busy the prospect may be by saying something like, "What is a less hectic time of day for you, mornings or afternoons (daytime or evenings)?" Early in the week may be better for some prospects; later in the week may be better for others. Many people work one or both days of the weekend and may have time off during the week. The key is to show that you value the prospect's time and therefore would like to meet when it is most convenient for him or her. As soon as the prospect chooses a broad time period, you've got an appointment! Then, together, you can decide on a specific day and time.

The following example shows a full script from part A to part G.

Example: "Hi, this is Jane Doe, an advisor with ABC Insurance Company here in Bryn Mawr. I just received your response card requesting information about funding your retirement. Because I share this information on a face-to-face basis, I'm simply calling this afternoon to schedule an appointment to get together and explore some strategies as they might apply to your situation. As a result of our time together, you might discover the need to improve your financial position, or at least we can confirm that you are already doing everything you can.

"I know you're extremely busy. What would be a less hectic time for us to talk—this week or next?"

There it is. Jane Doe has just made a professional call to schedule an appointment to a direct mail replier. And it all took less than 40 seconds. Bear in mind that the call will last longer if the prospect either shows some interest or expresses a concern you must address. Even so, the entire call should take no more than 2 minutes. Calls that take longer are usually unsuccessful (as we will discuss in the section titled "Prospect Responses").

End your call by confirming the appointment.

Example: ". . . Okay, I just want to confirm that we will be meeting at 7 o'clock next Thursday evening at your home. I will call for directions to your house prior to the appointment."

> **Planning in advance what words you are going to use in your phone call will always result in a smoother, more organized presentation than ad-libbing.**

How Not to Sound Canned

Planning in advance what words you are going to use in your phone call will always result in a smoother, more organized presentation than ad-libbing. If you sound like you are reading, however, you will defeat the purpose of having a structured script in the first place.

The best way to sound natural but prepared is to compose a script that includes all the necessary components. After reading the script out loud and approving the exact words, you should rewrite the script in short, bulleted phrases. That way, you will "talk" the key points in the proper order and remember the important words. For instance, an advisor who is new to the financial services business and calling a prospect in his or her natural market might use the script in Example 1. The advisor's script in the abbreviated bulleted format might look like Example 2.

Example 1:	"Hi, this is Paul Harris, and I'm sure that you've heard about my new career with the XYZ Company. I'm really excited about it. The reason I'm calling is that I would like to position myself as a financial resource to you. I'd like to set up a time when we could get together so that I can share with you the scope of the work I do. That way, you can use me, my expertise, and the resources of my company any way that makes you feel the most comfortable. With that in mind, what is a less hectic time for us to get together—days or evenings?"
Example 2:	• heard about my new career • excited about it • reason I'm calling • position myself • financial resource • get together • share with you • scope of the work I do • me, expertise, resources • any way that makes you feel the most comfortable • less hectic

PROSPECT RESPONSES

Four Categories of Responses

If you are prepared with the right script and the right approach, your offering to the prospect is not likely to result in a flat "No." But there might not be a wholly positive response either. Responses generally come in four categories. The prospect

- agrees to an appointment
- says either "I'm not interested" or "I'm all set"
- asks you a question
- states a problem or raises an objection

Prospect Agrees to an Appointment

In some situations, the prospect will say "Yes" and readily agree to appointment. If you give the prospect a choice of times, however, the response is more likely to be, "Well, mornings are kind of busy, but afternoons later in the week look pretty good." Notice the value of not asking anything that can be answered with a simple "Yes" or "No." For example, avoid a weak question like, "Would you be interested in getting together sometime?" What are you going to do if the prospect says "No"? It would be better to say, "Let's get together when things quiet down next week. Are mornings or afternoons better?"

Prospect Says, "I'm Not Interested" or "I'm All Set"

In this category, the prospect gives a reflexive response that he or she has learned usually works with most phone solicitations. The two especially notorious responses in the financial services industry are "I'm not interested" and "I'm all set."

Both are essentially knee-jerk reactions and are frequently followed by the prospect's hanging up. The reason the responses are hard to deal with is that they are noncommunicative. Neither gives a specific reason or answers the question, "Why?" Think how much easier it would be if the prospect responded, "I'm not interested right now (or "I'm all set right now"); I already have life insurance." You can deal with that response effectively using the formula discussed later in this text.

You need to react correctly to these two reflexive responses. If the prospect says, "I'm not interested," you can simply say, "May I ask why not?" Only two things can happen. Either the prospect repeats, "I'm not interested," because he or she truly is not interested, or the prospect gives a reason why he or she is not interested—for example, "because I'm very busy right now giving my baby his bath" or "because this is my inventory season and I'm shorthanded." It may be best to hang up politely and move on to the next prospect. Such responses occur mostly in situations where you are cold calling strangers with whom you have no connection. The way to avoid these negative responses is to call warmer leads or referrals. Or you can always offer to call back at a future date when the prospect has more time and may be in a better frame of mind.

If the prospect says, "I'm all set," your response should be, "What exactly do you mean?" You need the prospect to share the real reason that he or she believes an appointment with you will not be useful.

> **"I'm not interested" and "I'm all set" are hard to deal with because neither gives a specific reason or answers the question, "Why?"**

rule of two

If the prospect continues to show a lack of interest, remember the *rule of two*: If the prospect responds negatively two times, let go of the call. Generally, if prospects have not offered a reason for their lack of interest or why they think they're "all set," nothing you say is likely to change their minds; you may simply antagonize them.

If you hear a legitimate reason why the prospect thinks an appointment is a bad idea or inconvenient—for example, "One of my children is sick," or "Soccer playoffs take place over the next few weeks"—it's usually best to politely end the call. It is more likely that you can keep this person in your prospecting file if you exit graciously and not increase the tension further.

Example: "I understand (whatever the problem is) and this probably is a bad time for us to talk, but I'd like to keep in touch with you. I will call you in the next couple of weeks. In the meantime, I'd like you to be on my mailing list. Is that okay with you?"

Prospect Asks You a Question

Often, a prospect will pose a question to you in response to your offering of an appointment—for example, "Can you send me something in the mail?" Your response should be brief, and it should be some variation of "Yes," "No," or "That depends." Although the answer to the can-you-mail-me-something question could surely be "Yes," what would you send? Your company probably has hundreds of brochures, but you can't send them all. However, you're certainly not going to say, "No" and possibly disappoint or anger the prospect. The best answer to the prospect's question in this case is a variation of "That depends."

Example: "I'll be glad to send you some information but it's hard to know specifically what type of information would be of most value to you right now. The best thing I can do for you is to get together with you for a quick cup of coffee so I can find out what financial planning concept interests you the most. Then, when you decide you might be interested in discussing it, you'll think of me. Would getting together be better in the morning or afternoon?"

Notice the phrase "the best thing I can do for you." It indicates that you have the prospect's interests at heart, which should make him or her feel more favorably inclined to see you.

Be careful about answering too many questions. It may seem that the more questions you answer, the more knowledgeable you'll appear, but actually, questions answered on the telephone can eliminate the need for an appointment.

It is imperative that you be brief in responding to questions, and remember to ask the prospect for an appointment after answering. After the prospect's third question, you should be able to set up an appointment. Why? Because three questions are a sign of interest. Use a response similar to the one in the example below if you find yourself caught in a question-and-answer session.

> **After the prospect's third question, you should be able to set up an appointment. Why? Because three questions are a sign of interest.**

Example: "You know, that's a good question that deserves a better answer than I can give over the phone. The best thing I can do for you is to meet you for a quick cup of coffee and get to understand your situation a little better. Would early next week be good, or would later in the week be better?"

You can find some of the most common questions you will hear from prospects in section 2 of appendix A. Each question is followed by a short answer that has worked for advisors in the field. Add your own answers as you encounter these questions. After answering a prospect's question on the phone, remember to close him or her for an appointment, just as you did in part G of your initial phone script.

Example: ". . . With that in mind, what is a better time to get together—daytime or evening?"

If the prospect's response is positive, say, "Good, _____ . I'll look forward to seeing you on (day) at (time) o'clock." (Be sure to repeat the date and time of the appointment.)

Prospect States Problem or Raises Objection

In the final type of response, the prospect states a problem or raises an objection. (See appendix A, section 3, for specific responses.) Here, the prospect is sharing a real or perceived situation that the prospect believes would make an appointment a mutual waste of time. There are typically four categories of responses:

- "I don't have any money."
- "I already have an advisor."
- "I'm too busy. Call me sometime in the future."
- "I'm all taken care of at my job."

Problems are not insurmountable. They are simply the prospect's rationale for what stands in the way of meeting with you.

Your job is to acknowledge the problem, show that you respect the prospect's opinion, and then address the problem so that it does not become an obstacle to the appointment. Although certain problems may eventually impede or prevent a sale, none of them should be a roadblock to making an appointment. Remember that your focus must be on giving the prospect a reason to meet with you, not on selling any financial product over the telephone. You are merely conveying the benefit of getting together for an appointment.

The "feel, felt, found" technique is one of the most persuasive methods for handling objections. These three statements follow immediately after the propsect states a problem or raises an objection:

Standard Response	Alternative Response
• I can appreciate that you feel that way.	• I can appreciate you telling me that.
• Other clients of mine initially felt that way too.	• Other people initially told me the same thing.
• But they found, after meeting with me that. . . .	• But they found, after meeting with me that. . . .

The first two sentences of the alternative response should be used when the first two sentences of the standard response do not appropriately acknowledge the problem stated by the prospect.

A response to the no-money objection, for example, might sound something like this.

Example: "I can understand that you feel things are very tight with two young children. Many of my clients initially felt the same way when I called. But they found after our meeting that my work is about prioritizing your financial needs, not necessarily spending your money."

The opening statement in the example above conveys the advisor's empathy. The prospect is likely to feel that the advisor is actually listening and not simply answering with a canned response. The second sentence links the prospect's problem to similar problems the advisor has encountered, which reassures the prospect that his or her problem is not unusual and therefore can be solved. Sentence three illustrates the benefit of meeting with the advisor. Following the phrase "But they found . . . that" in the third sentence, you must add a benefit that relates to the stated problem. (Section 3 of appendix A gives

---

The "feel, felt, found" technique is one of the most persuasive methods for handling objections.

you some excellent benefits that you can use in answer to some of the most common problems or objections.)

The feel-felt-found technique works well for three reasons: It is easy to memorize, it is short, and it moves the prospect past his or her initial objection. The response acknowledges your understanding of the problem, shows that it should not be an obstacle to at least meeting, and asks for an appointment. Perhaps most of all, this technique tends to diffuse the tension that naturally seems to occur during a sales situation.

After stating the three sentences above followed by a benefit, you should then add:

- "I'd like to see if I can do the same for you."
- "Sound fair enough?" (This is a rhetorical question; don't wait for an answer.)
- "With that in mind. . . ." (Close the call by making an appointment.)

The example above, therefore, might conclude like this.

Example:	". . . I'd like to see if I can do the same for you. Sound fair enough? With that in mind and given your hectic schedule, would mornings or afternoons be better for us to meet?"

It takes practice to train yourself to use the feel-felt-found technique. But the results are worth the effort. Skillful use of the technique reduces the prospect's tension, establishes your credibility, and clarifies why an appointment with you makes sense.

TRACKING YOUR RESULTS

Need for a System

> A system to track your sales call results is vital to determine what you need in order to achieve your prospecting goals.

A system to track your sales call results is vital to determine what you need in order to achieve your prospecting goals. Without a system, you will not be able to reach your goals because there will be no quantifiable measure of your progress. The system discussed below for tracking your telephoning results includes some statistics in the financial services industry so that you can gauge your level of success against the national averages.

The first question you must answer is how many face-to-face appointments you need per week. (The average for a new financial advisor is 10.) You will have to make three gross appointments (unconfirmed) in order to net two face-to-face appointments—that is, you must multiply the desired number of face-to-face appointments (net) by 1.5 to determine the number of gross appointments you must make, as follows: number of net appointments x 1.5 = number of gross appointments, or

10 net appointments x 1.5 = 15 gross appointments

Next, you must determine how much time you need to spend on the phone to acquire your desired number of gross appointments for a given week. According to records compiled by telephone skills trainer Gail B. Goodman while training thousands of financial advisors on how to use the phone, the national average of appointments acquired per hour is 2 to 2.5. To calculate your required time on the phone, divide your desired number of gross appointments per week by your average number of appointments per hour. The result is the number of hours you must spend on the phone to obtain these appointments: number of gross appointments per week ÷ average number of appointments per hour = time on phone, or

15 gross appointments ÷ 2.5 per hour = 6 hours

Thus, if you're willing to spend 6 hours on the phone per week and your average appointment rate per hour is 2.5, you can expect to get 15 unconfirmed appointments per week. If you acquire 15 gross appointments per week and your rate of reschedules and cancellations is in line with the national averages (discussed below), you can expect to net 10 face-to-face appointments per week. (Confirmations, reschedules, and cancellations will be discussed later in this chapter.)

Counting Calls

It is necessary to use a counting system to track your personal appointments per hour. The counting system should track the following six elements:

dials
- *dials*. This tracks every single time you dial the phone.

contacts
- *contacts*. This is a record of the times you reach the proper party.

presentations
- *presentations*. Presentations are the number of times that you complete your script from beginning to end without interruption.

appointments
- *appointments*. You must have a record of each person's name, date, time of appointment, and the source of this lead. You must keep track of where your appointments came from to formulate your marketing plans. If you attend a trade show and book many appointments as a result, this tells you to attend more shows. On the other hand, if the seminars you do result in few or no appointments, you'll know to avoid them or to change how you conduct them.

dialing time
- *dialing time*. This tracks your actual time on the phone in hours and fractions of hours. (For example, 1 hour and 15 minutes is 1.25 hours; 1 hour and 20 minutes is 1.30 hours; 1 hour and 30 minutes is 1.50 hours.)

- phone session time of day and day of the week. It is essential to keep track of what time of day and day of the week you are making calls so that you can analyze what times and days yield the most (and least) effective results for you. For instance, if you find that you're able to consistently book appointments on Tuesdays and Thursdays between 10:30 a.m. and 12:00 p.m., obviously you should set aside these days and times for phoning.

Effectiveness Ratio Averages

telephone effectiveness ratios

Using statistics compiled for dials, contacts, presentations, appointments, and dialing time, you can quantify and track four *telephone effectiveness ratios*. These ratios allow you to measure how productive your time on the phone is during each session. Your effectiveness will vary based on the geographic area where you live, the time of day you make calls, the markets you are calling, and the source of the leads. The ratios will help you determine how much time you will need to schedule each week to accomplish your appointment-setting goals when prospecting on the telephone.

> **Phone effectiveness varies based on the geographic area, time of day, markets you are calling, and source of leads.**

Let's take a look at some of the national averages for each of the four effectiveness ratios.[1]

contact percentage

Contact Percentage = Contacts ÷ Dials. The national average for this effectiveness ratio is 30 percent. In other words, if you make 10 dials, you can expect to contact an average of three of the prospects you are trying to reach. The percentage varies, as shown below, depending on the type of prospects you are calling, and the time of day you are making the calls:

- daytime business calling—22 percent to 29 percent
- evening residential calling—25 percent to 35 percent

For businesses, the lower end of the average is usually in larger urban areas, and the higher end is in smaller towns and rural areas. The average contact percentage for evening residential calls can exceed 40 percent if more people are in their homes because of bad weather in the area. (These averages assume that there are no cold calls or calls to individuals on the DNC list.)

presentation rate

Presentation Rate = Presentations ÷ Contacts. Overall, the national average for this ratio is 75 to 80 percent. Thus, of all the contacts you reach, 75 to 80 percent will result in your making your full telephone presentation. This rate has risen since the DNC Law went into effect because the contacts you reach these days are more inclined to listen to what you have to say.

appointment-setting rate

Appointment-Setting Rate = Appointment ÷ Presentations. The national average for this effectiveness ratio hovers at around 50 percent. That is, you can expect to set appointments with about 50 percent of the prospects to whom you have made a full presentation. The following variations exist, depending on the source of leads you are calling:

> **You can expect to set appointments with about 50 percent of the prospects to whom you have made a full presentation.**

- natural markets—70 percent to 80 percent
- referrals—50 percent
- seminar follow-ups—80 percent to 90 percent
- cold calls—17 percent
- face-to-face canvassing to businesses—50 percent

appointments per hour

Appointments per Hour = Appointments Made ÷ Dialing Time. The national average is 2 to 2.75 appointments acquired per hour. The average number of appointments varies as follows, depending on the source of the leads and the markets you are calling:

- natural market—3 to 5 (sometimes even more)
- referrals—2 to 3 (sometimes more)
- cold calls—.5
- seminar follow-ups—3 to 5

Many successful advisors "work backwards" and calculate the dollar value of each call, whether it is productive or not. This encourages them to focus on their income goals and keep making calls in the face of rejection. (This concept will be discussed in greater detail in a later chapter of this text.)

Confirmations, Reschedules, Cancellations, and No-Shows

You need to know which appointments to confirm, which to reschedule, and which will never happen because the prospect has canceled. This will enable you to determine your rate of success in meeting with prospects with whom you have made appointments. You also need a system to control the process of scheduling appointments made over the phone in order to effectively manage your time when meeting with prospects.

There are four possible outcomes of appointments made over the phone:

confirmed appointment

- *confirmed appointment*—one that is scheduled with a prospect for a specific date, time, and place, to which the prospect (within 24 hours before it is set to begin) reiterates his or her commitment to attend

rescheduled appointment

- *rescheduled appointment*—one that the prospect wishes to change. Rescheduling can occur either at the time you call to confirm or in the near future.

cancellation

- *cancellation*—an appointment the prospect refuses to attend even though you have offered to reschedule. Generally referred to as dead leads, they are relatively rare because it is unlikely that a prospect who has agreed to meet with you would completely reverse his or her decision when you call to confirm.

no-show

- *no-show*—a scheduled and confirmed appointment that the prospect fails to attend, often referred to as a stand-up

Procedure to Confirm and Reschedule Appointments

> **Professionals always call to confirm their appointments.**

Professionals always call to confirm their appointments. After all, it is in the advisor's best interest to confirm appointments rather than waste time attempting to see prospects who are not interested.

You should call to confirm 24 hours before the appointment. If you get the prospect's voice mail, leave a message saying, "Hi, this is _____. We have an appointment tomorrow at 10 o'clock. I'm calling to remind you about it and just to let you know I will be there."

If you are calling to confirm an appointment at the prospect's house, be sure to get directions. This is important because it encourages the prospect to visualize your coming to his or her home, which strengthens the prospect's commitment to seeing you. For business owners, on the other hand, it is better to use an Internet service to get directions; business owners usually prefer not to spend time on the phone to give you directions.

Be sure to use the word "remind," not the word "confirm," when you are calling. It is generally more effective and typically better received by prospects.

Example: "I'm calling to remind you about my appointment with you (or you and your partner, husband, or wife) tomorrow at _____."

With reschedules, try to get a new appointment for the same time in the next week. Sometimes reschedules can be challenging; you may keep getting voice mail and wind up calling the prospect repeatedly to set a new time. Twice a year the number of reschedules you encounter will increase dramatically: during the holiday period at end of year and during the week before elementary schools open, which is a somewhat chaotic week for parents.

> **Wait a maximum of 20 minutes for the prospect to arrive before you leave a no-show appointment.**

When you arrive at an appointment and the prospect is not there, call on your cell phone; leave a message saying that you are concerned about him or her, and ask the prospect to let you know that he or she is okay. Despite the tendency to get angry, don't convey that anger to the prospect. There is a greater likelihood that you will hear back from a no-show prospect if you remain calm. Wait a maximum of 20 minutes for the prospect to arrive before you leave a no-show appointment.

Statistical Averages for Appointments

- Confirmations   70% to 75% of all appointments you make monthly
- Reschedules   20% to 25% monthly
- Cancellations   7% to 10% yearly*
- No-shows   2% to 3% a yearly*

* The reason that average rates for cancellations and no-shows are listed yearly is that overall they are so statistically insignificant for most advisors that some months will have no cancellations, while others may have several. You may not encounter any no-shows; it depends on the markets in which you work.

ADDITIONAL PHONING CONCEPTS

Human Gatekeepers

When you make business calls, you are going to confront human gate-keepers. One of the gatekeeper's jobs is to protect his or her boss from unnecessary phone calls. You must be polite to the gatekeeper, but you must also let the person know why he or she should take your phone call and why your call is worth the boss's time. Here is a sample script for use with a gatekeeper.

Example: "Hi, this is _____. Who am I speaking with?"
("Mary.")
"Mary, you're in the best postiton to help me. I need to speak to _____. Is he busy—you know, on the phone, in a meeting—or can you please put me through right now?"

If the prospect is busy, you should ask whether you should wait on the phone or call back when it might be a better time to talk. Try to get the gatekeeper to help you, but don't become rude or aggressive.

If the gatekeeper asks, "What is this about?" be prepared with a response that states a benefit that will directly affect the gatekeeper. For instance, if you are calling a benefits coordinator who has a gatekeeper to talk about employee benefits, you may say something like this.

Example: "I'd like to speak to your boss about a program that adds to your employee benefits but also allows you to keep this benefit if you decide to leave the company."

If the concept is intriguing enough to the gatekeeper, he or she will be motivated to deliver your message to the boss.

If the concept is intriguing enough to the gatekeeper, he or she will be motivated to deliver your message to the boss, which increases the possibility that you will receive a call-back.

Automated Call-Screening Devices

Voice Mail

Voice mail efficiently manages phone messages from callers. Its primary use is for the call recipient to give the caller a way to leave a message in the recipient's absence. Another use for voice mail is to screen incoming calls so that the recipient can decide whether to take the call immediately, return the call later, or not call back at all.

Advisors have two main questions concerning voice mail:

- For whom (what types of prospects) should I leave a voice mail message?
- What should the content of the message be?

Recall the three types of connectors discussed earlier in this chapter. You should leave a message for D1s (memory jogs) and D2s (referrals and orphan leads) but not for D3-type prospects (direct mail leads and cold calls).

Also recall the earlier discussion on the sequence of script components. When leaving a voice mail message, use parts A and B; then skip to part D. Do not state the reason for your call or the name of your company in the message. Save that information for when you actually reach the prospect. Make your message brief, and at the end, state your phone number two times slowly.

Caller ID

Caller ID is a fact of life. Increasingly, individuals are using it to protect themselves from unwanted calls, especially from telemarketers.

You have no way of controlling a prospect's use of caller ID. Therefore, you should simply ignore it, working around it in your everyday prospecting activities. If the prospect answers your call, you have the opportunity to use the phone techniques you have learned in this chapter. If not, there is nothing you can do about it.

Asking Prequalifying Questions

Many successful advisors prequalify prospects once they have agreed to an appointment. Whether you prequalify your prospects depends on your type of practice and personal views on this matter. For example, if you market multiple financial products, prequalification before the initial interview may not be crucial because you have other products to which you can turn to satisfy other needs prospects may have. Nevertheless, prequalifying before the initial interview will allow you to prepare a smooth transition to these other needs and products.

If you do wish to prequalify before the initial interview, the next step is to decide what information you need to know and build a script. Choosing what questions to ask is based on your philosophy of prequalification.

The transition from getting the appointment to asking the prequalifying questions needs to be smooth—that is why a script is important. Here is one commonly used transition.

Example: ". . . Okay, Richard, that's going to be 7 o'clock next Thursday night at your home with you and your wife. Before I hang up, however, I just want to let you know that there are certain financial products that you can't have just because you want them. You have to qualify for them. Richard, I know your time is valuable. For our appointment to be more productive, therefore, I would like to briefly ask you several critical questions about your health and medical history."

Once you have made the transition, keep the prequalifying questions short and general. For example, here are a few different questions that successful advisors use for life or long-term care insurance prospects:

- What is your date of birth?
- How is your health?
- What medications are you taking?
- What surgeries have you had?
- Can you tell me a little about your medical history?
- Are there any medical problems you manage daily?

After your prospect answers these questions, refer to your company's underwriting rules to determine how and whether to proceed. Prepare a short script you can use to let the person down gently if you know he or she will not qualify. This requires the utmost care and sensitivity. You do not want the prospect to feel bad.

Example: "Stan, I am really sorry, but because of your health, we will not be able to help you at this time. I highly recommend that you consider seeing an eldercare attorney who specializes in alternative strategies for meeting your long-term care planning needs. I can provide the names of several attorneys if you like."

Summary

Expect to encounter a variety of problems (objections) that the prospect considers obstacles to an appointment. Deal with each one quickly and in a friendly but professional tone, and always end with a request for the appointment. If you hear more than two reasons why the prospect thinks an appointment is a bad idea or simply not acceptable, it's usually best to politely end the call.

Never argue with the prospect. Maintain a pleasant tone of voice at all times. Resist any temptation to engage in a discussion about the benefit of your products. Your only discussion should be about the benefit of your meeting.

Have realistic expectations. If you are cold calling, you can expect to obtain less than one appointment per hour. That's why successful advisors make few, if any, cold calls. They find their prospects through referred leads, personal observations, some direct mailing, canvassing to businesses, and business and social networking contacts.

Remember that the "No's," even the abrupt or rude ones that can be so annoying, are not directed at you personally. You may hear "No" before your prospect even knows why you called. This indicates that the response is nothing more than an automatic reflex. It may result from a preconceived notion that you are an ordinary telemarketer selling some commodity, or it may be because the prospect has no knowledge of what your work involves. More likely, such responses simply reflect the natural human tendency to take the path of least resistance and have nothing to do with a stranger who calls, regardless of the product or service being offered.

| Planning and preparation are essential to success on the telephone. |

Planning and preparation are essential to success on the telephone. Use the planning and preparation steps discussed in this chapter to make phoning sessions a routine and inviolable part of your daily sales activity.

It is advisable to keep a phone log or tally sheet. Over a fairly short time, you will develop consistent ratios of dials to contacts, contacts to presentations, and presentations to appointments. Your personal telephone effectiveness ratios serve both as a benchmark to track your progress toward reaching your telephone prospecting goals and as a way to keep your individual phone sessions in perspective.

CHAPTER FIVE REVIEW

Key Terms and Concepts are explained in the Glossary. Answers to the Review Questions and Self-Test Questions are found in the back of the textbook in the Answers to Questions section.

Key Terms and Concepts

elevator speech	telephone effectiveness
10-second commercial	ratios
Do Not Call Registry	contact percentage
Do Not Call Law	presentation rate
rule of two	appointment-setting rate
dials	appointments per hour
contacts	confirmed appointment
presentations	rescheduled appointment
appointments	cancellation
dialing time	no-show

Review Questions

5-1.   Explain the difference between an elevator speech and a 10-second commercial.

5-2.   Define the criteria for contacting prospects who fall within the exceptions under the current Do Not Call Law.

5-3.   Identify the four principles of phoning.

5-4.   Briefly describe the three main elements in preparing for a phoning session.

5-5.   Identify the seven basic components in a telephone script.

5-6.   Describe the variations of part D (the connector) of the telephone script when approaching each of the three categories of prospects.

5-7.   Describe how an advisor can avoid sounding canned on the phone when using a written script to prospects.

5-8.   Explain the procedure for handling a prospect's questions on the telelphone.

5-9.   Explain the procedure for handling a prospect's concerns or objections to meeting with the advisor.

5-10.   Identify the four telephone effectiveness ratios, and explain their importance in tracking phoning session results.

5-11.   Describe the procedures to handle appointment confirmations, reschedules, cancellations, and no-shows.

5-12.   Describe the rationale and procedure for using prequalifying questions prior to meeting with a prospect.

Self-Test Questions

Instructions: Read chapter 5 first, then answer the following questions to test your knowledge. There are 10 questions; circle the correct answer, then check your answers with the answer key in the back of the book.

5-1.   Which of the following best describes the purpose of the telephone call to a prospect?

(A)   to establish a relationship with the prospect
(B)   to qualify the prospect prior to meeting with him or her
(C)   to get an appointment for a face-to-face meeting
(D)   to set the stage for making a sale

5-2.   Which of the following is the most important concept regarding the consistent and effective use of the telephone as a prospecting mechanism?

(A)   enthusiasm
(B)   calmness
(C)   discipline
(D)   attitude

5-3.   Which of the following statements concerning the handling of a prospect's questions during the phone call is correct?

(A)   The advisor should not be afraid of answering too many questions.
(B)   It is imperative to be brief in responding to questions.
(C)   Answering the prospect's question thoroughly is more important than making an an appointment.
(D)   Just because the prospect asks three or more questions, it does not indicate interest.

5-4.   Which of the following statements concerning the "I'm not interested"/"I'm all set" response from a prospect is (are) correct?

    I.   Both are essentially knee-jerk reactions, frequently followed by the prospect's hanging up.

    II.   The advisor's reaction to these two responses should be to ask the prospect to elaborate.

    (A)   I only
    (B)   II only
    (C)   Both I and II
    (D)   Neither I nor II

5-5.   Which of the following statements concerning component D of the phone call, the connector, is (are) correct?

    I.   It is the most imortant component in a phone script.

    II.   The advisor should always state his or her company affiliation before the connector.

    (A)   I only
    (B)   II only
    (C)   Both I and II
    (D)   Neither I nor II

5-6.   Which of the following statements concerning use of the phone to obtain the desired number of appointments in a given week is (are) correct?

    I.   To determine the advisor's required time on the phone, divide the average appointments per hour into the desired gross number of appointments needed.

    II.   To determine the gross number of appointments to schedule, multiply the number of net appointments the advisor wants per week by 1.5.

    (A)   I only
    (B)   II only
    (C)   Both I and II
    (D)   Neither I nor II

5-7. All the following statements concerning preparation for phone calls are correct EXCEPT:

(A) It is essential for the advisor to manage and plan the physical phoning environment.

(B) There will never be a second chance for the advisor to make a first impression.

(C) Attitude affects phoning because the way the advisor feels is the way he or she sounds.

(D) All the advisor needs to make calls is a telephone and an appointment book.

5-8. The advisor can contact easier leads within the parameters of the national Do Not Call Law in all the following ways EXCEPT by

(A) calling friends and family within the advisor's natural market

(B) making follow-up calls to business owners the advisor has canvassed

(C) phoning new homeowners from a recent mortgage list

(D) calling recent attendees from a seminar the advisor conducted

5-9. Each of the following is a recommended alternative-choice close when asking the prospect for an appointment EXCEPT:

(A) "Are mornings or evenings more convenient for you?"

(B) "What is better for you, this week or next?"

(C) "Is Tuesday at 6 p.m. or Wednesday at 7 p.m. better for you?"

(D) "What's less hectic for you—mornings or evenings?"

5-10. When using a counting system to track personal-appointments-per-hour statistics, the advisor should keep records of all the following EXCEPT

(A) dials

(B) contacts

(C) presentations

(D) rejections

NOTE

1. The national averages for each of the four effectiveness ratios were provided by Gail B. Goodman and based on empirical data gathered from advisors in the financial services industry who have participated in her telephone training workshops.

6

Goal Setting and Time Management

Learning Objectives

An understanding of the material in this chapter should enable the student to

6-1.   Describe the importance of goal setting in achieving personal success.

6-2.   Explain the role of discipline in attaining goals.

6-3.   Describe techniques for using perseverance to break through "the wall" and keep from quitting.

6-4.   Define structure and explain the impact it can have on behavior.

6-5.   Describe the difference between problem solving and creating as they relate to obtaining the desired results.

6-6.   Identify the four steps in creating the structure that will help to achieve goals.

6-7.   Explain how to inventory and analyze the use of time.

6-8.   Describe the green, red, and yellow time management zones for apportioning daily business and social activities.

6-9.   Identify strategies and techniques to schedule, prioritize, and manage time effectively.

Chapter Outline

INTRODUCTION

This chapter examines the concepts of goal setting and time management. These concepts will be discussed generally, as two important elements of an individual's overall self-improvement efforts, and specifically, as they relate to achieving success for the career financial advisor. The discussion will include some of the broader abstract psychosocial aspects of goal achievement, as well as practical applications of goal setting that are specific to the financial services industry.

Likewise, time management principles will be also be considered generally, as they relate to improving any individual's efficient use of time, and specifically, as they concern the daily planning and performance of effective prospecting techniques and sales activities by financial advisors.

PRINCIPLES OF GOAL SETTING

Goals Defined

goal

A *goal* (or objective) is a desired end point in personal or organizational development. The intention is usually to reach the goal in a finite period of time. Goals can have short (less than 1 year), intermediate (1 to 5 years), and long-term (greater than 5 years) time periods for their targeted completion.

Goals can also encompass many areas of a person's life. Life is multi-faceted and consists of many types of activities. Likewise, goals often reflect an individual's ambition to accomplish tangible results not only personally, but also for friends, family, and, in the broader perspective, society as a whole. Personal goals entail the pursuit of success in various categories of a well-rounded individual's life, including family, career, education, social involvement, recreation, philanthropy, spirituality, self-development, and financial planning.

Importance of Determining Goals

goal setting

Goal setting is the process of establishing clearly definable, measurable, achievable, and challenging objectives toward which an individual can target his or her effort, resources, plans, and actions. On a personal level, goal setting is a process that allows people to identify and then work toward realizing their own objectives such as financial or career-based goals.

> Without a time frame attached to it, your goal is merely a frivolous wish.

To be most effective, all goals should have deadlines. Without a time frame attached to it, your goal is merely a frivolous wish. Goals should also be in writing so that they act as a motivating reminder to the goal setter and represent an explicit commitment to their completion. There must be realistic plans to achieve the intended goal.

goal prioritization

Another important aspect of goal setting is *goal prioritization*—the process of ranking goals according to their importance to determine the order for their achievement. Accordingly, it is helpful to establish flexible plans for achieving long-term goals in the event that they are blocked by or incompatible with other priorities. For example, if you set the acquisition of a luxury automobile as a goal, you may find that your current monthly budget will not accommodate the additional expense, so you may have to delay the purchase date.

> To consistently perform at peak effectiveness, the most successful people begin with the end in mind.

To achieve success, ask yourself, "What are my goals in life?" This is usually a difficult question to answer because you are busy and it's not easy to think about goals when you have so much to do in your daily life. Nevertheless, to consistently perform at peak effectiveness, the most successful people begin with the end in mind. "Knowing why you're doing what you're doing when you're doing it" goes a long way toward helping you motivate yourself in your daily routine. Having goals enables you to maintain your focus. Your focus can help drive you to greater personal achievement.

In order to establish and attain your goals, it is helpful to understand the concepts of discipline, perseverance, determination, and structure. These concepts are discussed in the sections that follow.

Three Keys to Strengthen Discipline

You may have heard that to get what you want, you cannot take any shortcuts when it comes to discipline.

Discipline helps people realize their full potential and get the results they want to achieve. It is not enough merely to have the skills and the talent. To succeed, you need the discipline to exercise those skills on a consistent basis.

> **It is not enough merely to have the skills and the talent. To succeed, you need the discipline to exercise those skills on a consistent basis.**

There are three keys to strengthen your discipline. To enjoy the freedom that comes from being disciplined, you must use them simultaneously. These three keys will bring you closer to your goals:

- Delay gratification.
- Make your choices before the "moment of choice" arrives.
- Know what you want.

Delay Gratification

Delaying gratification acknowledges that life has both pleasure and pain. As author M. Scott Peck, MD, observes in *The Road Less Traveled*, "Life is difficult."

The process of writing a book is a great lesson in delayed gratification. It is similar to writing a college term paper every night for a week, a month, or even a year before experiencing the satisfaction of a completed project.

Financial advisors are like coaches overseeing a group of athletes. The late Tom Landry, former coach of the Dallas Cowboys, described his job in an interesting way: "I have a job to do that isn't that complicated, but it is often very difficult—to get a group of men to do what they don't want to do so that they can achieve what they've wanted all of their lives."

Delaying gratification helps you build the discipline you need to succeed. It enables you not only to earn more income, but also to become more responsible with the money you earn, to plan sensibly for the future, and to live within your means.

Make Choices before the "Moment of Choice" Arrives

Life is full of choices. Your choices determine the outcome of your career —and your life. You are responsible for the choices you make.

To ensure that you make the right choices, you must know exactly what you want so that you can organize your options. You must have a clear understanding of your priorities.

When you establish clear priorities for your goals, making the correct choices proactively becomes much easier than making choices reactively; your priorities serve as guidelines for your decisions. However, clarifying your priorities requires one very important prerequisite—and that's the third key to strengthening discipline.

Know What You Want

Everything you do must be for a reason. Knowing what you want gives direction to your goal and answers the question, "Why am I doing this?"

Discipline, then, is not restrictive, Indeed, it is quite liberating. Furthermore, it is a responsible use of freedom.

Many people define freedom as the ability to do whatever they want, whenever they want. In essence, they believe that freedom is the ability to make choices based on immediate gratification. Few people who use their freedom recklessly for immediate gratification, however, really achieve happiness or truly get what they want out of life. Ironically, they usually end up feeling trapped by their own short-sightedness and frustrated because they cannot achieve more meaningful long-term satisfaction.

On the other hand, those who recognize the true potential of their freedom to make choices are able to establish their goals and live disciplined lives in pursuit of them. In doing so, they are the ones who are truly liberated.

The use of discipline must come from the passionate motivation to achieve your goals. This passion will give you strength in moments when you feel weak. Practicing the three keys to strengthening discipline will enhance your perseverance in the pursuit of your goals.

The Power of Perseverance

Discipline will get you started on your way to achieving your goals. But it will take perseverance to actually reach them.

Marathons, as you know, are long-distance races, and the most notable of them attract the best competitive runners from around the world. Everyday people, however, also compete in these events, and their biggest challenge is simply to finish. The vast majority of them do not expect to come in first, or even 50th. They are there to compete with themselves. They are there to finish.

Many runners consider the marathon two races in one: the first 20 miles and the last 10 kilometers. In the last 10 kilometers, runners exert the most effort, their legs hurt, their bodies have run out of fuel, and their heads feel cloudy. The pain seems too much to bear. They are tempted to quit. Marathoners have an expression for this—hitting the wall.

Hitting the wall is when you want to give up. You don't have to be a marathon runner to experience this. It's when the realization that you can quit

---

Sidebar: When you establish clear priorities for your goals, making the correct choices becomes easier; your priorities serve as guidelines for your decisions.

Sidebar: Discipline will get you started on your way to achieving your goals. But it will take perseverance to actually achieve them.

and put an end to your pain overshadows your positive thoughts. You rationalize that everything will still be okay; the world won't end because you quit. It is here that your good habits are in conflict with your bad habits. You get yourself going, but you don't see things through. This is why discipline and perseverance go hand-in-hand.

Discipline isn't any good without the wherewithal to stick to it. To see your goals through—to finish—you need to have perseverance.

Five Techniques to Break through "the Wall"

Advisors encounter many "walls" throughout their career. These walls are often the excuses they give themselves—the reasons for not doing the things they know they should do, including prospecting. To sustain perseverance, you must identify these walls and be ready not to hit them and stop trying, but instead to break through them and continue on your way.

The following five techniques will help you adopt a lifelong habit of breaking through walls you may encounter:

- Have a goal.
- Divide each goal into manageable parts.
- Anticipate obstacles and overcome them.
- Stay objective and focus on the facts.
- Surround yourself with winners.

> If you truly believe in what you do and how you serve your clients' interests and improve their lives, you are less inclined to quit.

Have a Goal. To persevere you must have a real, meaningful goal. If you truly believe in what you do and how you serve your clients' interests and improve their lives, you are less inclined to quit.

You need to maintain goals that make sense—goals that enable you to relate your present activity to ultimate and worthy ends. All the mundane parts of daily life can drain your energy if you aren't vigilant. You have to keep the big picture in mind.

Important, meaningful goals—for your clients, your family, and your income—make for a wonderful finish line. Because they are often so far away, however, and there are so many excuses that could keep you from reaching them, you need to use a second technique to help you persevere.

Divide Each Goal into Manageable Parts. Marathon runners often refer to running the course one block or one mile at a time. Authors write books one page or one chapter at a time.

Goals in every area of your life can be broken into manageable parts. Manageable parts are good for your career and your relationships. Perseverance is far more achievable when you run your race through life at a tolerable pace.

Shortly, this chapter will discuss further the technique of dividing goals into manageable parts. This technique will be examined in the context of helping advisors get from where they are today to where they want to be concerning their prospecting, sales, and income objectives.

Anticipate Obstacles and Overcome Them. Perhaps you know someone who simply cannot handle a crisis. He or she does not anticipate the occasional headaches or make allowances for the inevitable setbacks that are a part of life. As a result, obstacles (many of them minor) wipe this person out instead of just tripping him or her up. The person hits a wall and crumbles.

It's worth repeating the wise words of M. Scott Peck: "Life is difficult." Once you accept this fact, you can go about the business of getting important things done.

How do you overcome obstacles? The answer involves facing and overcoming your fears. As stated in chapter 1, a big "wall" in prospecting is the fear of rejection. No one likes to be rejected. In chapter 1, you were asked to consider the most common fears in prospecting such as approaching prospects on the telephone or face-to-face and asking for an appointment. Then you were challenged to view those fears objectively and ask yourself, "What's the worst that could happen?"

stress

Understanding that these are not near-death experiences will help you to put them in perspective, and it will reduce stress. *Stress* is defined as mental tension resulting from the gap between your expectations and reality. Close this gap by expecting obstacles and challenges, and you will find that your stress is reduced and your perseverance stays intact.

Stay Objective and Focus on the Facts. When you encounter an obstacle, your mind fills with excuses and rationalizations for quitting. It is at these critical junctures—when you hit that emotional, intellectual, physical, or spiritual wall—that you need consciously to weigh your options. You must put emotion aside, review the facts, and tell yourself the following:

- I have a goal—I know what I want.
- I've broken my goal into manageable parts.
- I expect there will be obstacles.
- Difficulties are a part of life that make me stronger.

Don't pretend that all the pain, hurt, and frustration will go away just by thinking positive thoughts. That is, do not rely overly on "positive thinking," at least in the traditional sense. Being objective and focusing on the facts will help you avoid the pitfalls of playing positive mind games. Perseverance requires you to face your obstacles honestly, acknowledge them for what they are, and work through them. You must aggressively attack, fight, and defeat the temptation to quit. That's perseverance.

Family, friends, and colleagues can assist at this point. They can help you to stay objective. That leads to the final step to strengthen perseverance.

> **Your behavior is strongly affected by the company you keep.**

Surround Yourself with Winners. Perseverance is a lot easier when it is the standard operating procedure for the people with whom you spend your time. Your behavior is strongly affected by the company you keep. Therefore, make sure that the significant others in your life are the types of people who embrace perseverance—people with a good track record for success.

Example: Imagine a game of tennis or chess. When you compete, you should always play someone who has skills superior to yours. You won't improve if you can easily beat your opponent. To improve your skills, you must play an opponent who makes you work hard, concentrate fully, and maximize your strategy. You want your opponent to do his or her best so you can, in turn, do your best.

Whether it's a game of tennis or chess, or a career in financial services, everyone does better when he or she has a circle of friends who make a habit of persevering. Without strong, encouraging friends, you're going to have a difficult time breaking through the walls.

AN INTRODUCTION TO STRUCTURE

The concepts of structure, vacillation, behavior, problem solving, and creating are derived from the writings of management consultants Tom Wentz, author of *Transformational Change*, and Robert Fritz, author of *The Path of Least Resistance*. These concepts, which are discussed below, will give you a deeper understanding of getting what you want instead of attempting to "problem solve" your way into the future.

Hitting the Wall: Vacillation

Hitting the wall is a natural part of facing any challenge. However, there are things you can do to help diffuse the situation so that you don't find yourself hitting the wall again and again, which causes what we call *vacillation*—bouncing back and forth between good habits and bad ones. Many advisors struggle when it comes to prospecting, which results in

vacillation

<table>
<tr><td>Many advisors struggle when it comes to prospecting, which results in frustration and production slumps—ultimately leading to attrition.</td></tr>
</table>

frustration and production slumps—ultimately leading to attrition. Advisors who vacillate when prospecting and selling typically do the following:

- Identify a number of prospects and call on them.
- Continue the selling/planning process focused only on those prospects (that is, the advisors stop prospecting). Gather facts and information for solutions only to those prospects' situations.
- Conduct sales presentations over the next 3 months (undertake no new prospecting).
- Make sales (but now have depleted the current supply of prospects).
- Start all over again!

The example below illustrates this frustrating cycle of activity over a typical calendar year.

Example:

Month	Activity
January	Prospecting
February	Fact finding, case preparation
March	Presentations and closings
April	Prospecting
May	Fact finding, case preparation
June	Presentations and closings
July	Prospecting
August	Fact finding, case preparation
September	Presentations and closings
October	Prospecting
November	Fact finding, case preparation
December	Presentations and closings

At the end of the year, the advisor in the example has no prospects and literally has to start all over again, because he or she concentrated on only one task at a time. Every day, you must focus on acquiring new prospects.

Life is filled with conflict. Whether you are attempting to prospect for clients, lose weight, achieve a personal goal, or solve a business problem, you cannot succumb to vacillation. It generates turmoil and frustration, and it results in your ending up back where you began. If you hope to achieve peace and satisfaction by moving from one immediate gratification to the next, keep this in mind: It seldom works.

Nevertheless, vacillation is common in most of the struggles people deal with in their lives. They may even be unaware that it is taking place. But the results can be far reaching. In some cases, it may take years to bounce back after hitting the wall repeatedly.

What Is Structure?

It has been said that undisciplined people lack willpower. Willpower is important, certainly, but there is a force that is far more powerful. It's called structure.

structure

Structure refers to the arrangement of the key elements in your life in relation to each other. These elements include such abstract things as ideas, desires, fears, beliefs, aspirations, and values. They also include the very tangible elements of your life such as the assets and property you own, where you live and work, what you do, and equally important, your interpersonal relationships.

Changing behavior without changing the underlying structure of how you work will result in failure. The structural approach has been well articulated by Robert Fritz, whose unique approach to planning and creating has been hailed as revolutionary. Fritz observed that people who produce results—regardless of their current problems, obstacles, or resources—are the ones who are responsible for the extraordinary advancements of the human race.

In *The Path of Least Resistance,* Fritz explains that energy in a system follows the path of least resistance, going where it is easiest to go. Water in streams, electricity in circuits, companies in the marketplace, and individuals in daily life all follow a path laid down by an unseen but underlying structure.

For instance, a stream bed interacts with the rate and volume of water flow to produce changing surface patterns. The surface runs smooth when the stream bed is smooth and/or water flow is high. Standing waves appear when the stream bed is irregular and water flow is moderate.

Similarly, in businesses, key components interrelate to give rise to patterns of behavior. Some structures support desired results. Others lead instead to conflict between elements.

Example: An investment firm that sells stocks and other investment products has as its official, executive-decreed rule, "Quality service is number one!" In the field and on the trading floor, however, the unspoken rule is, "Daily commissions are number one!" And this rule drives the day-to-day actions of the people who work in the firm. Despite its official claim, the company structure of advancement, recognition, and compensation rewards short-term transactions over long-term client relationships.

In such a structure, quality service is not really number one. In spite of the official rule, the unspoken rule takes precedence. This structure—the relationship between formal and informal rules—causes considerable conflict within the organization. As a result, quality suffers and revenues decline. The firm has compromised a core value and relinquished its commitment to quality service in a highly competitive industry. If the firm doesn't do something to change the momentum, financial insolvency and even bankruptcy are likely.

What the company in the example above can do to remedy this conflict depends on how the leaders view their current situation. If they focus on the short-term problems—increasing sales revenues and decreasing expenses—and merely cut jobs and costs to maintain the status quo, they are probably not going to be able to maintain a viable, healthy organization.

If, however, they look at the root of the problem, the executives will recognize that they can accomplish the long-term results they want only by creating a hierarchy of values in which they acknowledge rules and establish priorities. Company leaders must make it clear to employees that quality service and long-term relationships take precedence over short-term transactions.

The company's reward structures should change to reflect the emphasis on establishing long-term client accounts as the primary goal. The new structure will ensure that quality service drives the behavior—and is, in fact, number one. As a result of this positive environment, quality service, employee morale, and company production should all increase significantly. Instead of breaking even, the company can grow rapidly; it can thrive. This would not be possible if company leaders merely attempted to "fix" a corporate structure that was profoundly flawed.

Structure Determines Behavior

Sometimes structure leads people toward what they want. Sometimes it leads them away from what they want. The challenge is consciously to establish structures that consistently support your highest and truest goals. To do so, you must understand the influence of structure on your own personal behavior.

In your professional and personal lives, most structures probably do not direct you toward your goals. Why? Because you are conforming to a structure you did not design for yourself with your goals in mind. Remember, however, that, although you must adhere to certain structures, you also have the power to create new structures that lead you toward your goals.

> **Sometimes structure leads people toward what they want. Sometimes it leads them away from what they want.**

Using Structure to Create a New Reality

Some time ago, there was a panel discussion on a local urban radio station about public transportation and the necessity of helping lower-income people find good jobs. The concerns were that people in poverty did not have cars or easy access to transportation. Nor were there many good-paying jobs within walking distance of where they lived. As a result, according to members of the panel, poor people could not work without buses to get them where they needed to go.

Kevin, a caller to the radio station, shared his story. Two years ago, he was on welfare, dwelling in a seedy apartment with roommates who stole anything of value from him, and living in a neighborhood that was not close to any meaningful employment. He did not own a car; nor did he have access to a bus or other public transportation.

To public transportation advocates, Kevin was a clear example of someone who was trapped. Without public transportation, they asserted, people like Kevin would not be able to get themselves off welfare.

Today, however, today Kevin has a new reality—a successful job, across town from where he lived before, in which he earns $65,000 a year. He created this new reality without relying on the bus. The host of the radio program asked Kevin how he did it.

"I managed to save up about $100 and keep it hidden from my roommates," Kevin answered. "I took that money and bought a bicycle and a good lock."

That bicycle changed Kevin's structure and, as a result, changed his life. Acquiring the bike gave him transportation to meaningful employment.

Instead of viewing himself as a victim of his circumstances, Kevin took personal responsibility for his situation and created his new reality.

The structure of your thinking affects your actions, attitudes, and emotions. The structures of your target market and your network have an impact on whom you are able to connect with and meet. The structure of your communication influences how well you can nurture those connections into client relationships. Last, the structure of your own board of advisors (see chapter 8) and how you manage it holds you accountable.

> Your goal is not merely to learn about prospecting. Ultimately, your goal should be to increase your service to your clients and thus increase your income.

The focus of this section is on goals. Structure is a means to help you achieve a goal. In the context of this textbook, your goal is not merely to learn about prospecting. Ultimately, your goal should be to increase your service to your clients and thus increase your income. Prospecting is simply a necessary means to that end.

Therefore, prospecting must become part of your structure.

GOAL SETTING VERSUS PROBLEM SOLVING

Worthy goals are not formulated simply to solve problems or make something undesirable go away. Problem solving has dominated many peoples' thinking and planning. If this is all you care about, then another problem will simply take its place. Instead, you need to find more positive approaches to goal setting.

**What Do You Have in Common with a Teacher
of Inner-City Youths?**

Years ago, Steve Mariotti, a New York City high school teacher, faced his first day on the job at Boys & Girls School. This inner-city high school had become known as the worst in the entire district: Seventy-two teachers preferred unemployment over going to work, the student dropout rate was 50 percent, and the NYC Board of Regents placed the *entire school* on probation.

Mariotti was a math teacher. His class had 59 students enrolled; there were 42 seats and 39 textbooks. The students lacked discipline and respect. They didn't even have pencils and paper. Initial diagnostic test scores were dismal at best.

When confronted with this seemingly hopeless situation, this was Mariotti's response: "The issue here is more than getting the students to learn math. I've decided that I am going to help them develop skills and experiences that will profoundly change the direction of their lives, increasing their performance in all subjects—not just math. They will want to go to school and they will want to learn. They will acquire personal life skills such as integrity, relationship building, and communication. Most important, they will believe in themselves and will believe that they can accomplish anything to which they set their minds.

"Oh, by the way," Mariotti added, "they will do all this within 1 school year."

Fortunately, if Mariotti heard any negative advice from colleagues, he ignored it. The reality is that Steve Mariotti changed those children's lives and accomplished all of his objectives—within 1 school year—by introducing them to the concept of entrepreneurship.

By teaching high school students the methods and rewards of starting and owning their own businesses, Steve Mariotti set out to achieve far more than he would have if he had focused only on solving the problem of getting his students to learn math. Every one of those children discovered how they could build their futures, how they could set goals and reach them. They did well in math and in all other subjects. Several went on to college.

Today, Mariotti's programs and methods serve low-income youth throughout the country. He is founder and president of the National Foundation for Teaching Entrepreneurship (NFTE.com).

> **Most people become so fixated on their problems that it limits how they establish their goals.**

Most people become so fixated on their problems that it limits how they establish their goals. People tend to look at what they know they can do and what their current capacity or structure will allow. With that information, they set goals to fit within the bounds of their existing situation.

The problem-solving approach to goal setting seldom works effectively, however, because it makes people fit their goals into the same structure that helped create the problems in the first place.

> **"fit" goal**

A *"fit" goal* rarely yields significant, lasting results. Moreover, it is also misleading because it sounds doable. And, as you will see shortly, "fit" goals limit the potential for success.

Four Fatal Flaws of "Fit" Goals

There are four fatal flaws in the limited "fit" approach to goal setting:

- First, a focus on getting rid of what you do not want rarely results in creating what you truly do want. This problem-solving attitude reflects a victim mentality.

Example: Many of your clients may have a desire to become debt free. But becoming debt free will not necessarily make their lives more fulfilling. Being debt free, although important, is not the equivalent of financial strength. An individual can be poor in terms of net worth and still be debt free.

- Second, small goals fail to bring out the highest and best in you. "Make no small goals," the old saying goes, "for they lack the power to stir our souls." They hardly ever inspire sustained action, let alone consistently outstanding results. Simply put, a small goal becomes just another thing you have to do.
- Third, problem solving rarely accounts for unpredictable changes. And unpredictable changes are an inevitable reality. Rather than viewing them as the enemy, you must harness them as a useful force to support your vision and end results.
- Fourth—perhaps most significant—conventional problem-solving approaches are based on what you know is feasible *today*. This limits you from the very beginning. Essentially, the approach prevents you from stretching for goals for which no conventional approach is currently available.

> **The kind of thinking that gets you into a predicament will not get you out of it.**

According to Albert Einstein, the kind of thinking that gets you into a predicament will not get you out of it. To produce real and lasting results, you must have a strategy that focuses on *creating*—bringing into being— what most matters to you and to your career. You need an approach that enables you to reach for the highest and best in yourself to achieve results that will benefit your family and your clients.

Doing the Improbable

stretch and leverage strategy

Rather than concentrating on solving your specific problem, you must envision what you want beyond all proportion to your currently perceived resources and capabilities. In business, this is referred to as a *stretch and leverage strategy*. Businesses set what appear to be impossible goals for their size and capacity. Then, relying on resourcefulness and innovation—by inventing, experimenting, copying, borrowing, or learning—they leverage available resources into outstanding results. You must learn to do the same as you prospect for new clients and expand your income base.

"Stretch" goals create a chasm between ambition and resources, instead of a nice, comfortable fit. They challenge you to improve dramatically—not by 10 or 20 percent, but by 500 or 1,000 percent—even higher perhaps! Fitting limits; stretching liberates.

Think about Kevin, the man you met earlier in this text, whose situation seemed hopeless. If he sought to "fit" his goals only to his current resources, he would likely still be on welfare.

What It Means to Create

creating

Creating is a widely misunderstood concept. Many people confuse being creative with creating. Being creative is traditionally viewed as applying existing methods a little differently to a problem for the purpose of becoming more productive. But creating goes far beyond what is conventionally regarded as being creative. "To create" means to bring something new into existence.

Creating is not about fixing what doesn't work, solving problems, or doing old things differently. It is not positive thinking, visualization, or brainstorming. Nor is it a trendy type of strategic planning designed simply to change the way you do things. The creating process, for the purpose of this discussion, is a fundamentally different way of approaching what you do. It is a reliable, step-by-step process that leads to tangible, recognizable results.

Although the processes that creators use may vary greatly, the end results are the same—a new creation that did not exist previously. Novelists end up with new novels, painters with new paintings. An architect sees a new building take shape as he or she envisioned it. A company creates the new products and services its customers value.

> **Even though creators may not know explicitly how to get there when they set a goal, they teach themselves to forge the path.**

Creating is a learning process in which creators teach themselves how to bridge the gap between their goals and their current capacity. Even though creators may not know explicitly how to get there when they set a goal, they teach themselves to forge the path. Through learning, they extend their capacity to create, and they establish the structure they need to get what they want.

Creators set up a dynamic framework—a structure—in which they see vision and reality simultaneously and through which daily decisions and actions consistently support the realization of desired results.

APPLYING THE STRUCTURE OF CREATING TO YOUR GOALS

The structure of creating is driven by your vision and focused on actions that consistently move you toward realizing your goal. In this structure, there is a dynamic tension between your vision—what you want—and your current situation, which generates energy for action.

When you apply the structure of creating to your goals, planning becomes truly strategic and is often crafted in a just-in-time fashion as the creation unfolds. Decisions and actions are aligned with primary values and visions. Small steps lead to success and generate momentum. Momentum creates a force that makes larger steps easier to take, increasing the capacity to create. An expanded capacity to create generates further successes and can eventually lead to outstanding results.

There is a four-step process to achieve your goal (vision). To illustrate the process, let's say that the goal in the discussion below is to increase annual income. The behaviors that relate to achieving that goal are prospecting activities and certain assumed sales results. The four steps in the process are as follows:

- Start with a clear, compelling picture of what you want (your vision).
- Assess your current situation.
- Prepare for creative tension.
- Use feedback and make adjustments.

Step 1: Start with a Clear, Compelling Picture of What You Want (Your Vision)

Your goal is to increase your annual income. How much income do you want to earn?

Caution: If you answer this question based on what you think you can earn, you are falling into the trap of setting a "fit" goal. You must not limit your income goal to what you perceive as your present earning capacity.

This is an extremely important aspect of goal formation. Learning what to want is one of the most radical, painful, and creative acts of life.

> **Leaning what to want is one of the most radical, painful, and creative acts of life.**

Unfortunately, most individuals and organizations are surprisingly vague about what they want—even those with elaborate mission and vision statements. Too many mission statements are full of generic easy-to-agree-with phrases like "to excel" or "to be the best in our industry." Furthermore, too many people and organizations focus on what they don't want or on what they can reasonably expect. Most are unwilling to specify what they truly want because, given current capacity, they don't believe they can achieve it.

If you regard yourself as a realist, you may keep your dreams and plans within the realm of what you consider possible. Unwilling—or afraid—to stretch for what they truly want, realists compromise, settling for reasonable, doable goals. To be sure, doable goals serve a valuable purpose: as action steps to higher goals. However, modest goals rarely stir anyone to sustained action.

An important mental hurdle to overcome is to separate what you want from what you believe is possible.

Visionaries are the opposite of realists. Visionaries are not afraid to set challenging goals. An important mental hurdle to overcome is to separate what you want from what you believe is possible. Creating is about learning and inventing. You must decide on the "what" and in doing so, not be too concerned about the "how." You simply need to know what you want to achieve. You'll learn to create the desired outcome as you go. The possibility of achievement is discovered by doing. As the saying goes, "Where there's a will, there's a way."

Example: John F. Kennedy's 1961 challenge to put a man on the moon by the end of the decade had far more power than if he had merely said, "Someday, we'll go to the moon."

Vision isn't merely words on paper. Kennedy's "man on the moon" speech was a vision. Martin Luther King's "I have a dream" speech was also a vision.

For the purpose of our discussion, let's assume that your goal is to earn $100,000 in first-year commissions (FYCs). When building a structure to reach this goal, you should start at the end and work backwards.

Step 2: Assess Your Current Situation

To begin to move toward your goal, you must have a clear understanding of where you are today, what resources you have or don't have, and the capacity of those resources you can use to reach that goal.

This may be intimidating for some of you, because the process may force you to confront issues that you have ignored for far too long. In doing so, you must be objective about what you have to work with. In addition, you must assess your current situation only after you have clearly defined your vision.

It is vitally important to identify goals that reach beyond what you deem possible or feasible.

As this text emphasizes, it is vitally important to identify goals that reach beyond what you deem possible or feasible. To dwell first on your current status only reinforces what you believe you can't do, not what you can do. Assessing your "stretch" goals first, on the other hand, ensures that problems, crises, shortages, and limitations in capacity do not hinder you in your vision-building process. Clarifying vision first ensures a stretch strategy, not a compromising fit.

Once you decide what you want to earn for a given year (or other time period), start with that objective and work backwards. This involves calculating how many of each type of prospecting and selling activities it will take to achieve your income goal for the desired time period. Concentrate on the prospecting activities; the sales results will fall into place.

Let us take a look at this process. First, you need to estimate (if you have kept accurate records, you can use your actual numbers) your prospecting and sales effectiveness ratios discussed in chapter 3. That is, how much FYC income do you derive from each sale on average? What is the ratio of closing interviews (total number including those that did not result in sales) to sales? Then calculate the ratio of fact-finding interviews to closing interviews, the ratio of qualified prospects to fact-finding interviews, the ratio of appointments to qualified prospects, and the ratio of total contacts to appointments. (See table 6-1.)

You must know how many prospects you need to identify and contact to ultimately reach your financial goal. If you are new to the financial services business, consult with your managers and colleagues to help you determine these ratios. Then, use these prospecting and sales effectiveness ratios to

TABLE 6-1
Prospecting and Sales Effectiveness Ratios

What is the average value of each sale?	**$1,000**—This is based on your current first-year commission income and production levels.
What is your ratio of closing interviews to sales?	**2 to 1**—This example assumes that it takes two closing interviews to result in one completed sale. This may take some research of your past sales results. (The closing interview should be the last meeting in which you ask the prospective client for the sale.)
What is your ratio of fact-finding interviews to closing interviews?	**2 to 1**—This example assumes it takes two fact-finding interviews to generate one closing interview. (The fact-finding interview is one in which you are successful in completing a fact finder on a qualified prospect.)
What is your ratio of qualified prospects to fact-finding interviews?	**1.5 to 1**—This example assumes that for every 1.5 qualified prospects you obtain, you will complete one fact finder. (A qualified prospect has a need, the money to buy, is suitable, and will see you on a favorable basis.)
What is your ratio of appointments to qualified prospects?	**1.25 to 1**—This example assumes that you need to acquire 1.25 appointments to identify and obtain one new qualified prospect. (An appointment is an initial formal meeting with a new prospect.)
What is your ratio of total phone and face-to-face contacts to appointments?	**4 to 1**—This example assumes that you need to make four phone or face-to-face contacts to acquire one appointment. (Calling easier leads like referrals and individuals in your natural market[s] will reduce this ratio.)

calculate how much prospecting and sales activity is necessary for you to generate your desired FYC income. Based on the goals and averages in this example, it will take 3,000 telephone and face-to-face contacts to generate 750 new appointments. Those 750 appointments will result in your obtaining 600 qualified prospects. Those 600 prospects will generate 400 fact-finding interviews; 200 will become closing interviews. Of those 200 closing interviews, 100 will result in sales. (See table 6-2.)

The averages above dictate that you must identify and contact 600 qualified prospects to close 100 sales. Therefore, based on a 50-week year, your first obligation is to secure 12 new qualified prospects per week.

The structure is full of variables, many of which you can control. For example, you make the decision about how much you want to earn. The markets you target and the products and services you offer will determine the amount of the commission you earn per sale.

The law of averages is inescapable. Nevertheless, it is within your power keep your success rates in setting appointments and closing sales at the high end of the probability curve. In other words, even if most advisors must identify 12 qualified prospects to set eight fact-finding interviews, your individual

TABLE 6-2
Required Activity Level

Objectives/Results/Activities (Time Frame: 1 Year)	Amount
FYC income goal	$100,000
Average FYC per sale	$1,000
Sales required to reach your goal (based on dividing your goal —$100,000—by your average FYC per sale—$1,000	100
Closing interviews (based on your ratio of closing interviews to sales: 2 to 1)	200
Fact-finding interviews (based on your ratio of fact-finding interviews to closing interviews: 2 to 1)	400
Qualified prospects needed (based on your ratio of qualified prospects to fact-finding interviews: 1.5 to 1)	600
New appointments needed (based on your ratio of appointments to generate one qualified prospect: 1.25 to 1)	750
Total contacts required (based on your ratio of total contacts to acquire one appointment: 4 to 1)	3,000

marketing methods and selling skills have a bearing on how close you must come to this number. By continually refining your prospecting skills, interviewing techniques, and market knowledge, you may be able to set nine or 10 fact-finding interviews for every 12 qualified prospects you identify. Conversely, if you are working in colder markets, you may get only three or four fact-finding interviews for every 12 qualified prospects identified.

Example:	If your average FYC per sale is $2,000 and you continue to acquire 12 new prospects per week, your FYC income would double from $100,000 to $200,000! On the other hand, if your average FYC per sale doubles, you can cut the number of prospects required to achieve your initial FYC income goal of $100,000 in half.

Although the prospecting and sales activity factors outlined above for establishing goals are quantifiable and predictable, they are based on the law of large numbers over long periods of time, as is an actuarial table. Accordingly, you should not make any value judgments regarding your progress toward achieving your goal for at least 1 year. This will allow a reasonable amount of time for you to assess the value of each activity factor toward the achievement of your goal.

Step 3: Prepare for Creative Tension

creative tension

> During the process of creating, there will always be a discrepancy—a gap —between vision and reality.

During the process of creating, there will always be a discrepancy—a gap — between vision and reality. This *creative tension* is a real, tangible force recognized by many experts. Earl Nightingale, pioneer in the personal development industry, referred to it as "constructive discontent." It is a pragmatic view of where you are as opposed to where you want to be. (Do not confuse creative tension with stress, defined earlier as mental tension resulting from the gap between expectation and reality.)

Example:	Consider the following scenario: You are dissatisfied with your current production level. You feel that your career growth is not what you would like it to be, and you do not want to become complacent. Given your level of discontent, you choose to channel your energy positively into a goal: to earn $150,000 next year. You realize how much work it will take and how many more new clients you will need to obtain. To reach your goal, you must iden-

tify 15 new prospects per week. With your family responsibilities, however, you wonder whether you are setting yourself up for disappointment by envisioning an unattainable goal.

This is the creative tension. It is a feeling of disequilibrium between where you are now (reality) and where you want to be (vision). It establishes a dynamic structure. Action within this structure can move you toward your vision or toward your reality.

There are three ways to resolve creative tension:

- *Give up.* Let go of your vision completely and allow reality to drive your actions.
- *Compromise.* Fit your vision to what seems possible.
- *Create.* Stretch for the vision; then change reality until you realize your goal.

Only the last strategy consistently produces real positive, lasting results. The key to success is maintaining creative tension, not just having a vision. This constructive discontent will move you to alter your current situation.

Your current situation changes in response to both successes and failures, as well as to outside events and issues beyond your control. You must continually reassess your current position in relation to these changes and make corrections in your course of action immediately. You can't foresee what all these changes will be, but you can anticipate that there will be some and that you will have to make adjustments. For instance, you may decide to start your own business. In designing your business plan, you cannot over-analyze every aspect to the point that it blocks movement toward your goal. It is important simply to get started and focus on what you will do next. Ultimately, the final plan will emerge out of the implementation process.

Example: Recall the sailing analogy in chapter 1. Creating is more like sailing a boat than taking a train. On a train, you know not only where you will start and end but also all the stops in between. There is almost no flexibility from the preestablished route to accommodate shifts in current status.

On a sailing trip, however, all you know for sure is where you start and where you want to end. The route you sail is not predictable. It is affected by external forces like winds, tides, currents, and storms. So you must make up the route as you go.

> Operating within the structure of creative tension allows creators to devise innovative, often elegant, paths that leverage even limited resources into extraordinary results.

Sailors and creators both use the forces they encounter to their advantage. They incorporate the energy of change. When blown off course, sailors immediately establish a new position and revise their course as needed.

Compare this to what happens when a train goes off the track!

Operating within the structure of creative tension allows creators to devise innovative, often elegant, paths that leverage even limited resources into extraordinary results.

Step 4: Use Feedback and Make Adjustments

Chapter 8 of this book discusses the concept of having your own board of advisors. This board will give you the valuable feedback that will enable you to update your current situation, augment successful actions, and stop or change the actions that don't consistently produce useful results.

Feedback leads to learning, increased capacity, and greater effectiveness. It also leads to expanded momentum. Momentum is a powerful force that can sustain action in the face of adversity, problems, and setbacks.

As the creative process brings you closer to your desired results, there is less creative tension. That means there is also less energy and momentum. Therefore, it is essential to rebuild that momentum to enable you to follow through fully, make the necessary adjustments, complete the action steps, and achieve the final results.

Creative Tension and Reaching Your Goal

If creative tension is the gap between vision and reality, what happens when you actually reach your goal? What happens when your vision becomes your reality? Some people fall into what can be called "the success trap."

When you attain a quantifiable goal, be careful not to shortchange yourself. For example, the goal in the earlier discussion was to earn an annual income of $100,000. Should that be your final objective? Why not $150,000, or even $200,000?

> No goal is an absolute end in itself—it is simply a means to something greater.

Whatever your goal may be for the next 12 months, keep it in perspective. No goal is an absolute end in itself—it is simply a means to something greater. If you decide you want to earn $100,000 this year, strive for it—but next year you must reset your goal to allow you to exercise your creative tension.

Your goal may not be to make more money. For example, you may wish to structure your business and client base so that you can earn a determined income level but from a smaller, more select group of clients. Working with this smaller group will enable you to give them more attention, which provides intrinsic rewards beyond a paycheck.

Ultimately, your goals will be tied to your quality of life. If you choose to earn $200,000, ask yourself, "Why?" What will that bring you that you do not presently have?

You are financial advisors. You may also be mothers or fathers, wives or husbands, and friends or lovers; you are certainly daughters or sons. Balance in life is important. Remember the maxim, "Money is only a means to an end." Remember, too, "You can't take it with you."

Summary

For almost all advisors, the missing link is a lack of prospects, not a lack of product knowledge or selling skills.

Recall that structure determines behavior. Once you establish your goal, the next step is to ascertain where you are in relation to what you want. For almost all advisors, the missing link is a lack of prospects, not a lack of product knowledge or selling skills.

Use the following procedure:

- Determine what you want to achieve.
- Assess where you are currently relative to what you want to achieve.
- Develop a structure to implement your plan.
- Periodically review and analyze your strategy to make certain you are progressing toward your goal.

As you work toward a successful career in financial services, you will see that the concepts discussed above will apply to more than just your career. Consider how what you have learned will affect other areas of your life—namely, your relationships with family and friends. As you move forward, do not forget that it is important to maintain a balance between your career and your personal life.

MANAGING YOUR TIME

Time has fascinated mankind throughout the ages. "Where has the time gone?" has been the lament of the young and the striving as well as the old and the dying. If you have ever wondered where time has gone, you may have been using your time ineffectively.

Today, thousands of people everywhere in the world tirelessly pursue that fleeting thing called time. Yet time is the one thing that is doled out in equal amounts—the rich and the poor all have the same number of minutes in an hour, hours in a day, and days in a year.

Not Seeing the Forest for the Trees

One of the best stories concerning the ineffective use of time is about a young lumberjack. The first day on his new job, this lad chopped down 10 large trees entirely by himself.

The next day, he seemed to work just as hard and just as long, but he chopped down eight trees. The rest of the week passed, and each day the young man worked as hard and as long, but each day he produced less.

"Sir," he eventually said to his boss, "I'm working so hard, but I'm afraid I'm a disappointment to you. I have yet to fell one tree today."

"Why are you accomplishing so little?" the boss asked.

"I'm really trying, sir," was the response.

"Have you taken the time to sharpen your ax, son?"

The young man answered, "No, sir, I really haven't had time because I have been so busy working."

The lesson: Work sharper, not harder.

What Is Time?

It has been said that your time is your life. If you waste your time, you waste your life. Think of time as a series of decisions, small and large, that gradually shape your life.

Not only is it difficult to define time, but there are also misconceptions about time. For example, people often say that time flies. However, time moves at a predetermined rate. Can you make up time? No. Once time is spent, it is irretrievable. Some say time is money, but it is valuable only if it is productive or enjoyable.

You may often feel that time and the clock are against you. But time can be on your side once you have the ability to organize your life to use it to your benefit. If the use of time—that is, a series of decisions—is ineffective, it can produce frustration, it can lower self-esteem, and it can increase stress levels.

Time and Productivity

Productivity can increase as you expend more time and energy—but only up to a point. Past this point, additional time and energy become counter-productive. Time management experts use a stress-productivity curve to demonstrate this finding. Along this curve, there is a critical point beyond which increasing time, energy, and stress lead to decreasing productivity. This is the point of diminishing returns.

Time can be your friend or your enemy. If you have enough time, the activities you do will bring you fulfillment, joy, and self-satisfaction. Too little time or too much stress, however, can result in wasted energy, which will cause dissatisfaction. Successful use of time management techniques will give you control over your life and will enable you to find solutions to various time wasters.

Time Wasters

An analysis of your time wasters helps you to build the foundation for good time management. A time waster is anything that prohibits you from reaching your objectives most effectively. Time is a perishable asset; if you are not careful, it can be stolen from you without your even noticing.

> **Time is a perishable asset; if you are not careful, it can be stolen from you without your noticing.**

Time wasters can be divided into two broad categories: major and minor:

- *Major time wasters* stand between you and what you want to accomplish. They include problems with your attitude, goals, objectives, priorities, plans, and ability to make basic decisions. These problems can surface in the form of procrastination, which is a fairly universal time management challenge that can be difficult to overcome if left unchecked.

- *Minor time wasters* are distractions that hinder you once you are on the way to accomplishing what you want to achieve. They include daily interruptions by coworkers, extended telephone calls, unexpected visitors, lengthy meetings of minor importance, unnecessary reports, excessive e-mails, annoying junk mail, and prolonged phone messages. Others include coffee breaks, surfing the Internet, and searching for or re-creating misplaced materials.

Personal Time Inventory

Before you can learn the steps to effective time management and techniques to control time wasters, you must take your personal time inventory. To do this, make a chart, breaking the day into 15-minute segments. You can also use an appointment book or hand-held personal digital assistant (PDA). Fill in the chart, book, or PDA as you go through your normal day's activities. Do not try to recall activities later or fill them in at the end of your day.

Taking a time inventory will enable you to separate and examine the various categories of time and how you use them. You can then determine whether you want to spend more or less time in particular activities.

Categorize the activities according to your business or daily routine—for example, socializing, routine tasks, low-priority work, productive work, meetings, and telephone calls. Categories away from work may include phone calls, television, the Internet, recreation, errands, commuting, shopping, household chores, eating, personal hygiene, and sleeping. Modify or change any categories that will help you understand how you use your time. The list of activities and abbreviations (codes) in table 6-3 may help you get started.

The next step is to determine your reaction to each 15-minute segment: satisfied (+), unsatisfied (−), or neutral (0). By doing so, you will learn which activities to increase and which to decrease, depending on your goals, your level of satisfaction, and the value of each activity—does it contribute to or

TABLE 6-3
Activities and Codes

Activity	Code	Activity	Code
Socializing	Soc	Passive recreation (reading, TV, radio, Internet)	PR
Telephone calls	Tel	Active recreation (exercise classes, golf)	AR
Meetings	Mtgs	Errands	Err
Low-priority work	LPW	Shopping	Shop
Medium-priority work	MPW	Household chores	HC
Productive work	PW	Eating	Eat
Commuting to work	Com	Personal hygiene	PH

detract from your productivity? You will also gain insight into which times of the day you are most and least productive so that you can plan your work and relaxation time accordingly.

Complete a time inventory for at least 1 week, and then add up the amount of time you spend in each category of activity. Table 6-4 is a sample portion of a blank time inventory that you can use to track your activities.

Time Analysis

Finally, group the activities according to each category and record the total time you spend on that activity. Indicate whether you are satisfied or dissatisfied with that amount of time. If you are dissatisfied, indicate in a column labeled "Adjustment" how much more or less time you want to devote to a particular activity. See table 6-5 for an example.

Time Spent Planning Leverages Time Spent Working

After you have analyzed how you spend your time, you need to decide how to use it most effectively. Remember that you have choices about how you manage this finite resource. Taking responsibility for the way you use your time is no more or no less important than taking responsibility for the way you live your life. Making wise time management decisions allows you to control how you choose to spend your time.

The best way to work effectively toward your goals and to complete your high-priority activities is to plan your day productively. Here is one of the most powerful rules of time management: 1 hour spent in effective planning

> **One of the most powerful rules of time management: 1 hour spent in effective planning saves 3 to 4 hours in execution.**

TABLE 6-4
Personal Time Inventory

Date _____

Time	Activity	Satisfaction (+, 0 –)	Time	Activity	Satisfaction (+, 0 –)
8:00 a.m.			10:00 a.m.		
8:15 a.m.			10:15 a.m.		
8:30 a.m.			10:30 a.m.		
8:45 a.m.			10:45 a.m.		
9:00 a.m.			11:00 a.m.		
9:15 a.m.			11:15 a.m.		
9:30 a.m.			11:30 a.m.		
9:45 a.m.			11:45 a.m.		

TABLE 6-5
Inventory Analysis

Activity	Time Spent	Satisfaction	Adjustment
Office gossip	1 hour	–	45 minutes less

saves 3 to 4 hours in execution. Often, you can avoid crises with proper planning. Proper planning will also ensure that the efforts you make during the day are directed at your main priorities—the major goals you have set in your life. Without a plan, without a road map, or without a chart for navigation, you will be left adrift in a sea of stress.

Making mistakes is part of being a fallible human being. But remember the saying, "If you don't have the time to do it right, when will you have the time to do it over?" Plan enough time to do quality work the first time.

Green, Red, and Yellow Time Management Zones

Because time is so precious, you must coordinate your time with your objectives. To use time optimally, you must prioritize your activities and then perform only the ones that will help you accomplish your objectives. Some of you may have adopted the habit of making simple to-do lists, while others may carefully plan and schedule each hour of the day. Whatever methods you use to manage your time, you can easily adapt the following system into your daily planning routine.

The system is called the green, red, and yellow time management zones. (See figure 6-1.) Using this system, divide your activities as follows:

- green zone (offense)
- red zone (defense)
- yellow zone (neutral)

The discussion below lists the basic types of business activities in the green and red zones and the recommended percentages of time you should typically allot to each activity. The nonwork activities that a well-rounded advisor who balances work with family, recreation, and community time are listed in the yellow zone.

FIGURE 6-1
Green, Red, and Yellow Time Management Zones

Time	Mon	Tues	Wed	Thurs	Fri	Sat	Sun
6:00 a.m.							
7:00 a.m.							
8:00 a.m.			Green				
9:00 a.m.							
10:00 a.m.							
11:00 a.m.							
Noon							
1:00 p.m.							
2:00 p.m.			Red				
3:00 p.m.							
4:00 p.m.							
5:00 p.m.							
6:00 p.m.							
7:00 p.m.							
8:00 p.m.			Yellow				
9:00 p.m.							
10:00 p.m.							

Sort the "zones" in your calendar in the proportions that work best for you.

Green Zone (Offense)

In the green zone—offense—you must be proactive about acquiring new business and finding new prospects.

As you analyze the usage of your time, you will soon recognize that there are distinct portions of your day during which the people in your market are more reachable and available, and distinct portions during which these individuals are more difficult to reach or meet with. These blocks of time could be continuous (for example, from 9 a.m. to 5 p.m.) or separate (for example, from 9 a.m. to 11 a.m. and then from 4 p.m. to 8 p.m.). Time periods will vary from advisor to advisor; yours will be unique to you.

The portion of your day during which the people in your market are more easily reachable and available is the time when you should be making appointments or conducting interviews. This is your *offense time*, your high-priority time. Ideally, you should allocate at least half (if not more) of your work time for the following income-generating activities.

offense time

> Allocate at least half of your work time to income-generating activities.

Prospecting (20%). Every day you must identify a steady flow of qualified prospects who need and want your services. You can send pre-approach letters, especially to referred leads, and then approach prospects on the telephone and/or face-to-face to secure at least two new appointments per day, to confirm appointments, and to prequalify the prospects. You should spend approximately 20 percent of your time on prospecting.

Fact-finding Interviews (15%). After obtaining appointments with prospects, you must meet with them in a favorable setting to uncover information about their current financial situation, future goals, and planning needs. You should devote about 15 percent of your time to fact-finding interviews.

Selling/Planning Presentations (15%). You should spend about 15 percent of your time on this activity. Here, you present your recommendations and solutions in a format that satisfies your company's compliance guidelines and meets your ethical standards to help clients reach their financial goals.

Red Zone (Defense)

In the red zone—defense—you need to hone your selling/planning, product knowledge, and service skills. Activities in the red zone include your ongoing service to your clients and building your existing client-advisor relationships. In the red zone, you are investing time and energy from which you will reap the benefits when you are in the green zone.

The portion of your day during which the people in your market are more difficult to reach and meet with—for example, between 11 a.m. and 3 p.m.—is the time during which you should be performing activities other than setting or

defense time

conducting appointments. This is your *defense time*, your low-priority time. The following activities belong in the red zone.

Case Preparation (10%). This activity entails the preparation of recommendations and product solutions that are suitable for your clients' budgets, risk tolerance, and financial priorities. It should involve about 10 percent of your time.

Case Processing and Client Follow-up (10%). These activities include processing applications and supplementary case requirements, sending thank-you letters, and establishing pending case files for tracking cases and follow-up. You should spend approximately 10 percent of your time on these activities.

Planning and Organization (10%). Another 10 percent of your time should be allotted to the daily, weekly, and monthly planning and organization activities that are essential to a successful career. Administrative office work is also in this category.

Training Meetings and Personal Development (20%). Enhancing your selling skills, improving your product knowledge, conducting individual performance reviews, reading professional journals and industry publications, and studying educational materials are in this category. Whether you are new to financial services or a veteran, keeping up with the latest sales techniques, tax law and product changes, compliance issues, continuing education requirements, and methods to upgrade your work habits is essential. The amount of time you spend on each activity will vary, depending on your markets, experience, and individual goals.

Advisors tend to spend too much time in the red zone. If that sounds like you, you may want to consider getting some help. You must dedicate your time and energy to activities that *only you* can do. Delegate the remainder to others.

| Formally defining your offense and defense times helps you to use your time wisely. |

Formally defining your offense time and your defense time enables you to use your time wisely. It forces you to recognize that performing a defense-time activity—such as sitting down at the computer to run illustrations—during your offense time doesn't make sense. That's when you should be making or keeping appointments. Conversely, if you are engaging in an offense-time activity—such as phoning for appointments—during your defense time, you probably are not maximizing your use of time because, as discussed earlier, prospects are not likely to be available during these hours of the day anyway.

Yellow Zone (Neutral)

The yellow zone—neutral—is for activities that are not necessarily directed toward your business goals but that give your life a sense of completeness. As you must know, it's not always about the money. Family and community give your life purpose. They are what individuals value most and work to nurture and protect. Activities in the yellow zone include the following.

Family Time. This time is as important—if not more important—as your work time. The enrichment your family provides enables you to recharge your batteries and become a better person.

Socializing. To really learn what makes people tick, you have to meet with them in social situations. Whether it is a game of bridge or a garden club meeting, this is your opportunity to interact with people in a nonwork environment.

Community Service Work or Volunteering. Advisors who are active in community services or volunteering are giving back some of their time to their local community (not to mention creating social mobility, a topic discussed in chapter 4). To really feel good about yourself, therefore, you should make volunteering one of your top priorities.

Benefits of the System

The green, red, and yellow time management zones allow you to accomplish the following:

- *Build structure.* As discussed earlier, structure determines behavior. If you structure your days to prospect in the mornings, then you need to do nothing except prospect in the mornings. During the time you have scheduled for prospecting, do not take phone calls (unless they are prospecting related), and do not deviate from the scheduled activity.

  The main point of a structured day is to enable you to manage your time, rather than letting time manage you.

- *Stay focused.* By not allowing other types of activities to intrude during your planned time zones, you will be able to stay focused on the scheduled activity. You will not let yourself get distracted by external stimuli or extraneous tasks. You will also prevent yourself from becoming obsessed with doing "urgent" work; if something must be done, there will be an appropriately scheduled time later to attend to it.

  There will, of course, be exceptions when you have to divert from your schedule for an unexpected emergency. Even in time management, therefore, you have to be flexible.

- *Maintain balance.* This system will help you to see in what type of activities you are spending your time. If you seek balance between your career, family, and social life, you can use the system as a

> **The main point of a structured day is to enable you to manage your time, rather than letting time manage you.**

reminder to spend your time in all three zones as planned. You may also want to look back over your list of important goals, revise it, and arrange to include family and personal goals.

As it says in the book of *Ecclesiastes,* "There is a time for everything, and a season for every activity under heaven."

Techniques to Schedule and Prioritize Your Time

> **Although you cannot control time itself, you can control how you use it.**

There are only so many hours in a day. Although you cannot control time itself, you can control how you use it. Allocate your time according to your priorities. Plan your time every day of your life. The idea is to make your time work for you, not against you. The only way you can "invest" your time wisely is to spend it wisely now.

You must take charge of the time you have. This means catching yourself in the act of wasting time. Periodically ask yourself, "Given the goals I have, is what I'm doing right now the best use of my time?"

The remaining sections of this chapter discuss some practical time management strategies, and they describe various techniques to schedule your time, prioritize your activities, and manage interruptions.

Combating Procrastination

One of the greatest obstacles to effective time management is procrastination. What makes people procrastinate? Procrastination is an avoidance tactic that controls people's behavior in certain situations, including the following:

- They want to put off doing something they perceive as unpleasant, overwhelming, or threatening—for example, prospecting for new clients.
- They believe that they are going to fail at the task, feel embarrassed, be unprepared, or have an unsatisfying experience.
- They have put the task off for so long that a mildly unpleasant (but important) activity has become even more unpleasant and is now a real problem. This generally provokes a lot of guilt and anxiety, but it stimulates no action.

Procrastination is a serious time management problem. What's more, it is self-perpetuating. The longer you postpone a task, the worse you feel about it, and the more undesirable it is to think about doing it. Often, advisors perform inconsequential tasks as a way of putting off what causes them anxiety—the important items such as telephoning prospects for appointments. You need to find ways to overcome your anxieties; try to tackle the tough and important jobs first, not last.

When you face an unpleasant task, one way to combat procrastination is to think about how guilty or how anxious you feel because the task is not getting done. Then imagine how relieved and energized you'll feel when the task is behind you.

Sometimes, getting started on a task is more difficult than the task itself. The next time you want to procrastinate, remind yourself that you could be doing the task now. Keep the importance of meeting your goals in the forefront of your mind, and maintain a sense of urgency about activities that will contribute to completing those goals successfully.

Understanding the 80/20 Rule

The *80/20 rule* is an important concept in time management. Also called the Pareto principle, it is named after Wilfredo Pareto, a nineteenth century economist and sociologist. According to this principle, 20 percent of causes are responsible for 80 percent of effects. Another way to put it is that most people spend 80 percent of their time on tasks that relate to only 20 percent of the total job results. Still another way to look at this principle is that addressing the most troublesome 20 percent of a problem will solve 80 percent of it.

You can apply the 80/20 rule to your practice by recognizing that 20 percent of your clients will command 80 percent of your time and that only 20 percent of your prospecting efforts or calls will produce 80 percent of your sales. Also recognize, with respect to time management, that you will spend 20 percent of your time on 80 percent of your most important activities and that 20 percent of your total activity will generate 80 percent of your production. Imagine the impact on your bottom line if you were able to make some of the less productive 80 percent of your time more productive!

Understanding the 80/20 rule can play a vital role in improving your productivity. First, in the area of overall activity, it is time efficient to group similar activities together, putting key activities in time slots that are most effective for accomplishing the particular type of task. For example, do all your calling to businesses between 9:30 and 11:00 in the morning; contact individual prospects at a given time—for instance, between 7:00 and 8:00 in the evening. Likewise, take note of your energy peaks and valleys. If you are a morning person, it is probably best to schedule your intense and demanding work for the morning and complete more mundane tasks in the afternoon.

Second, in the time management area, the 80/20 rule will help you concentrate your efforts on the 20 percent of the tasks that will generate 80 percent of the benefits. Spending too much time doing defense activities will probably not yield the results you want. Concentrating on your offense activities and their related tasks is the best way to reach your ultimate productivity goals.

In summary: Don't waste your most precious time.

> **Sometimes getting started on a task is more difficult than the task itself.**

> **80/20 rule**

> **Only 20 percent of your prospecting efforts will produce 80 percent of your sales.**

The solution to the problem of managing your time is not in learning to do everything more efficiently but in learning to spend less time to work more effectively. Using time more resourcefully will enable you to boost your production.

Delegating Wisely

When you have too many defense items to do, delegate them. Before you delegate, however, determine the time necessary to do the task and the time required to explain, instruct, and coordinate. Delegate tasks for which the time needed to complete the work is long in relation to the time required to instruct and supervise. Avoid doing tasks you should delegate, but be sure not to delegate jobs you should do yourself.

> **Avoid doing tasks you should delegate, but be sure not to delegate jobs you should do yourself.**

Managing Interruptions

It has been reported that, on average, the typical manager in the United States is interrupted once every 8 minutes. The interruptions may be from visitors, colleagues, or service-related phone calls. Your productivity is threatened whenever you work in an environment that permits clients or coworkers to contact you at any time. Time inventories generally reveal that individuals are involved in only 2 hours of essential work in an 8-hour day. This can result in low productivity, having to work overtime hours, and a reduction in the amount of time you can spend with your family and on recreation.

Telephone Interruptions

Phone calls can break your concentration, and many of them are not important. To block the interruptions from phone calls, you can make yourself unavailable during selected parts of the day. Use voice mail systems and assistants to screen incoming calls during these time periods. (Remember to return all calls later in the day to maintain your credibility.)

Check your personal time inventory to see how much time you spend on the phone when you return, accept, or initiate telephone calls. Perhaps you spend more time returning or accepting calls than you do initiating them. Analyze the reasons. You may want to set a time limit for each type of phone call and use a call timer. When you reach your allotted time or complete your intended discussion, close the conversation politely.

Scheduled Interruptions and Quiet Periods

Build time into your daily schedule for interruptions and unforeseen problems. If you schedule times for them, you can avoid rushing and feeling pressured.

It is also necessary to estimate a reasonable amount of time for each activity. If you schedule only 30-minute lunch breaks between important appointments, for example, you will probably have difficulty keeping within the time boundary and enjoying your lunch.

You should also set aside periods each day for relaxing and unwinding. During these quiet times, you should be interrupted only if there is an emergency. Be sure to schedule quiet times in your appointment book, or you may not take advantage of them.

Prioritizing Your Time

To-Do Lists

One technique to help you maintain focus is to keep a daily to-do list. Break your larger goals into small, short-term goals, and list the ones you would like to accomplish on a particular day. Attaining these short-term goals can be revitalizing. It gives you a real sense of accomplishment and creates momentum.

> **If you find yourself performing a defense activity and you haven't finished an offense activity for the day, you can be sure that you are wasting time.**

Compose your to-do list for the next day during a quiet time in the last half hour of your workday or after your last appointment at the end your day. Review what you have accomplished that day, and decide on five to 10 important jobs that you want to do the next day, carrying over any unfinished items from today. Define each item as an offense or defense item. If you find yourself performing a defense activity, such as cleaning out your desk, and you haven't finished an offense activity for the day, such as prospecting, you can be sure that you are wasting time. Take each day's list home with you, and review it before you go to bed.

In his book *Psycho Cybernetics*, Dr. Maxwell Maltz states that the sub-conscious mind has an amazing capacity to work and plan while people sleep. Thus, your subconscious will start to work on the next day's tasks, practicing and rehearsing their completion during your sleep. Each night, therefore, you can mentally prepare yourself for the day ahead by knowing what your goals are and planning what you want to do to accomplish them.

To-do lists are not for work days only. Remember to plan your play time and your weekends to make the most of your leisure time. Early in the week, write down your weekend plans and begin the necessary preparations. Resting your mind and rejuvenating your energy level are important activities. On Friday, review your weekend plans just before you go to sleep.

10-Minute Priority-Setting Conference

If you have set your priorities for the next day, composed your to-do list of the five to 10 most important things to do, and allowed your subconscious mind to review those tasks while you sleep, you will be ready in the morning to schedule time to execute your tasks.

Spend a maximum of 10 minutes in a priority-setting conference at the beginning of the day. Review your plans with the people with whom you work you so that they can also plan their days. This review enables both you and your colleagues or assistants to decide which task needs to be done first. Do not turn this priority-setting conference into a staff meeting. It should be a work-oriented and planning-oriented session with a numbered agenda to keep everyone focused.

Setting your priorities at the beginning of the day is just as important if you work alone. Take about 10 minutes to review your list and allocate time to each of the activities. Make any revisions to the list at this time so that you won't have to waste time later thinking about what you need to accomplish.

Your Daily Calendar

> **Your appointment book or daily calendar can be a powerful time management tool if you expand its usage.**

Your appointment book or daily calendar can be a powerful time management tool if you expand its usage. Consolidate all miscellaneous notes, phone messages, and memos in your book or calendar. The information will then be readily available for use now and retrieval later. List your activities and quiet periods for the following week.

Part of your appointment book can serve as a place for your to-do lists, relaxation cues, and meeting agendas. Schedule reminders to make appointments and to follow up on delegated tasks. Include yearly maintenance schedules for your car, home appliances, and office equipment. Enter anniversaries, birthdays, and other important dates with notes a week before each occasion to purchase gifts or cards.

If you use a PDA as your daily calendar, be careful not to spend so much time making entries that it becomes a time waster instead of a time manager for you.

Summary

The emphasis in this section was on how to manage your time more effectively. Manage your time as carefully as you manage any other valuable asset. Having a plan is a key aspect of time management.

Time management is not a skill that you will acquire simply from reading this chapter. You must practice, which may take 2 months, 6 months, or longer. Remember, however, that as long as you are striving toward an objective and practicing the appropriate techniques, you will accomplish your goal.

CHAPTER SIX REVIEW

Key Terms and Concepts are explained in the Glossary. Answers to the Review Questions and Self-Test Questions are found in the back of the textbook in the Answers to Questions section.

Key Terms and Concepts

goal	stretch and leverage
goal setting	strategy
goal prioritization	creating
stress	creative tension
vacillation	offense time
structure	defense time
"fit" goal	80/20 rule

Review Questions

6-1. Explain the relevance of goal setting in achieving personal success.

6-2. Identify the three keys to strengthening discipline.

6-3. Identify the five techniques that will enable the advisor to break through "the wall."

6-4. Explain the impact of structure on behavior and goal achievement.

6-5. Identify the four fatal flaws of "fit" goals.

6-6. Explain the difference between being creative and creating.

6-7. Identify the four-step process for achieving goals.

6-8. Identify major and minor time wasters; give examples of each type.

6.9. Briefly describe how the green, red, and yellow time management zones can help financial advisors use their time more effectively.

6.10. Identify several techniques and strategies to schedule and prioritize time and to manage interruptions.

Self-Test Questions

Instructions: Read chapter 6 first, then answer the following questions to test your knowledge. There are 10 questions; circle the correct answer, then check your answers with the answer key in the back of the book.

6-1. Which of the following statements concerning the principles of goal setting is correct?

   (A)   Once established, long-term goals should not be changed.
   (B)   To be most effective, goals should be tangible, specific, challenging, and have a targeted time for completion.
   (C)   Goal setting is the process of ranking goals according to their importance to determine the order for their achievement.
   (D)   Personal goals are usually confined to financial or career-based objectives.

6-2. Which of the following statements concerning the green, red, and yellow time management zones is correct?

   (A)   Green zone activities are those that hone the advisor's selling/planning, product knowledge, and service skills.
   (B)   Red zone activities are proactive—acquiring new business and finding new prospects.
   (C)   Yellow zone activities are not directed toward business goals but give the advisor's life a sense of completeness.
   (D)   In the red zone, advisors will reap the benefits of having invested time and energy in activities in the green zone.

6-3. Which of the following statements concerning creating and the creative process is correct?

   (A)   Creating is about fixing what doesn't work, solving problems, or doing old things differently.
   (B)   The creative process is a fundamentally different way for an individual to approach what he or she does.
   (C)   In the process of creating, there are strict rules for bringing into being what truly matters, which is the creation of something new.
   (D)   Creating is traditionally viewed as applying existing methods differently to a problem for the purpose of being more productive.

6-4.   Which of the following statements concerning the Pareto Principle is (are) correct?

   I.   Twenty percent of causes are responsible for 80 percent of effects.
   II.   Most people spend 80 percent of their time on tasks that are related to only 20 percent of the total result.

   (A)   I only
   (B)   II only
   (C)   Both I and II
   (D)   Neither I nor II

6-5.   Which of the following statements concerning goal setting is (are) correct?

   I.   The problem-solving approach to goal setting does not work because it keeps individuals in the same structure that helped create the problem in the first place.
   II.   The stretch and leverage strategy fails because it requires the individual to envision a goal beyond his or her currently perceived resources and capabilities.

   (A)   I only
   (B)   II only
   (C)   Both I and II
   (D)   Neither I nor II

6-6.   Which of the following statements concerning time and productivity is (are) correct?

   I.   There is a critical point beyond which increasing time, energy, and stress leads to decreasing productivity.
   II.   Too little time or too much stress can result in wasted time, which will cause frustration.

   (A)   I only
   (B)   II only
   (C)   Both I and II
   (D)   Neither I nor II

6-7. An advisor should implement all the following techniques to break through the wall and keep striving to reach his or her goals EXCEPT:

(A) Divide goals into manageable parts.
(B) Ignore obstacles.
(C) Stay objective and focused on the facts.
(D) Surround himself or herself with winners.

6-8. All the following are keys the advisor can use to strengthen discipline EXCEPT

(A) knowing when to use shortcuts
(B) delaying gratification
(C) making choices before "the moment of choice" arrives
(D) knowing what he or she wants

6-9. All the following are steps to create the structure that will enable the advisor to achieve his or her goals EXCEPT

(A) starting with a clear, compelling picture of what he or she wants (the advisor's vision)
(B) assessing his or her future resources
(C) preparing for creative tension
(D) using feedback and making adjustments

6-10. All the following statements concerning the concept of structure are correct EXCEPT:

(A) Structure refers to how key elements in a person's life are arranged in relation to each other.
(B) Although an individual must adhere to certain structures, he or she also has the power to create new structures that lead the individual toward his or her goals.
(C) The structure of a person's thinking affects his or her actions, attitudes, and emotions.
(D) Structure includes only such abstract intangible elements as ideas, desires, fears, beliefs, aspirations, and values.

7

Prospecting through Service

Learning Objectives

An understanding of the material in this chapter should enable the student to

7-1. Discuss the relationship between service and prospecting.

. 7-2. Identify the difference between client satisfaction and client loyalty.

7-3. Explain what "moments of truth" are and how to recognize and respond to these moments.

7-4. Discuss the differences between customers and clients.

7-5. Identify the eight laws of extraordinary service.

7-6. Explain what should be included in monitoring and servicing a financial or insurance plan.

7-7. Identify methods to provide extraordinary service to clients and how this can enhance the advisor's business.

7-8. Describe the process of categorizing clients and delivering service on a class basis.

Chapter Outline

INTRODUCTION

Have you ever been so satisfied or so well served by another person or business that you were actually eager to share that experience with a friend? Wouldn't you like your own clients to feel that way about the service they receive from you? This is the result of "extraordinary" service. It is also what creates the "tipping point" between passive referrals and personal introductions. In this chapter you will learn some traditional ideas for being extraordinary, as well as some ideas and methods that can give your service and prospecting that creative edge.

Long ago, milk was delivered to your home and doctors made house calls. That may have been the heyday of customer service—a time of great customer loyalty and satisfaction based on excellent service to the customer. Times are different. Today, it is common to hear complaints rather than compliments from customers. What has happened to genuine customer service? How does it affect your role as a financial advisor?

Many experts agree that cost cutting in corporate America is the primary reason for the decline in customer service. From the company's point of view, good service is expensive because it requires training staff and maintaining customer service employees. Ironically, in today's competitive, price-driven environment, some companies have difficulty seeing how they can strengthen their competitive edge by enhancing client service and creating customer loyalty.

Many advisors in the financial services industry are overly focused on selling clients, rather than servicing them.

But that's no excuse for you—regardless of what your company's policy may be. Many advisors in the financial services industry—especially those whose incomes are derived primarily from commission-based sales—are overly focused on *selling* clients, rather than *servicing* them. Once the commission check has cleared, the client does not hear from the advisor—until he or she has something else to sell.

As a financial advisor, it is your job to service your clients, not just to sell them products. People don't like to be sold—but they do like to be served. A client-focused sales approach emphasizes helping clients by providing solutions to their needs and goals. The client's satisfaction with the process and the results is crucial to maintaining existing business and procuring new business.

RELATIONSHIP BETWEEN SERVICE AND PROSPECTING

The selling/planning process focuses on the advisor's role in helping people recognize and understand their financial needs, providing them with the necessary information, facilitating the removal of obstacles to their financial success, and guiding them to take positive actions to achieve their goals. Your selling/planning process should concentrate on helping prospects see how you can assist them in solving their financial problems.

"I agree," you may say, "but what does client service have to do with prospecting for *new clients*?" Everything! Think back to a time when you ate in a great restaurant or discovered a new store and received what we will call extraordinary service. What did that service involve?

- The people went above and beyond what you would normally expect—just to please you.
- They were polite, respectful, and you could tell that they cared.
- The service was outstanding, delivering what you wanted, and then some.
- The products were top quality.

And how did you react to that positive experience? You probably told someone. You recommended the restaurant or store to your friends. You were that business's advocate. You became its salesperson, and you weren't even on the payroll!

Extraordinary service is part of building client loyalty—and client loyalty is essential for repeat business and referrals to new prospects.

That is why providing extraordinary service is more than just doing the right thing by your clients. Providing extraordinary service is part of building client loyalty—and client loyalty is essential for repeat business and referrals to new prospects. This is where service and prospecting connect.

Good client service separates the best businesses from those that fail. The successful advisor recognizes that each service opportunity is a marketing opportunity, a chance to demonstrate to clients that he or she is reliable, responsive, trustworthy, ethical, and empathetic to their needs. When the advisor sees servicing opportunities as marketing opportunities, referrals and more sales follow.

As your career as an advisor progresses, you will have more and more opportunities to provide service to your clients. Some advisors mistakenly view client service as something that takes time away from their selling activities. Successful advisors, however, know that providing good client service increases both client retention and sales productivity.

In every industry, businesses that have demonstrated the greatest success are those that deliver the best customer service. People return year after year to Disney World because the grounds are clean, the employees are courteous to all guests, the entertainment is excellent, and the staff manages the large crowds expertly. The sale occurs when the customer purchases the first ticket, but it is the high-quality service that brings that customer to the gate again and again thereafter.

The same is true in the financial services business. The best advisors realize that in the client's eyes, they are the company. They also realize, just like the employees at Disney World, that every service request is an opportunity to build a reputation for caring about clients. The sale is made when the check is signed, but it is the service that leads to repeat sales and referrals to new prospects.

There are many ways to build trust and rapport throughout your relationship with clients, as will be discussed later in this chapter. You can express your interest in a new client by establishing a relationship with him or her. Send a letter regarding benefits you provide through your service. Return phone calls and letters promptly. Contact the client on a regular basis. Acknowledge birthdays, promotions, and other events and dates that matter to the client. Schedule annual reviews. Be there to provide the service you promised. Always exceed the client's expectations of service.

How Would You Rate Your Service?

> **Take a moment to imagine you are your own client. Would you hire you?**

Take a moment to imagine you are your own client. Would you hire you?

Think about this and decide how you would rate your service. Put yourself in your client's position. Are you providing the kind of service that you would want to receive? Are you embodying the Golden Rule? This is a tough question—you must answer it honestly.

Platinum Rule

Perhaps an even better way to look at the service and client relationship question is to consider the *Platinum Rule*, as described by Dr. Tony Alessandra.

The Platinum Rule

An indisputable fact is that people do business with people they like. It makes sense, therefore, to like and be liked by as many people as possible. The ability to create rapport with a large number of people is a fundamental skill in sales, management, personal relationships, and everyday life.

We have all heard of the Golden Rule—and many people aspire to live by it. The Golden Rule is not a panacea. Think about it: "Do unto others as you would have them do unto you." The Golden Rule assumes that other people would like to be treated the way that you would like to be treated.

The counterpart to the Golden Rule is even more productive. I call it the Platinum Rule: "Treat others the way they want to be treated." Ah hah! What a difference! The Platinum Rule accommodates the feelings of others. The focus of relationships shifts from "This is what I want, so I'll give everyone the same thing" to "Let me first understand what the person wants and then I'll give it to that person."

Reprinted with permission from Dr. Tony Alessandra.

Questions to Ask

Let us explore this issue further with some specific questions.

> Are you focused only on the current paycheck, or are you keeping in mind the total profitability of each client?

Do I Recognize the Total Profitability of My Client? In commission-driven industries such as financial services, this question may seem self-evident. Of course, you know where your paycheck comes from. But are you focused only on the current paycheck, or are you keeping in mind the total profitability of each client? This value can be determined not only from what the individual client buys, but also from the additional new business that client sends your way.

The worksheet in figure 7-1 demonstrates how you can determine the total profitability of a client. You need to calculate the income items of that client over a period of time, and then subtract the expense items created for the client over the same period. The net result is the client's total profitability.

> The best way to measure the impact of service is by looking at whether or not it produces a measurable level of sales revenue.

The best way to measure the impact of service is by looking at whether or not it produces a measurable level of sales revenue. Monitoring and servicing will have predictable costs, including overhead, staffing, and supplies, as well as marketing and sales costs attributable to its operation. The amount of profit is determined by a simple formula:

$$\text{Profit} = \text{Income} - \text{Costs}$$

You can determine how profitable your client-building and service efforts are by identifying the income and expense items associated with these activities for any given period.

The income items in figure 7-1 should be the actual commissions or bonuses paid in each category for the period. Some of the expense items entered, such as policy wallets, business cards, and client lunches, represent costs that are fully attributable to service activity. Other general office and staff expenses, including computer, telephone, and secretarial costs must be estimated and apportioned.

FIGURE 7-1
How Profitable Is Your Service Work?

Period: From _____ to _____

Income Items		Expense Items	
Renewal commissions	$ _____	Travel	$ _____
Persistency bonuses	$ _____	Office	$ _____
New sales (first-year commissions)	$ _____	Computer costs	$ _____
		Supplies	$ _____
Number of prospect leads from service calls	$ _____	Postage	$ _____
		Lunch with clients	$ _____
Number of sales from prospect leads	$ _____	Policy wallets and portfolios	$ _____
		Calendars	$ _____
		Newsletters	$ _____
		Others (list)	$ _____
			$ _____
			$ _____
Total income from service work			$ _____
Total expenses attributable to service work			$ _____
Profit from service work			$ _____

With This Knowledge in Hand, Do I Make My Clients Feel As Important As They Truly Are? Many advisors follow the formula above to satisfy clients' needs. But who today *isn't* focusing on satisfying clients? In today's ultracompetitive marketplace, if you're doing what everyone else is doing, you'll never get to where you want to be. You need to set yourself apart from the competition. If you want to be a leader and top producer, you must develop customer loyalty.

The most effective way to ensure growth is to turn existing clients into loyal clients—advocates who will "sing your praises" to other qualified prospects.

You need to cultivate skills that will establish long-term relationships with your best clients. Too often, the constant push for sales leads drives advisors away from their current clients toward finding new ones. The most effective way to ensure growth, however, is to turn existing clients into loyal clients—advocates who will "preach your message" and "sing your praises" to other qualified prospects.

You can do this by selecting the right prospects through identifying your ideal client and using other target marketing skills developed in chapter 2. Recall, for example, from chapter 6, that according to the 80/20 rule, 20 percent of your clients will produce 80 percent of your income. By profiling the top 20 percent of your clients, therefore, you can identify characteristics of your best, most potentially loyal customers.

You can also identify the characteristics of clients in the bottom 20 percent of your book of business. Understanding the traits of that percentage of your client base can help you avoid the wrong prospects and instead direct your resources to the upper 20 percent.

Do I Know Why My Clients Buy? Everyone has his or her own personal reasons for the choices he or she makes, and those reasons will never have anything to do with *your* needs. You may want your prospect to buy because you need to pay your mortgage, but your reasons do not matter. The only things that matter to the prospect are his or her reasons and concerns.

You must understand your prospect's needs, wants, and goals. This will enable you to address these concerns and thus motivate the prospect to buy.

Do I Truly Care about My Clients? True professionals take a very close and almost intimate view of their clients; it is through these people that the professional carries out a big part of his or her life's mission. Just as you value your relationships with your family, so too must you make your clients near and dear to you. In many cases, your clients will assist you in your career far more than your families are able to do.

Do I Deliver What I Promise—and Then Go beyond Expectations? Not only is it vitally important to follow through on your word, but it is also crucial to develop the habit of going above and beyond the call—to surprise your clients—by doing more than is expected. We will discuss some specific ideas shortly.

Do you base your choices on your client's interest, or on what is most convenient for you? This is a personal question and one that only you can answer. Your choices and actions should be driven by what is in your client's best interest and not by what is easiest for you.

> **Your choices and actions should be driven by what is in your client's best interest and not by what is easiest for you.**

Do I Take Responsibility for My Client's Happiness with My Company's Products and Services? If you are an advisor for a larger company, it is likely that you will not be the only person with whom your client interacts. You may be, however, the primary advisor—the "face" on what is normally considered a faceless bureaucracy. This means that if there is any problem, regardless of who is at fault, it is your responsibility to resolve it and ensure that your client's needs are satisfied. If a client calls with a concern or a complaint, your response is not to assign blame by deciding whose fault it is. Rather, your response must be to determine what must be done to resolve the problem.

These are just a few of the questions that you can explore to evaluate your own level of commitment in serving your clients. To ascertain where you want to go and how you will get there, it is important first to determine where you are. By honestly examining your own actions, attitudes, and priorities as they relate to your clients, you will be in a stronger position to identify and correct weaknesses. You will be open to new opportunities to go beyond the expected.

Service Is Your Business

When customers buy your product or service, they receive an extra bonus—you. You have a professional obligation to ensure your client's satisfaction. Your value to the customer takes many forms:

- You are a person, not a company, for the client to contact. People prefer to relate to individuals instead of large customer service departments.
- You are an open, concerned person who listens and understands. Your client can reach you for even the smallest questions.
- You have a depth of knowledge that can be helpful in solving specific problems. You also have the means to obtain further information inside and outside your company to save your client time.
- You are accessible and make your company accessible. You act as a conduit through which clients can reach other people in your company. If a client wishes to consult a lawyer, for example, he or she can arrange to do so through you.
- You represent security. Clients know you have a vested interest in their business. They also know you are reliable and consistent. You are a known entity, whereas a competitive financial advisor is not.

Develop an attitude of friendship, respect, and concern. The more you give of these qualities, the more you receive from your clients. The working relationship that grows from this foundation is a strong one that resists destruction. Clients who return these feelings are hesitant to change to a competitive product or service.

Tony Alessandra, "Service is your Business." Reprinted with permission.

Client Satisfaction versus Client Loyalty

In his best-selling book *Customer Satisfaction is Worthless; Customer Loyalty is Priceless,* author Jeffrey Gitomer defines the key differences between satisfied customers and loyal customers.

client satisfaction

In essence, *client satisfaction* is too low a standard upon which to measure effectiveness in customer service, according to Gitomer. After all, you know that it is far more cost effective to keep a client than to find a new one. Yet this concept still has not penetrated the core of American business. If customers are satisfied, many believe, that is good enough. They are wrong. Your goal should be to go beyond satisfaction. Your goal should be *client loyalty.*

client loyalty

Figure 7-2 is a summary of Gitomer's levels of client satisfaction.[1] Note how much of an advocate a loyal client can be for you.

FIGURE 7-2
Levels of Client Satisfaction

When Your Client Is	Tells Others	Refers Others	Buys Again
Loyal	Everyone, all the time	At every opportunity	Always, for ever and ever
Satisfied	If asked	If asked	If convenient
Apathetic	No	If asked	Maybe, maybe not
Unhappy	10 people	No	Maybe, after several years
Wronged	25 people	No	Never
Angry	Everyone, all the time	Never	Never

In his book, Gitomer actually adds a lower level of client dissatisfaction, which he calls *lawsuit.* Guess how many people are told at this level? *The whole city!*

Have you ever been so upset with a business that you cannot help but spread the word about your substandard experience? This is called giving "reverse referrals," and the dynamics that can do wonders for spreading the good news about your service can also work against you. Beware!

> **The dynamics that can do wonders for spreading the good word about your service can also work against you. Beware!**

The Moment of Truth

moment of truth

In customer service, there is a common expression known as the *moment of truth.* This is defined as a time of client contact, during which the advisor is confronted with a situation in which he or she must make a choice about how he or she will respond to the client.

This concept can be broken down further to reveal the types of experiences your clients may have that shape their view of you and your service. This, of course, will affect your future relationship with that client and the amount of referrals and repeat business you may obtain from him or her. These client contacts include the following:

- *moment of magic*—interaction that exceeds the expectations of your client and leaves him or her with a positive experience

- *moment of misery*—interaction that has negative connotations and leaves the client dissatisfied
- *moment of mediocrity*—routine, uninspired service that leaves neither a strong positive impression nor a strong negative impression

Your objective is to achieve as many moments of magic as possible—interactions that consistently exceed the client's service expectations. They require that you make a commitment to put the client first, working hard to maintain a proactive attitude to meet client needs. Clients whose expectations are exceeded consistently (experience moments of magic) are those who will become advocates and centers of influence for your business.

The choices that you make reveal the truth about how deeply you care about your clients. In the selling/planning process, there are many different moments of truth. Each of these moments—and their outcomes—can be divided into three parts:

- the moment (the stimulus)
- the choice (your response)
- the truth revealed (the message you send to the client)

Moments of Truth from the Client's Perspective

A Prospective Client Calls the Company for the First Time. Take a look at the following example. Although it is not specific to the financial services industry, it illustrates how important that initial contact with a prospect is.

Example: Suppose you want to have your kitchen remodeled, so you phone four companies that specialize in kitchen remodeling.

When you call Company A, a polite representative answers the phone. You ask for an estimate to have your kitchen remodeled. The receptionist connects you with a sales representative who schedules a time that afternoon to come take some measurements and ask a few questions. He arrives on time, is very polite, listens to your concerns, and asks intelligent questions. He assures you that he will have a detailed estimate for you by noon the next day. As promised, a typed estimate arrives by fax the next morning at 10:45.

You call Company B. The initial experience is similar to your experience with Company A, except the sales rep begins to ask questions about your budget—as if he wants to know how much money you plan to spend before he will schedule an appointment. You don't give him a specific amount, but you make it clear that you are interested in quality—and you know that quality costs money. "The soonest I can fit you in is next Monday," he responds. Obviously, he's pretty busy. Good for him.

You call Company C and get an answering machine. You leave your inquiry on the machine. No one calls back for 5 days.

When you call Company D, the owner answers the phone. The conversation goes well; the owner is polite and interested in what you have to say. She asks if tomorrow afternoon is okay to meet, and that's fine with you. She arrives on time, takes notes, asks questions, and agrees to get an estimate to you in a day or two. It arrives 3 days later, handwritten on a sheet of notebook paper.

Four companies, four moments, four choices, four truths. What truth does each company reveal about its attitude toward you, the prospect? What subconscious message do you, the prospect, receive based on the behaviors of the people you call?

Truth of Company A: You're important to me, and I value your business. I'll do whatever it takes to earn your business, your loyalty, and your trust.

Truth of Company B: How important you are depends on how much money you want to spend. I am busy enough right now that I can live without your business.

Truth of Company C: I'll call back when I get around to it.

Truth of Company D: When I say I'll get something to you, it really depends on how other things play out.

> **The way you initially respond to a prospect has a tremendous impact on whether or not he or she chooses to do business with you.**

Notice that in all of the above, the amount of the quote is irrelevant. Therefore, the way you initially respond to a prospect has a tremendous impact on whether or not he or she chooses to do business with you at all. If he or she does not like you or if your actions and attitudes reveal a "truth" that says you do not value the prospect, you have seriously damaged your chances of acquiring a client, let alone building a relationship.

A Client Is Upset, Angry, or Has a Complaint. There will always be times when problems arise. If a client calls you and is obviously upset—even angry—how you respond at that very moment will affect your ability to resolve the situation and maintain the client-advisor relationship.

Example:	An advisor received a call from one of his clients because the company he represented mistakenly withdrew the incorrect amount from her checking account. When he answered the phone, he could tell she was upset. He needed to understand fully what had happened.
	Even though the advisor was not directly involved, he apologized. "We screwed up," he said.
	"I know you weren't to blame," the client said.
	"No," said the advisor. "We screwed up," he repeated. And he apologized again. "I will correct the situation immediately. How soon can we meet for lunch to ensure that all your needs are being met?"

Although the advisor in the example above was not directly responsible for the mistake, it is still a reflection on him. He is her advisor. He is the company. He must find out what went wrong and make sure it doesn't happen again.

If a client is particularly irate—even hostile—you have two choices:

- You can become defensive.
- You can empathize and seek greater understanding.

Which response will create a more positive result? Let's consider it further: What "truth" does the client glean from each of these options?

Becoming defensive sends a message, or reveals a truth, of selfishness. It says to the client, "You have a problem, but my first concern is to protect myself, regardless of any level of direct responsibility I may have for the situation." Remember this, however: Shifting the blame for mistakes never works.

Empathy and understanding, on the other hand, send the message that you are caring, responsible, and professional—even in situations where the client may be angry or hostile. Because the client loses his or her cool does not give you an excuse to do likewise. The truth revealed by empathizing and understanding: "I care about you. My priority is to make certain that your issue is resolved and that we can move forward with a positive relationship. I am equally responsible for what goes wrong as I am for what goes right."

Note: Whatever conversations the advisor has with the company to resolve a conflict with a client should never involve the client directly. The client should not part of these conversations. Nor does it help the advisor's cause (or the company's) to subject the client to petty excuses or defensiveness. What the client wants to hear is that resolving the problem is the advisor's top priority. No excuses. Just fix the problem and move ahead.

You Discover a Mistake, but Resolving It Will Inconvenience You. There will be times when you must rise to the occasion even when resolving a problem will be inconvenient—perhaps greatly inconvenient—for you.

> There are times when you must rise to the occasion, even when resolving a problem will be inconvenient—perhaps greatly inconvenient—for you.

Example:

Your client lives about 5 1/2 hours away. She absolutely has to have the final paperwork for a new policy by Friday morning for a meeting at 7:00 a.m. The paperwork is scheduled to mail to her overnight on Wednesday. Wednesday comes and goes, and the underwriting department confirms that the policy is completed and will be sent that night. Great, you think. Your client will receive it Thursday—in time for her deadline.

Thursday arrives; around noon you call your client to make sure she has received the paperwork. She indicates that it has not yet arrived. You contact the underwriting department and discover that a clerk in the office mistakenly sent the documents for second-day delivery. Your client will not get the package until Friday afternoon. Too late.

The moment of truth is upon you. Should you contact the client with the bad news? "I'm sorry—I wish I could do more," you could tell her. Or should you have a solution to her problem that shows her the level of service you deliver to your clients?

You obtain a duplicate copy of the documents. That golf game you scheduled for that afternoon will have to be postponed, because you are about to make a long drive. Once on the road, you call your client to tell her you are delivering the package yourself.

Your client realizes that the trip is more than 5 hours each way. "You're driving all that way?" she asks. She is incredulous.

Ten years from now, she will still be sharing the story about how you drove nearly 12 hours to deliver a package to her. That's service. That's something to talk about!

Servicing versus Prospecting

The concept of valued service gets a lot of publicity, but not all advisors actually deliver. You have heard slogans such as "Service is our middle name," but what you hear may not be what you get. "Talking the talk" is not enough. In fact, service is a sore point for many consumers today.

What is service? According to *Webster's Ninth New Collegiate Dictionary*, service is "contributing to the welfare of others."

If the prospect does not actually do business with you, what service have you rendered? Do you have the luxury of being able to provide service and time to people who never become your clients?

This is a business decision you must constantly make as you decide how to allocate your precious marketing and selling hours and dollars. Although service can play a tremendous role in nurturing a prospect, true service begins with the sale of your product—when the prospect becomes your client. Otherwise, what you are really doing is prospecting rather than servicing clients.

Orphans—The Abandoned Clients

An excellent example of this dilemma for an advisor is an orphan lead. Recall from chapter 3 that an orphan is an existing account that you did not personally produce and for which there is no active advisor assigned. Orphans are created when advisors in your company retire, die, or are reassigned. Other advisors are told to service them—traditionally, by letter, e-mail, fax, or phone.

Large insurance companies have thousands of orphan clients; no one calls them. The usual complaint of these abandoned clients: "I haven't talked to anyone from your company in 5 years or more." Sound familiar?

Given the number of orphan clients, you would think that advisors must have enough people to call. Their future is set, right? Wrong! The number one reason for leaving the financial services business is a lack of prospects.

What has happened is that because so many orphan clients have been given no service, they are not receptive to talking to any advisor of that company. Thus, they transfer their business to another company. The amount of business, including life insurance, annuities, investment accounts, and retirement plans that is transferred from one company to another is staggering. Often, it comes down to one common factor: lack of service.

Example: A few years ago, an advisor was assigned 900 orphan clients who were acquired through a merger of two companies. His question: What do I do with 900 orphans? He sorted through the list and focused first only on 30 names. It was the holiday season, so instead of calling or sending information, he purchased poinsettia plants and hand-delivered them to each client's home.

When he called a few days later, each of the recipients was thrilled that he had thought so much about them. In fact, 27 of the 30 orphan clients adopted him as their new trusted advisor. The referrals and recommendations from these 27 clients created a flood of new clients.

Can it be that simple? Yes, it can.

> There are about 1,100 insurance companies. There are thousands of mutual funds. Why should prospects choose you? It is probably not because you are good looking!

Your goal should not be merely to give just good service; it should be to give extraordinary service. Remember, your clients and prospects can purchase any product or commodity you are selling from someone else. There are about 1,100 insurance companies. There are thousands of mutual funds. Why should prospects choose you? It is probably not because you are good looking!

THE EIGHT LAWS OF EXTRAORDINARY SERVICE

extraordinary service

There are eight laws of *extraordinary service*. Together, they help you determine the most effective ways to serve your clients, stand out from the competition, and generate valuable referrals.

Law #1: Serve for the Joy of It

Think back to a time when you came to someone's rescue. Maybe it was not too long ago. How did that make you feel? Most likely, it gave you a sense of joy in being able to help someone.

One advisor's first experience of joy in helping others was when he was 10 years old and started his paper route. He never realized how much fun it was to get to know his customers and become part of their lives. When customers were ill, he would visit them or leave something extra with the paper, like a box of candy. Did he do it for the tips? No. He did it because he cared—*because he wanted to*. Of course, the tips didn't hurt, either.

Being an advisor is not something you should do just for the money. If it is, you will burn out quickly.

Law #2: Never Substitute Convenience for Service

Remember the earlier example in which the client's package was shipped incorrectly? In that situation, the proper response was getting in the car and making the long drive. Forget the golf game. Forget convenience.

Had the advisor been unwilling to be inconvenienced for the sake of his client, he would not have deserved his commission. Too many people on the customer service front line, especially companies that do not reward great service, make customers feel like they are actually a bother. "I'll get to you when I get around to it" is a common unspoken message. Serving should never be an inconvenience—it should always be an honor and a privilege.

Law #3: Every Complaint Is a Cry for Service

> When a client expresses a problem or a concern, it is a request for service at point-blank range.

You have heard that other people's problems are your opportunities. When a client expresses a problem or a concern, it is a request for service at point-blank range. In many cases, it is the ultimate moment of truth, because your commitment and character are revealed in the difficult times, not the easy ones.

What about those annoying clients who are continually complaining? If you have a client who is always finding something to gripe about, that person is most likely attempting to fill a void in his or her life. The complaints are requests for service. Your reaction should be to stop looking at that client as a heckler and start seeing him or her as a human being who has needs. Be proactive. Schedule some extra time to review his or her entire situation. In the process, listen and understand; you may discover that the frequent complaining is merely a symptom of some deeper issues. Help your client uncover the real issues, and you may increase that client's loyalty to you as his or her advisor.

Law #4: Seek Out Moments of Truth and Embrace Them

Earlier, this textbook illustrated some examples of moments of truth. You will encounter them everywhere. Too many advisors avoid such situations. That is no way to provide extraordinary service. To do so, you must seek out moments of truth and embrace them.

Moments of truth are more than just responding to complaints or correcting mistakes (even if they are not yours). You find moments of truth in the following situations:

- *A client (or prospect) has a special need or request.* The more you can customize your service to meet the specific needs of your client, the more you will distinguish yourself among your colleagues. Go the extra mile, take the road less traveled, and become exemplary.
- *A prospect cannot make up his or her mind.* Many prospects do not know what they need; many of them do not even know what they want. If you run into this situation, do not put them off or tell them to

call you when they are ready. Instead, evaluate the prospect's situation and the desired results; then leverage your own knowledge (and that of your colleagues) to determine possible solutions. If the prospect is stuck, you need to be proactive and get his or her mental gears going.

- *A prospect is resistant to buying.* Various levels of financial products often meet corresponding levels of resistance or hesitation from prospects. If a prospect has objections, listen and understand them. Reflect your own understanding of the prospect's particular situation, and educate him or her from that perspective. Walk your prospect through each concern and address it. Doing so solidifies the prospect's commitment to his or her purchase—and enhances the prospect's loyalty to you as his or her advisor.

- *The prospect is ready to buy.* In his classic book *How to Win Customers and Keep Them for Life,* author Michael LeBoeuf "retells" an experience from Samuel Clemens (aka Mark Twain), who listened in awe to a preacher at a mission gathering. Clemens was so inspired by the speaker, he was ready to contribute $5.00 to the collection plate—up from his usual $1.00 donation. As time went by, however, and the preacher continued talking—apparently enjoying the sound of his own voice—Clemens' inspiration soon gave way to frustration. When the collection plate came by, he kept his $5.00 and instead removed a dime!

  What can advisors learn from this story? Look for signs that the prospect is ready to purchase or agree to an interview. Then stop selling and close.

- *The client buys.* Is this the time to relax? Never! Once the client buys, the real service begins. Now is your opportunity to give your client what he or she doesn't expect—extraordinary service. Continually show the client that you appreciate and value his or her business by sending him reminders that you are thinking of him or her.

- *The prospect refuses to buy.* If you do not make the sale today, is that the time to quit? Do you remove that prospect from your file? Many of the best advisors will tell you that some of their best clients were their longest and toughest acquisitions. One advisor, for example, pursued a prospect for more than a year, without ever getting an interview. Today, that person is one of the advisor's best clients.

> Once the client buys, the real service begins. Now is the time to give your client what he or she doesn't expect—extraordinary service.

Law #5: Remember Whom You Serve

Supervisors and sales managers play a specific role in the client-advisor relationship. The good ones will be the first to state, "You do not work for me. You work for your client, but you report to me."

There are two models that exist officially or unofficially in many work and service environments—the inverted pyramid and the chain of command. In the inverted pyramid, or customer-driven, model the client is on top; the needs of the customer come first. Everything else flows downward. In this model, everyone works for the client, and this thinking is what drives everyday actions and decisions. The company may even bend its own rules to please its customers. Notice that the higher up in the corporate structure, the lower the person is in the pyramid. This model cultivates moments of magic. (See figure 7-3.)

The chain of command, or operations-driven, model is more antiquated. In this model, the focus is not on what the client wants, but on what the boss wants. The model puts the needs and procedures of the company first. Keep the boss happy, and you keep your job. If you're lucky, you won't have any irritating clients to get in your way. This model is destined to cultivate moments of misery. Following it will disappoint customers by triggering this exasperating excuse: "I'm sorry, it's against company policy." (See figure 7-4.)

Law #6: No Person Is an Island

In both models in Law #5, you will see that the advisor, even though he or she may be the lifeline to the client, is still only part of a greater team. In many cases, for example, you may be managing a group of associates who help you serve your clients' needs and interests. In essence, your clients are their clients. When your associates contribute to a job well done, be sure to provide positive reinforcement that will ensure that their high-quality help and assistance continue. You cannot provide extraordinary service all by yourself.

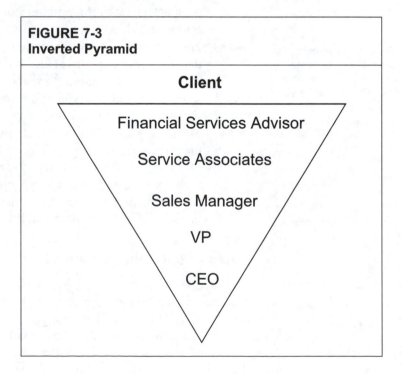

FIGURE 7-3
Inverted Pyramid

Client

Financial Services Advisor

Service Associates

Sales Manager

VP

CEO

> **Show the same courtesy and respect to your colleagues and associates that you show to your prospects and clients.**

Show the same courtesy and respect to your colleagues and associates that you show to your prospects and clients. Make your colleagues and associates happy, and they will help you make your prospects and clients happy. It's as simple as that.

Law #7: Serve More; Sell More

Your first goal is to serve. Selling is a means to this end. It is a service to sell a financial product or insurance policy to a client who really needs it and can understand the benefits that he or she receives. Is it a service to oversell someone, or get the client to buy something he or she does not need just so you can get the commission? Of course not. That is not service. It is exploitation and it is unethical.

To serve is to care. It is to ensure, to the best of your ability, that your clients are receiving the benefits of what you provide. As you know, people do not buy products or services—they buy the benefits of those products or services. Focus on the benefits, and you will succeed and thus become more valuable to your clients.

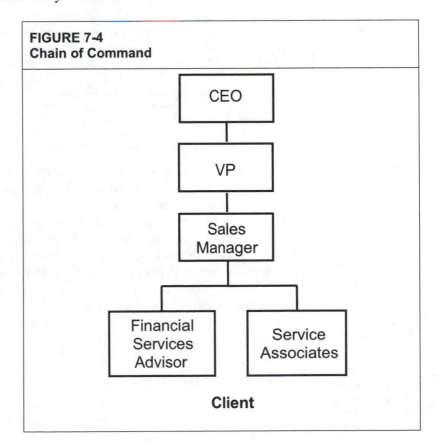

FIGURE 7-4
Chain of Command

Law #8: What Goes Around Comes Around

What does extraordinary service have to do with prospecting? The answer is law #8. What goes around comes around.

Laws #1 through #7 have all been focused outward—directing you to keep your attentions on your prospects and clients. They say very little about what prospects and clients can do for you—it has all been about what you can do for them.

Law #8 is all about you. When you focus on serving, you allow yourself to be served. When you offer and deliver help, you receive help in return.

Example: Al is an advisor. When he wants referrals from his client, Monica, what is the first question he asks her? Is it for names of her friends? No. He first asks her whether or not he has provided extraordinary service to her. If she says, "Yes," Al asks Monica if she believes that his services would be equally beneficial to others. If Monica answers, "Yes," to that question, Al asks her for referrals and introductions. Monica is happy to help Al because Al has helped her.

SERVICING THE PLAN

Proactive versus Reactive Service

reactive service

Service is reactive when it is provided at the client's initiative. *Reactive service* is like a reflex. It corresponds to the minimum level of required service.

proactive service

Service is proactive when it is provided at the advisor's initiative. With *proactive service*, the advisor cultivates client service requests in order to create client contact and opportunities to demonstrate reliability and responsiveness. Proactive service sets you apart and is the key to additional sales and a regular stream of referrals. It requires a service plan, hard work, and consistency.

This section reviews the purpose and benefits of providing excellent service to a client. It then examines how to monitor the client's plan through annual reviews and how to make the most of servicing opportunities and staying in touch with your clients.

The Objectives of Service

> Service is the key to client building, which is the key to long-term success.

Service is the key to client building, which is the key to long-term success. Service actually means two things: The first is monitoring the progress of the plan. The second is delivering promised services to the client. Together, they work to satisfy your present clientele and provide you with referrals, an essential ingredient in building a practice.

Your success at client building through monitoring and servicing will accomplish the following:

- Maintain client persistency.
- Invite repeat sales and referrals.
- Lower expenses.

Maintain Client Persistency. A strong client relationship can help prevent competitors from replacing your business. In this competitive climate, service is clearly a necessary defensive strategy. Maintaining a high profile with clients through your service activities and other contacts will build client loyalty and commitment to you because of the business you have done together. A client who feels no such loyalty or commitment is not likely to think twice about accepting the next attractive proposal that comes along. Your persistency will suffer unless you take steps to keep your existing clients.

> **Experienced advisors will tell you that as much as 75 percent of their new business comes from existing clients or referrals these clients provide.**

Invite Repeat Sales and Referrals. Client building is also part of a smart proactive marketing and sales strategy. Experienced advisors will tell you that as much as 75 percent or more of their new business comes from existing clients or referrals these clients provide. If you are relatively new to the business, preoccupied with generating production and first-year premiums, or struggling to find a market, consider the difference it would make if most of your sales were to clients who had already bought from you or to individuals they referred. Remember that people have a tendency to refer those like themselves. If the client values your products and services, there is a good chance that people to whom the client refers you will as well.

Lower Expenses. Another consideration is the cost of developing clients. What are your marketing and sales costs for finding one qualified prospect and going through a multiple-interview and sales process? Factor them all in: the cost of the lead; sales promotion; telephone, administrative and mailing expenses; other overhead expenses; automobile costs; meals; computer time; and presentation materials. Multiply that amount by the number of prospects it takes to make a sale. By selling primarily to clients, however, you will be able to lower your sales and marketing costs.

Most of you work hard to bring in new customer. Once that customer becomes a client, you can begin to build a barrier through which your competitors cannot penetrate.

> **Monitoring is the servicing aspect that separates a financial product from a financial plan.**

Monitoring the Plan

Monitoring is the servicing aspect that separates a financial product from a financial plan.

A customer buys from the advisor, throws the policy or other financial product into a drawer, and never reviews it. More often than it should be, the advisor doesn't follow up on the client-advisor relationship, so the client-building process is nonexistent. A financial plan that is monitored and revised ensures that it is doing what the client intends it to do. Monitoring the client's plan is the backbone of client-building service activity.

Monitoring responsibilities include the following:

- identifying changes in conditions that would affect the current plan
- obtaining information from the client to determine changes in his or her personal circumstances
- evaluating the client's progress toward achieving his or her financial goals
- reviewing work that was done for the client by other professionals or providers

---

Orphan Clients:
An Overlooked Service Opportunity

You have a real opportunity to differentiate yourself from competitors if you are working with an orphan client who has been neglected by other advisors or companies. Although this client may be understandably dubious about your good intentions at first, he or she will soon appreciate your follow through and professionalism.

The Annual Review

One of the most important services you offer clients is the annual planning review. Make a commitment to clients to review their financial and life insurance programs at least once a year (more often if they want, or as needed). Explain what that process might entail and inform them that you will call in advance to arrange a time when you can sit down together. Remind clients that you will not recommend anything to them that you would not recommend to yourself in the same situation.

There are several key points to consider regarding the annual review. Let's examine each of these key points.

Set the Expectation. Build your client's expectations about the value of your proactive service. Lay the groundwork early on in the process, as early as the initial interview when you explain how you work and the services you offer. Position yourself as an advisor who helps people uncover their insurance and financial goals, creates and implements plans to achieve them, and continues to monitor their plans and make adjustments as needed. Make

> One of the most important services you offer clients is the annual planning review.

sure clients understand the importance you place on helping them realize their financial planning goals.

Set the Appointment. In the case of a life insurance sale, you should set a date for a follow-up appointment at policy delivery. For other financial plans, your goal should be to set a date after step 7 of the selling/planning process (implement the plan) to meet for step 8 of the process (service the plan). (See chapter 1 for the eight steps of the selling/planning process.) Be aware, however, that you should not try to force someone to agree to a date; different clients have different needs and expectations. Rather, you should determine a mutually agreeable schedule for reviews.

Keep Good Records. Record keeping is extremely important. An updated master folder and computer record will help you prepare for the review. Keeping good records along the way will shorten your preparation time, enable you to present a more professional image, and ensure that you have relevant topics to review. Make sure that your records include information about the client's family (for example, children's names and ages, and the names and ages of grandchildren), the client's personal interests, and performance of his or her financial plan(s). You should record personal interests and events to help reestablish rapport and for future servicing and marketing opportunities. Be sure to follow your company's compliance requirements regarding record keeping.

Confirm the Appointment. Call to confirm the appointment. Let clients know in advance if they will need to bring any documentation to the review. For example, if you are following up with a disability income plan, you will need the client to bring a W-2 form and Social Security or employee benefits statement to the meeting.

If you have not set the appointment in advance, you can treat this the same way you would treat a prospecting call except, of course, that you already have a relationship with the client. You may send a preapproach letter to remind the client of the review and then follow up with a phone call to set the date for the appointment.

Prepare for the Appointment. If you have kept good records, preparing will be much easier. Refresh your memory by reviewing the needs analysis and the financial information that the client provided you. Reexamine the information you gathered on your client's attitudes, values, and goals. Review plan recommendations, and pay attention to which recommendations were implemented and which were not implemented. Put together a game plan of areas where you feel client needs may exist.

Conduct the Review. Review the client's insurance and/or financial plan, and evaluate its progress in relation to the client's needs and goals. If appropriate, recalculate needs and note any shortfalls. Inquire about any changes in the client's current or future financial situation. Listen carefully and note any planning opportunities. Implement any plan changes and set any necessary follow-up appointments.

Ask for Referrals. An annual review is a perfect time to ask your client if he or she knows anyone who might share the same values and goals. There is a high probability that your client values and trusts you; otherwise, the client would not have let you review his or her financial plan. Thus, the client is likely to want to refer you to family members and friends.

CUSTOMERS VERSUS CLIENTS

> **All clients are customers, but not all customers are clients.**

All clients are customers, but not all customers are clients. Clients are people who have bought from you, but they are also people with whom you have developed, or are developing, a strong personal business relationship. Because of this relationship, if these clients have any problems in areas relevant to your expertise, they will look to you for advice.

How does a customer become a client? Most consumers never develop a personal relationship with the salesperson from whom they buy a product. They might not even recognize the salesperson after they make the purchase. They are customers but they are not clients. For advisors who represent products and services that depend on client-advisor relationships, however, it is essential to convert customers to clients. To build a career based on client relationships, time spent to develop personal relationships is time well spent.

Developing client relationships is mutually beneficial for the advisor and the client: The client knows that the advisor will deliver promised services and will be available when he or she is needed. The advisor understands that the relationship is a building block in the development of a long-term career.

As noted earlier, in client building, the sale is not the end; it is the beginning. You must fulfill all the promises of service you make during the selling/planning process. You should, as emphasized previously, exceed your client's expectations. Make frequent contact and updates. This entails more work than merely servicing a customer, but it is also more rewarding. As the client-advisor relationship grows, the advisor is able to meet more of the client's financial needs; more doors open through referrals.

What can you do to cement that relationship and protect your customers from the competition? Here are some ideas:

- Try to broaden the product portfolio you offer to customers, and develop strategic alliances (discussed in chapter 8) with other producers and professionals who offer products and services you do not.

- Expand your areas of expertise. Pursue professional designations, add additional licenses, and generally augment the knowledge and skills that make you more valuable to clients.
- Be sure to tell your prospects and clients frequently, through mail and personal contact, about all the products and services you offer. Highlight those you may not have previously addressed.

Your clients are your most valuable business asset. You can achieve great success in the financial services business when your established client base provides enough referrals and repeat business to enable you to attain your personal sales goals.

THE A-B-C METHOD FOR LEVELS OF SERVICE

When it comes to client service, if you are just winging it, your clients are more susceptible to your competition.

Are you winging it when it comes to client service, or do you have a plan to follow? Do you think your clients can tell the difference?

Here is something to consider: If you are just winging service, your clients are more susceptible to your competition. In addition, you are not likely to realize their full value in more business or referrals.

Unless you have decided to work with only a small number of clients, you will not be able to deliver the same level of service to all your clients as your book of business grows. There are a number of criteria you can use to segment your clients.

Typical quantitative criteria are as follows:

- net worth
- household income
- current and potential (movable) assets
- age range
- revenue generated
- occupation or industry

Qualitative criteria include the following:

- long-term goals
- service requirements
- probability of following your advice—loyalty potential
- additional sales potential
- source (or potential source) of referrals
- likability

A combination of quantitative and qualitative characteristics may move a client up or down in the rating groupings. This is largely a subjective call based on your ideal client profile.

A-B-C method

customer

A traditional way to classify clients based on these quantitative and qualitative factors is the *A-B-C method* in which you segment your clients into three categories or classes:

- *customer*. This is someone who has bought a product from you, but that's all he or she wants—or that's all you want. Customers have demonstrated that they do not wish to enter into an ongoing client-advisor relationship with you, or for some reason, you do not want to do more business with them. These are your "C" clients to whom you offer only basic services (for example, basic annual reviews and follow-ups on information and service requests).
- *customer/potential client*. This individual is a customer you wish to turn into a client, or a customer who has purchased a product from you but has not yet committed to a full client-advisor relationship. Customers/potential clients are your "B" clients. You will offer them a broader range of services than the "C" group. Your goal is eventually to lower them to the "C" group or raise them to client status.
- *client*. This is an individual who believes in you, the products you sell, and the services you provide. He or she is a source for repeat business and referrals. Your long-term goal is to deal only with clients. They are in your "A" group. Clients merit your very best service.

client

> Your "A" clients are a source for repeat business and referrals. They merit your very best service.

Why Classify?

In the same way you would not put most of your money in low-yielding investments, you should not put most of your time, energy, and money in the low-profit segment of your book of business.

Exercise some caution here. Do not focus solely on income or on how well you get along with an individual. Recall the total profitability of your client, as discussed earlier in this chapter. Look at the impact on your business. You can measure this in three ways:

- Is the customer or client a source or potential source for repeat business?

Example: Robert is a young man who does not have a lot of business with you right now. But he has demonstrated an appreciation for the financial plan you recommended, and he has the motivation and ability to increase his income over the years. It is likely, therefore, that Robert will be a source of repeat business for you.

- Is the customer or client a source (or potential source) for referrals?

Example: Robert is a plumber who is a part of a trade union. The union meets regularly for programs on interesting topics. This could be an opportunity for you to present a seminar to union members at one of their meetings.

- Is the client a good client? Is the client easy to get along with and not high maintenance? (Stay away from high-maintenance, low value-clients. You will expend a lot of time and energy and receive a negative net return.)

Example: Robert has many personality traits that match those in your ideal client profile. He also doesn't call repeatedly with irrelevant and time-consuming questions. Although his net worth is lower than others in your "A" group, Robert has the potential to be a high-value client.

How Classifying Works

> **Identifying your "A" clients, "B" clients, and "C" clients allows you to treat each category according to profitability criteria.**

Identifying your "A" clients, "B" clients, and "C" clients allows you to treat each category according to profitability criteria. You can distinguish between basic and discretionary services and select which services to offer each category of client. Clearly, you have an obligation to all customers and clients to process policy changes and handle other requests for information or service; you may also want to send every one of your customers and clients a calendar each year. You can be selective, however, about how you spend your service budget for discretionary services, such as greeting cards, lunches, receptions, and other gifts. You are not obligated to include all customers and clients on your newsletter mailing list or to send all of them special clippings from the local newspaper with personal notes. These can be expensive and time-consuming gestures that would simply be wasted on some customers. For example, you may take your "A" clients out to lunch or host receptions for them as a way to show your appreciation for their business. One advisor who owns a sailboat invites "A" clients for evening or weekend sails. This has been an excellent way to build loyalty and strengthen the client-advisor relationship. Golf outings, tennis matches, or jogging dates, for example, are other nonbusiness interests you can share with your clients.

Your discretionary service activities are important because of the impact they can have on your client. Both you and your clients should see the service you provide as timely, professional, and top quality. In addition, you should see it as an opportunity to differentiate yourself from others in the industry.

Let us now discuss the three service packages for the three categories of clients:

- standard service package—"C" clients
- take care plus package—"B" clients
- extra frills package—"A" clients

Standard Service Package

> **Excellent customer service is now the price of admission. It is the expected, not the exception.**

Excellent customer service is now the price of admission. It is the expected, not the exception. This means that your standard package must provide high-quality service to all customers and clients, regardless of their value to your business. Follow the three steps below to determine what yours will encompass.

Define It. Define your standard service package. What services will you provide to everyone and at what level? Your standard services may include the following, for example:

- annual or periodic reviews
- prompt handling of plan changes and inquiries
- claims assistance

Also define the level of service clients can expect. For example:

- accessibility. When can clients reach you? What after-hours services do you have?
- response time. How quickly will clients receive a response to their phone calls, e-mails, or letters?

Communicate It. Once you have defined your standard service package, communicate it to your customers and clients. Tell them what they can expect. The best time to do this is when you establish a new account. Many advisors include a flyer that explains servicing information.

Deliver It. The next step is to follow through and hold yourself and any staff to what you have committed. Some basic tips for delivering service:

- When you receive a request, let the client know how long it will take you to satisfy it. When will you get back to the client?

- Telephone your clients after service requests to confirm that the information received or action taken was what they requested. Make sure that the service was prompt and efficient. Take the opportunity to discuss any issues suggested by the service request that were left unresolved.

- If your client moves to another city or state, try to arrange to keep in contact and to retain the business. If you are unable to do that, offer to review the plan with his or her new advisor.

Take Care Plus Package

Excellent service puts you in a position to inquire about the client's other relevant needs. In other words, you are able to *take care, plus* market other products and services you provide. The keys to recognizing opportunities are listening carefully and asking relevant questions. Here are a couple examples to get you started:

- Beneficiary changes may reveal a need to revisit the client's plan. For example, the insured may be changing the beneficiary because of a new spouse or baby.

- Disbursement requests (loans, dividend withdrawals, partial surrenders, and so on) indicate that a service call is needed. They may reveal that the client is in financial trouble, opening a new business, or paying for a wedding. You won't know unless you ask what prompted the request.

As always, if the client is pleased with your service, ask him or her for referrals.

Extra Frills Package

You will want to give a premium service package to your best clients. This means you must identify what services you will provide as frills and what clients will receive this premier service.

The purpose of the extra frills is to turn every contact with the client into an opportunity to market. By this we mean that you should strive to accomplish at least one of the following three things:

- *Create visibility.* You want your clients to think of you when they think of financial services.

- *Create awareness of the products and services you provide.* You want your clients to associate your name with any product you sell and any service you provide.

- *Create a sense of high touch.* You want your clients not only to associate you with financial services, but also to think of you as a person who cares about them. You want them to see you as someone who provides what they need, when they need it, and how they how it.

Setting Up Your Service Plan

After you have segmented your client base, you need to establish your service plan. Select the services you want to provide as basic and discretionary services to your A-B-C client groups. Once you determine the services, automate the process as much as possible, creating a to-do list and calendar to trigger reminders for each activity. Computer contact management systems (discussed in chapter 8) allow you to manage these activities with precision. The more you plan ahead and systematize, the more likely you will actually perform the services.

Commit yourself to a service plan by including it as part of your business plan. Prepare a list of services, and let prospects and clients know that these are the services you provide to the people you value as clients.

Figure 7-5 illustrates how you can segment service activities according to client category to add value to your client-advisor relationship and build business friendships. Consider what you would like to include, and then create your own plan that you can commit to and implement.

Commitment to Service

One highly successful advisor who has made a commitment to his service plan reports that approximately 90 percent of new business comes directly from clients and 10 percent from people referred by clients. This advisor works exclusively with young professionals, meets with most clients in his office, and employs a full-time staff of three to handle the service work. He also states that he has not had to prospect for new business in more than 2 years.

This advisor is convinced that if you want to build a continuing business among people who look to you as their advisor, good service is key to helping you accomplish that objective.

HOW TO PROVIDE EXTRAORDINARY SERVICE

Virtually all client service strategies are designed to address two key objectives: adding value and building business relationships.

Remember that when it comes to service, what works for one advisor may not work for you. It depends on you; it depends on the client. What one client finds interesting, another may find annoying. Let's explore some ideas you may want to consider regarding your service activities. Virtually all client service strategies are designed to address two key objectives: adding value and building business relationships. But do not bite off more than you can chew. It is better to do a few things consistently and well than to do many things inconsistently and poorly.

FIGURE 7-5
Sample Client Service Delivery Plan

Level "A" Clients	Level "B" Clients	Level "C" Clients
Monthly call	Bimonthly call	Quarterly call
Annual formal review	Annual formal review	Annual formal review
Quarterly performance review	Semiannual performance review	—
Invitation to all company events	Invitation to two company events	—
Quarterly newsletter	Quarterly newsletter	Quarterly newsletter
Monthly article of interest	Quarterly article of interest	—
Books of interest	—	—
Birthday cards and phone calls	Birthday cards and phone calls	Birthday cards or phone calls
Holiday cards	Holiday cards	Holiday cards
Anniversary cards	Anniversary cards	—
Anniversary reminder call	—	—
Invitation to all client appreciation events	Invitation to one client appreciation event	Invitation to one client appreciation event
Invitation to select referral events	Invitation to select referral events	—
Notes on sports interests	—	—
Notes on other interests	—	—
Contribution to favorite charity	—	—

The more of these services you want to pursue, the more important a competent staff becomes. View a good staff not as an expense but as an investment in developing your business that frees you to do what you do best—selling/planning and building relationships. Take the time to clearly explain to your staff what frills you will offer. Make sure that you can deliver them and that they are cost effective from a monetary and time standpoint.

Service Activities and Ideas

Study your "A" client list to determine what you can do to provide extraordinary service to these clients. Here are several ideas.

Remember Your Client's Interests

Hobbies Rather Than Occupation. Think about your clients' hobbies rather than their occupations. There are lots of possibilities. Look for items that correlate with your clients' interests at retail stores or on the Internet. Clients will also likely appreciate tickets to hobby-related events or classes such as cooking, design, music, and so forth.

Example:	Recently, Henry, a financial advisor, had lunch with his client Mark, who has a passion for trains. Although Mark works as an auditor, he did not talk to Henry about how he conducts an annual corporate audit. What Mark wanted to talk about was trains. Knowing this in advance, Henry stopped at a bookstore to purchase an illustrated book on the most noteworthy trains over the last 100 years. Mark was thrilled that Henry had taken an interest in him and had given him the book. They spent the entire lunch discussing trains—that is, Mark talked and Henry listened. After lunch, Mark asked Henry how business was going and if he could do anything to help out. Henry asked Mark for some introductions, and Mark gave Henry the names of five people. Each person eventually resulted in a sizable sale.

Magazine or Newspaper Subscriptions. If you know your clients' interests, you can give them annual subscriptions to their favorite magazine or book-of-the-month club. Clubs that send their members a different item every month (food, books, music, wine, and so forth) are a great way to remind clients regularly of your ongoing relationship—without having to call them.

Educational Meetings and Seminars. Conducting an educational meeting or seminar is a potential marketing opportunity to a small group of your best clients who share a common interest. In addition, it offers an excellent occasion to ask for referrals. But you can also use educational meetings and seminars conducted by others as a way to offer extraordinary service.

Example:	Assume you have a client who loves gardening. You read in the newspaper about a 3-day symposium on gardening and landscaping. You give your client a call and ask her if she would

like to attend. You provide the ticket. You can be pretty sure you will receive a thank-you call from the client.

Tickets to Sporting and Other Events. Many people enjoy seeing a football, basketball, baseball, or soccer game, or events such as plays, flower shows, and movies—there is virtually no limit to the activities you could offer your best clients. Purchase tickets and ask if your clients would like to attend individually or as a group. With the appropriate fact-gathering process, you will learn about their interests and be able to provide suitable tickets.

Offer a Complimentary Service

Providing a complimentary service can be an unexpected bonus to your client that enhances the client-advisor relationship.

One dentist, for example, came up with a creative complimentary service that created goodwill, built his clientele, and separated him from the pack. While his patients were having their teeth cleaned, he arranged for a well-dressed, clean-cut teenager to shine their shoes at no charge. The dentist's patients praised his services. "Great dentist!" they said. "Plus you can get your shoes shined at the same time." Referrals poured in. Business boomed. (The teenager, who always received a large tip, was also a winner.)

What can you learn from this creative dentist?

Think of a complimentary service you can provide that your client will value and talk about to others.

Think of a complimentary service you can provide that your client will value and talk about to others. Because people today are so pressed for time, you might want to consider a time-saving gift for your most valuable clients such as a car wash, dry cleaning services, or a home-delivered meal, for example. Use you imagination.

Example: One advisor offers to drive her clients to the airport—especially for early morning flights. According to the advisor, "You can't really trust that a cab will always show up on time." The advisor says that she receives more referrals from this group of clients than she does from any other group.

Get Personal

Just as you record information about your client's hobbies, keep notes about your client's personal life that will show him or her that you care. It is so much nicer and more personal to say, for instance, "How is Bob?" than

"How's your husband?" This helps to develop your credibility with the client, establish rapport and trust, and enhance the relationship-building process.

Personal service to the client will reinforce his or her loyalty to you.

Birthday, Holiday, and Special Occasion Cards. Send birthday and holiday cards to your clients and their families. Cards can be imprinted with your name and address; the selection of messages is wide enough so that you will be able to find one or more cards that fit your style. Ask your manager if greeting card catalogs are available. The cost is deductible as a business expense, so keep track of the postage and number of cards sent. There are also Internet sites that allow you to personalize electronic greetings.

Instead of a commercial card, why not create your own? Make the design distinctive so that clients will soon recognize the cards as your "trademark." Perhaps the card could invite the client to call you so that you can buy him or her the beverage of the client's choice. If you want to take it up a notch, you could personally deliver a birthday cake.

Contact clients at other significant dates such as weddings, christenings, bar mitzvahs, births, home purchases, new jobs, or promotions. Do not limit your contacts only to happy occasions. Show your sympathy and support when there are deaths in your clients' families or if your client or client's spouse loses a job. Clients will appreciate your concern—and perhaps even more important, your advice and assistance.

> **In this era of high technology and fast-paced living, a handwritten note is an especially thoughtful gesture and adds a genuine personal touch.**

Handwritten Notes. In this era of high technology and fast-paced living, a handwritten note is an especially thoughtful gesture and adds a genuine personal touch. Be sure to hand address the envelope, and use a commemorative postage stamp, if possible. Your note will certainly stand out from the onslaught of mass-produced computer-generated envelopes, flyers, magazines, and newspapers. Consider having your note cards or stationery imprinted with your company's name and logo, your name, and business information.

Work on developing the message you want to convey. For example, when writing a note to a new client, you can communicate much more than a simple "thank you."

Example: "Thank you for giving me the opportunity to do business with you. My goal now is to continue to offer you excellent follow-up service so that you'll have no reservations about referring others who have similar needs."

Newspaper Articles. Scan the local newspaper every day, and copy articles that might interest one of your clients. Mail the article to him or her with a personal note that simply says, "I was thinking of you when I came across this article," or "Here's an idea I thought you'd enjoy." (Check with your compliance department before sending articles.)

Similarly, when you see an announcement about a client's family or business, send a copy of it with a brief personal note. Be sure to include the publication name and date. The announcement may also indicate a new sales or service opportunity. For example, you may find out that the client has been named a member of a community club or professional organization that may be a great target market for you.

Care Packages to Family Members in the Armed Services. If your clients have sons or daughters serving in the military, send care packages to them. Cookies, cakes, or anything else from home is a great morale builder. Your clients are sure to appreciate your thoughtfulness.

Make Regular Phone Contacts

You want to stay in touch with your clients periodically, and regular phones call can ensure that you do. The appropriate interval between calls is typically based on the A-B-C client categories. Often, these stay-in-touch calls are intentionally unrelated to business. Some advisors actually refuse to discuss business. This is all a part of creating a relationship with your clients.

In addition, you should always contact clients to let them know the results of any meetings you had with referrals they gave you. Be sure to thank your clients for their support.

> **Always contact clients to let them know the results of any meetings you had with referrals they gave you. Thank them for their support.**

Hold Client-Appreciation Events

A client-appreciation event is different from a referral event, where the purpose is to bring people together to meet you and generate new business. The purpose of a client-appreciation event is to thank your clients for their business and make them feel important. It is designed to reward loyal clients.

There are many ways you can hold a client-appreciation event, depending on the interests of your clientele. It's up to you. For example, you can have a formal dinner party, a family picnic, a holiday gathering, a trip to a ballgame, or a wine-and-cheese-tasting party. Whatever the event, it is an opportunity for you and your clients to get to know each other in new and more personal ways.

You can also bring together people who you think will benefit from meeting each other. This adds value to the event for your clients by providing them with personal and business connections, which, in turn, makes you more valuable and hence more referable.

Keep Your Name in Front of Clients

Newsletters. Have you considered a newsletter? It is an excellent way to keep your name in front of clients on a regular basis and build your prestige

> **Have you considered a newsletter? It is an excellent way to keep your name in front of clients on a regular basis and build your prestige as a professional advisor.**

as a professional advisor. It can be a newsletter you purchase through your company or marketing outlet or one you create yourself. Providing interesting and informative articles about tax laws and planning, insurance and other financial matters, as well as information about yourself and your practice will educate your clients about the services and products you provide and the needs that your clients may want to address with you.

Commercial (as always with company approval) or company-sponsored newsletters offer a wide selection of information that you can send to your clients either monthly or quarterly. In many cases, the letters can be personalized with your name, address, logo, and photograph. You may also be able to enclose a business return card that the client can mail to request more information about the topics covered in that issue. Ask your manager about the availability of these newsletter services and their costs.

Customized Letters and Gifts. To keep their names in front of clients, many advisors give them imprinted calendars and other customized items. Shop around. Look for package offers and quantity discounts on larger purchases. Clients appreciate receiving pens, refrigerator magnets, and other items customized with your information, if they are attractive and useful.

You can purchase individualized gifts for select clients or for those with whom you want to strengthen your relationship or show your appreciation. Here, especially, you want to choose a gift that complements the client's interests or has a special meaning. For example, a baseball fan may appreciate tickets to a game, a book on the sport, photos of baseball legends or famous stadiums, reviews of memorable games, and so on. Be creative.

Another possibility is a charitable donation. A gift to a client's favorite charity is both thoughtful and personal.

Food and Beverage Baskets. Whether it's a food basket, pizza, or bagels for an office or family, good food gets noticed and appreciated—not only by your clients, but also by their colleagues and/or families. "Who sent this?" they are sure to ask. Your name and company are likely to be the topic of discussion. How about the makings for a backyard picnic? That sounds like fun.

CONCLUSION

This chapter explains the relationship between service and prospecting. Providing extraordinary service is essential to building client loyalty, and client loyalty is key to repeat business and referrals to new prospects. This is where service and prospecting intersect.

Imagine that you are your own client. How would you rate your service? Do you embody the Golden Rule? Perhaps more important, do you embody the Platinum Rule? Do you, in fact, treat others the way they want to be treated?

It is much more cost effective to keep a client than to find a new one. Thus, the chapter explains why mere client satisfaction is not good enough. The advisor's goal must be to exceed satisfaction and develop client loyalty.

In addition, the chapter examines "moments of truth"—times when an advisor is faced with choices about how to respond to a client. These choices shape the client's perception of both the advisor and the company, which affects the future relationship with that client and the amount of any repeat business or referrals.

The chapter discusses the eight laws of extraordinary service. By following these laws, advisors can determine the most effective ways to serve clients and distinguish themselves from the competition. The chapter also explains the difference between reactive and proactive service, identifies the primary objectives of service, describes aspects involved in monitoring the client's plan, and reviews the key points to consider in the annual planning review.

Because all clients are customers, but not all customers are clients. Therefore, the chapter describes the A-B-C method of categorizing customers and clients into segments according to profitability criteria. The advisor can then determine which basic and which discretionary services to offer each level of customer or client. The chapter presents a sample client service delivery plan from which advisors can develop their own service packages.

Finally, the chapter explores some service ideas and activities that exemplify extraordinary service.

CHAPTER SEVEN REVIEW

Key Terms and Concepts are explained in the Glossary. Answers to the Review Questions and Self-Test Questions are found in the back of the textbook in the Answers to Questions section.

Key Terms and Concepts

Platinum Rule	reactive service
client satisfaction	proactive service
client loyalty	A-B-C method
moment of truth	customer
extraordinary service	client

Review Questions

7-1. Explain the importance of extraordinary service to prospecting, and discuss how are prospecting and service are related.

7-2. Discuss how financial advisors should evaluate their own service to clients.

7-3. Define extraordinary service.

7-4. Identify a moment of truth and explain how it affects service.

7-5. Explain why an orphan client may be reluctant to meet with a financial advisor who has been assigned to him or her.

7-6. Identify the eight laws of extraordinary service.

7-7. Discuss the A-B-C method to categorize levels of clients and the reasoning behind it.

7-8. Identify four ways an advisor can provide extraordinary service to clients.

Self-Test Questions

Instructions: Read chapter 7 first, then answer the following questions to test your knowledge. There are 10 questions; circle the correct answer, then check your answers with the answer key in the back of the book.

7-1. Which of the following statements concerning the relationship between service and prospecting is correct?

(A) Client service takes time away from the advisor's selling activities.
(B) Providing extraordinary service is part of building client loyalty, which is essential for referrals to new prospects.
(C) Most financial advisors focus on servicing rather than selling.
(D) A financial advisor's job is to sell products and services, not to service clients.

7-2. Which of the following statements concerning the Platinum Rule is correct?

(A) People like to be treated the way you like to be treated.
(B) Act in a manner that makes your behavior the universal standard.
(C) Treat others as they would like to be treated.
(D) Do unto others as you would have them do unto you.

7-3. Which of the following statements concerning the determination of the total profitability of a client is correct?

(A) It is the income items for that client subtracted from the expense items received from that client over time.
(B) It is limited to the commissions related to that client received over time.
(C) It is the commissions received over time and the new business attributed to that client, minus expenses created for the client over the same time.
(D) It is best not to measure the impact of total profitability by looking at whether it produces a measurable level of sales revenue.

7-4. Which of the following identifies a moment of magic?

(A) when a client interaction that has negative connotations leaves the client dissatisfied
(B) when an interaction exceeds client expectations and leaves him or her with a positive experience
(C) when uninspired service has neither positive nor negative connotations
(D) when an advisor must make a choice about how he or she will respond to a client

7-5. Which of the following statements concerning orphan clients is (are) correct?

I. They are clients who have no advisor with that particular company.
II. Advisors do not call orphans because they already have enough people to call.

(A) I only
(B) II only
(C) Both I and II
(D) Neither I nor II

7-6. Service and prospecting are related in which of the following ways?

I. Each servicing opportunity may be a marketing opportunity.
II. Providing extraordinary service is crucial to building client loyalty.

(A) I only
(B) II only
(C) Both I and II
(D) Neither I nor II

7-7. Which of the following statements regarding the laws of extraordinary service is (are) correct?

I. The chain of command, or operations-driven, service model, puts the needs of the company first.
II. If an individual offers and delivers help, he or she will receive help in return.

(A) I only
(B) II only
(C) Both I and II
(D) Neither I nor II

7-8.   All the following statements regarding the laws of extraordinary service are correct EXCEPT:

(A)   The advisor should seek out moments of truth and embrace them.
(B)   The inverted pyramid, or customer-driven, service model puts the needs of the client first.
(C)   Service should never be an inconvenience for the advisor.
(D)   It is a service to sell a financial product or plan to a client who really does not need it.

7-9.   All the following statements regarding clients and customers are correct EXCEPT:

(A)   All clients are customers, but not all customers are clients.
(B)   Not all customers want to become clients.
(C)   To build a career in financial services, it is essential to convert customers to clients.
(D)   "Client" is a title the advisor imposes on customers so they will think of themselves as clients.

7-10.   All the following statements regarding the A-B-C method of categorizing clients is correct EXCEPT:

(A)   "C" clients will never be "A" clients, so the advisor can just ignore them.
(B)   The advisor must give a basic level of service to all customers and clients.
(C)   Advisors should devote most of their time and money to "A" clients.
(D)   "B" clients are those the advisor wants to move up to "A" or down to "C" status.

NOTE

1.   Gitomer, Jeffrey, *Customer Satisfaction is Worthless, Customer Loyalty is Priceless* (Austin, TX: Bard Press, 1998), p. 50.

8

Professional Practice Management

Learning Objectives

An understanding of the material in this chapter should enable the student to

8-1.   Identify the four phases of a mentoring relationship.

8-2.   Describe the value of strategic alliances with other professional advisors.

8-3.   Explain the importance of the formation of a personal board of advisors.

8-4.   Identify the components of a contact management system.

8-5.   Describe the prospecting and service functions that advisors perform using a contact management system.

8-6.   Identify the typical sections in both an annual and monthly business planner.

8-7.   Outline the steps in the annual and monthly planning processes.

8-8.   Review state regulation of insurance companies and advisors.

8-9.   Explain the role of compliance, ethics, and professionalism in the financial advisor's selling/planning activities.

Chapter Outline

INTRODUCTION

This chapter discusses practice management concepts that will help you execute your prospecting and sales activities effectively and professionally. The chapter begins with a look at mentoring relationships. It then explores strategic alliances with other professionals and the importance of establishing a personal board of advisors. It provides a systematized approach to manage contacts with your prospects, followed by a discussion of techniques for planning all phases of your business activities. Finally, the chapter examines the regulation of advisors at the state and federal levels and presents concepts of ethical behavior and professionalism.

STRATEGIES FOR SUCCESSFUL MENTORING RELATIONSHIPS

mentoring

Mentoring is a process through which a more experienced person—the mentor—transfers knowledge and information to a less experienced person—the aspirant—who acquires, applies, and critically reflects upon that knowledge and information. It is a reciprocal and collaborative learning partnership between two (or more) individuals who share responsibility and accountability for reaching learning goals defined by both the mentor and the aspirant.

In her book *The Mentor's Guide: Facilitating Effective Learning Relationships*, Lois J. Zachary discusses the four phases of mentoring relationships.[1] We will examine them below, as well as several success strategies you can use to enhance the mentoring partnership experience.

Four Phases of Mentoring Relationships

All mentoring relationships are composed of four sequential phases: preparing, negotiating, enabling, and coming to closure. Each phase builds on the previous phase to form a predictable developmental sequence. The phases vary in length from one relationship to another. Taking any phase for granted or skipping a phase has a negative impact on the mentoring relationship. Being aware of the potential pitfalls may be enough to ward off the negative consequences.

Preparing

Each mentoring relationship is unique; it doesn't matter how often you've been engaged in a mentoring relationship. Facilitating successful mentoring relationships requires preparation (1) of the participants and (2) for the relationship. Preparation adds value to the mentoring partnership for both the mentor and the aspirant.

During the preparing phase of a mentoring relationship, several processes take place simultaneously. Mentors explore their personal motivation and readiness to be a mentor. They assess their individual mentoring skills to identify areas in which they need additional learning and development. Clarity about each person's expectations and role helps to define the parameters for a productive and healthy mentoring relationship.

Success Strategy #1

Before an initial meeting with a prospective mentoring partner, answer the following questions:

- Am I ready for a mentoring relationship?
- Am I clear about my role?
- Is this particular relationship right for me?
- Do I have the time to do justice to this relationship?
- Is mentoring compatible with my learning style?
- Am I comfortable with my mentoring partner?
- What are my personal developmental goals from the mentoring relationship?

Negotiating

Negotiating is the contracting phase of the mentoring relationship. This is when details of the relationship get spelled out: when and how to meet, who has what responsibilities, what the criteria are for success, how accountability is determined, and how and when to bring the relationship to closure. The outcome of this phase should be a partnership work plan consisting of well-defined goals, criteria and measurement for success, mutual responsibility, accountability mechanisms, and protocols for dealing with stumbling blocks.

Failure of a mentoring relationship is often attributed to lack of time. Mentors frequently underestimate the necessary time commitment they must make to establish and build this relationship. Regular contact (be it daily, weekly, or monthly) is vital. Be aware, however, that there must be a consensus about the meaning of "regular" and a decision to adhere to that agreement. Discuss what you and your mentoring partner need to do individually and together to maximize the time you do have.

> **Mentors frequently underestimate the time commitment they must make to establish and build the mentoring relationship.**

Success Strategy #2

Before you move into the enabling phase, answer the following questions. The answers should be crystal clear to you and your partner.

- What are the learning goals?
- What are the learning needs?
- Is there a mutual understanding of roles?
- What are the responsibilities of each partner?
- What are the norms of the relationship?
- How often will we meet?
- Who will initiate contact?
- What are the boundaries and limits of this relationship?
- What is our work plan?

Enabling

The enabling phase is when most of the learning between mentoring partners takes place. Each mentoring relationship is different and finds its own path during this phase. The longest of the four, this phase offers a tremendous opportunity to nurture learning and development. It is also the phase during which mentoring partners are most vulnerable to relationship derailment.

The mentor's role during this phase is to nurture the aspirant's growth by maintaining an open and affirming learning climate, by asking the right questions at the right time, and by providing thoughtful, timely, candid, and constructive feedback. It is difficult to create a learning environment, build and maintain the relationship, monitor and evaluate progress, and encourage reflection and assess learning outcomes without feedback. When feedback is given

---

Success Strategy #3

Brainstorm a list of learning opportunities with your mentoring partner. Consider the following:

- What kinds of opportunities exist to get exposure to new learning?
- What kinds of opportunities exist to reinforce new learning?
- What kinds of opportunities might accelerate learning?

> **When feedback is given or received in the wrong way, it can undermine the mentoring relationship.**

and received in the right way, it supports the mentoring relationship. When it is given or received in the wrong way, it can undermine the relationship. Mentors can evaluate the process when feedback is provided in a climate of readiness and expectation. Providing feedback without establishing this climate can be a frustrating and negative experience for aspirants and mentors.

Success Strategy #4

Regular feedback provides support to meet learning challenges as they occur. Some guidelines:

- Set clear expectations about the feedback you provide.
- Acknowledge the limits of that feedback.
- Be authentic and candid.
- Focus on behavior, not personality.
- Consider the timing.
- Be constructive.
- Ask for feedback on your feedback.

Coming to Closure

Although it is difficult to plan for closure, closure is necessary in any mentoring relationship. Coming to closure is the time when mentors and aspirants process their learning and move on, regardless of whether or not the mentoring relationship has been positive. It is a time to assess the experience, acknowledge progress, and celebrate achievement. When closure is seen as an opportunity to evaluate personal learning and take that learning to the next level, mentors and aspirants are able to leverage their own learning and growth.

What to Look for in a Mentor

When seeking mentors, consider individuals with these characteristics:

- *Seek out possible mentors whose success you would like to emulate,* and ask them what they like about their work. Beyond the financial rewards, what do they receive? Why did they choose a career in the financial services business? Why not something else? Ask them to share stories of their favorite clients and about their relationships. Their commitment probably goes much deeper than their involvement in financial services. Capture their passion!

- *Look for the qualities of honesty and integrity* when searching for a mentor. Because someone is successful in the financial services business doesn't automatically qualify that person to be your mentor. How a person conducts himself or herself in other roles may be much more important than how professionally and financially successful the person is. In the words of Gandhi: "Life is one indivisible whole." If you do not respect the individual, neither will you respect his or her counsel.

> **Because someone is successful in the financial services business doesn't automatically qualify that person to be your mentor.**

- *Search for people who share similar values to your own* and whom you admire. You needn't agree on everything or do everything the same way. But if your prospective mentor has qualities you admire, it makes him or her a good role model for you.

- *Look for someone who can provide you with guidance and leads* through his or her own network of clients, colleagues, and friends. As your relationship strengthens, your mentor will most likely introduce you to people who can make a profound difference in your future.

CREATING ALLIANCES

Team Approach to Planning

It is virtually impossible to be successful in the financial services marketplace without working with other professionals. These professionals may include, for example, attorneys, accountants, insurance product specialists, bankers/trust officers, and investment advisors. Coordinating the efforts of these other professional advisors in a comprehensive financial planning approach can yield the greatest benefits for you and produce the best results for your clients.

> **By combining efforts and expertise, advisors can provide their clients with competent assistance in all the major financial planning disciplines.**

The team approach to planning works well: By combining professionals' efforts and expertise, advisors can provide their clients with competent assistance in all of the major financial planning disciplines.

The financial advisor is usually regarded as the quarterback of the financial planning team, who takes responsibility for implementing and coordinating the specialties of the various team members. It is frequently the financial advisor who makes the first contact with the client, enlightens the client to the need for planning, and motivates the client to become involved in the planning process. Thus, the financial advisor may be the prime motivator in getting a client to take action on plan recommendations.

Working with Other Professionals

Joint work with specialists who are more knowledgeable in other disciplines is an excellent way to gain experience, learn from experts, and provide competent, professional advice to clients who have needs and concerns outside of your area of expertise. These specialists can serve as mentors to you to enhance your practice. They can be from your own financial services area or from another area.

Therefore, if you have not already done so, you should establish working relationships with other professionals in your community who can enhance your practice. This should be a continual process that allows you to expand these relationships and give your clients the best professional services in all disciplines to help them reach their financial and life goals.

Strategic Alliances

strategic alliance

A *strategic alliance* is a relationship that you, as a financial advisor, develop with another professional. These relationships enable you to have at your disposal an established team with knowledge in complementary professional disciplines to whom you can refer a client when his or her financial situation requires their expertise and services.

> **Strategic alliances are beneficial for both clients and advisors.**

Strategic alliances are beneficial for both clients and advisors. The client benefits from the services of competent, cooperative, and trustworthy professionals. The advisor benefits from the additional clients that result from the alliance.

In the course of providing your clients with a team of advisors whose financial services complement yours, you can also form mutually beneficial strategic alliances with professionals, which can further your prospecting efforts. You need to introduce your clients to these professionals when it is appropriate to do so. In return, they will introduce you to their clients.

Financial Planning Team

Below is a brief description of the services of some of the professional advisors with whom you should be working, along with ideas of how you can help them better serve their clients.

Attorneys. Attorneys draft all the legal documents that are necessary to implement the client's financial, business, and estate plans. These documents virtually always include wills and may include trusts, buy-sell agreements, and other documents if a more sophisticated estate plan is elected. When funding is needed for these legal instruments you can be the resource to provide attorneys' clients with the appropriate financial products.

Accountants. Accountants have knowledge of clients' financial transactions, income tax returns, and net worth. They may provide valuations for assets in clients' estates or in buy-sell agreements to facilitate transfers of a business interests upon clients' death or disability.

Accountants often recommend financial products to protect business owners and their families from the loss or devaluation of property, assets, income, and life. You can work with accountants to provide their clients with the financial products that will accomplish their planning needs and objectives.

Insurance Product Specialists. Insurance product specialists understand the unique features of specific types of insurance contracts and the many ways these contracts can provide solutions to a wide range of personal planning needs.

The life insurance specialist understands the ways that life insurance, annuities, and income and asset protection contracts provide solutions to financial, retirement, and estate planning needs.

Property and casualty insurance (P & C) specialists understand the ways that auto, homeowners, and commercial insurance products protect a client's property from specified perils, and how liability insurance protects assets from certain types of lawsuits.

If you work and are licensed exclusively in either the life or P & C insurance fields, you may be able to cultivate a potentially lucrative exchange of leads with an insurance specialist who does not directly compete with you. Thus, you may be able to provide each other's clients with expertise in complementary fields of insurance specialty.

Trust Officers/Bankers. The trust officer may be the person to whom the client initially turned for information and estate planning services if professional management is desired in the administration of an inter vivos (living) or testamentary trust. Trust officers can help your clients execute their estate planning objectives, and you can help them provide funding for the trust instruments they administer.

The banker often represents the link between a lender and a borrower. Where there is debt, there is a need for financial products to protect the purchased assets for which the debt is incurred. There may be cases when the bank has no investment or insurance representative. This can give you, the advisor, the opportunity to receive referrals for your services. Thus, bankers can be great allies for an exchange of prospects.

Investment Advisors. The investment advisor is a registered representative who earns fees for providing clients with advice regarding investments. These advisors usually provide asset-allocation strategies and may actively manage securities portfolios for their clients. As such, many of them have no interest in or knowledge of insurance products, which leaves the potential for you to provide their clients with these asset-protecting financial products.

Fee-only Financial Planners. Fee-only financial planners typically know more about their clients' net worth, cash flow, dreams, and financial goals than any other member of the financial planning team. Their method of working with clients is usually to evaluate the client's current financial plan, uncover goals and needs, and recommend solutions for the client to implement on his or her own. This can represent a tremendous opportunity for you to offer insurance and investment products to the planner's clients.

Summary

Plan to sit down with other professionals and discuss your businesses, your services, and the value you can bring to their clients. It's never too late to build mutually beneficial strategic alliances with fellow professionals.

> It's never too late to build mutually beneficial strategic alliances with fellow professionals.

YOUR PERSONAL BOARD OF ADVISORS

This section discusses the importance of establishing your personal board of advisors and being accountable to its members for your success. As with strategic alliances, this also relies on a team effort. In addition to you, the board includes an elite group of your most important centers of influence (COIs).

Companies, universities, and nonprofit organizations have boards of advisors, directors, or trustees. It's required by law. Why? The answer is simple: accountability. After all, leaders in positions of power cannot have absolute authority. They must answer to someone—especially in situations where the leaders are responsible for managing assets that belong to others (such as publicly held companies and nonprofit organizations). There is a process of checks and balances in effective boards; this process creates the accountability that guides senior management's actions.

> **personal board of advisors**

This is the value in establishing a *personal board of advisors*: three to five (or more) centers of influence who share your vision and want to see you succeed. Your board should provide you with advice, counseling, encouragement, and the drive to follow through on your commitments.

To begin, ask a group of your most loyal, trustworthy, and influential clients to serve as your board of advisors. Why clients? Consider this: Who better understands the importance of the services you offer? Board members may also be suppliers or work associates. They may or may not have a professional relationship with you. One thing to remember is that busy, prominent people are typically invited by many organizations to join their boards, so being asked by you to serve as a board member will be nothing new to them.

> Why should clients be on your personal board of advisors? Consider this: Who better understands the importance of the services you offer?

Share your mission statement and business plan with your board of advisors. Reinforce with them your personal values and business ethics. Describe your vision of the future from your perspective as an increasingly successful financial advisor. Elicit their commitment to meet with you as a group once or twice a year, if possible.

Establishing a personal board of advisors is not a new concept. In his book *Love the Work You're With,* author Richard Whiteley identifies six key "jobs" that, together, comprise an effective board of advisors:[2]

1. *mentor.* We have already discussed the significant impact that mentoring can have on your career. In many ways, your board is a group of mentors; however, some members may take this role more seriously than others. Your mentor should care about your success and have ownership in your future development.

2. *strategist.* What makes a good strategist? It's a person who can anticipate future challenges and opportunities and then alert you to them. It's a person who can teach you to do the same—to look beyond the next step to the next five or six steps and evaluate all the possible outcomes. The strategist helps you to prepare for business contingencies that may impede your success.

3. *problem solver.* This individual is more focused on the present, helping you confront immediate challenges so that you can move forward to achieve your goals. He or she provides you with useful ideas to overcome the current stumbling blocks to your success.

4. *coach.* Everyone needs a coach from time to time. Your coach will be someone who helps you stay focused on what you want to achieve and develop a strategy to achieve it. A coach is immensely valuable in situations when you begin to feel as if the world is bringing you down.

5. *motivator.* This is the individual who can inspire you to challenge yourself in ways you might never have imagined. It is also the person you do not want to irritate—if you let the motivator down, you will either have to listen to a lecture about staying the course or, worse, he or she will sever ties with you.

6. *cheerleader.* Your cheerleader reinforces a positive attitude—not only in a "Pollyanna" sense but in ways that will help you see problems as opportunities. He or she will encourage you to look for the positive side of every obstacle you may face in your career.

How to Hold Effective Board Meetings

Serving on your board of advisors should be a rewarding experience for members, not something that they regret. To obtain the maximum benefit from your board meeting, you must provide the maximum input—you must make the meeting a big event.

> Serving on your board of advisors should be a rewarding experience for members, not something they regret.

What does that mean? It's pretty straightforward: Make your meeting, the activities, and the advice your members share special. Your board meeting brings together some of the people you admire and respect the most—such a gathering should be given the attention and consideration it deserves.

If you make the board of advisors meeting a big event, members will recognize its importance. They will realize the magnitude of your work, your intentions, and your inquiries; they will not hesitate to give you guidance; they will refer you to the best prospects.

> **The more seriously you treat board meetings, the more seriously your board members will treat their roles as active participants.**

The more seriously you treat board meetings, the more seriously your board members will regard their roles as active participants. They need to see that you mean business. You must demonstrate that you respect them, their time, and their counsel, that you respect your mission and your business, and that you respect yourself.

At each meeting, share your sales production numbers. The more forthright you are, the better the advice you will receive. Many successful financial advisors have followed this procedure, and in almost every case, board members have acted as true COIs and given the advisors many introductions to prospective clients.

There are three key steps in holding a successful meeting of your personal board of advisors: scheduling, preparation, and administration of the meeting. Let's examine them one at a time.

Scheduling

Give Advance Notice. Schedule your meeting at least 2 to 3 months in advance. Everyone knows how challenging it can be to assemble a group of busy professional people in a single place at a single time. Choose at least three options, and find out which options are the best for *all* your board members (ask them to indicate to you which ones work for them—and emphasize that it is important that they choose more than one, if possible). The more choices you offer, the better your chances of having full attendance.

Choose a Business Setting. Hold your board meeting in a business environment. Meeting rooms are not difficult to obtain, whether you rent a meeting room in a hotel or simply use one in your corporate offices. Remember, this is business. Choose a first-class location that matches the impression you want to create.

Confirm Meeting Date and Time. Once you set the time and place, send out letters to confirm. This doesn't mean e-mail confirmations. It means letters sent through the postal system or a courier service. Personalized letters on your company stationery convey the message that you appreciate board members' time and that you respect them.

Preparation

Preparation begins with scheduling and continues until the board meeting itself.

Prepare Information Packets. Package your talking points into a handout or "facts" packet. A complete information packet should include the following:

> Your statement of objectives should follow the SMART formula: sensible, measurable, attainable, realistic, and time specific.

- *personal brochure.* Describe your services, qualifications and how you meet the needs of your targeted market. This may be a single page, a double-sided page, or an 11″ x 17″ half-fold newsletter format. It should have a clean design and be two- or four-color, if possible.
- *statement of objectives.* Your statement of objectives should follow the SMART formula: sensible, measurable, attainable, realistic, and time specific. Be concise.
- *meeting agenda.* Indicate specifically what results you wish to achieve through the meeting. Is it feedback on a sales and marketing plan? Is it business intelligence about what's happening in the community? The more tangible ideas you offer for feedback, the more effectively you will elicit responses from your board members.
- *one-sheet handout stating time and place of meeting.* Clearly state the time and location of the meeting, and include a map with directions.
- *presentation folder.* Presentation is important. Neatly arrange all of the above items—plus anything else you choose to include—in a crisp, attractive 9″ x 12″ presentation folder.

Deliver Information Packets. If possible, hand-deliver packets to board members. They should know what to expect at the meeting and how long it will last. Deliver the packets 1 to 2 weeks prior to your meeting. This will not guarantee that members will actually review your information—although an attractive presentation will improve your chances of its happening.

Confirm Receipt of the Packet. Call members personally to verify that they received the information package and confirm their plans to attend. It never hurts to make an extra phone call, and many of your members will appreciate the reminder. It also is the best way to ensure the highest attendance possible.

Administering the Meeting

Arrive Early to Set Up the Room. You do not want to arrive at the same time as everyone else. You need to make sure the meeting room is ready and that all the materials you need are on hand. When board members arrive, they should find you there to receive and greet them.

Have the following items available for your meeting:

- extra information packets for members who inevitably leave their packets behind
- pens and notepads

- refreshments that you are confident your members will enjoy and will not interfere with the progress of the meeting (nothing alcoholic)
- thank-you gifts such as a personalized pen or gift certificate to a restaurant (Gifts do not need to be elaborate or expensive, but they should convey a sincere message of thank you.)

Lead the Meeting. Using the meeting agenda in your information packet, outline where you are now, where you've been, and where you want to go. Use your materials to back up your information. Record your meeting if it's okay with your members. Get productive feedback, and stay focused on the agenda.

Schedule Next Meeting. While all board members are in the same room, arrange the next meeting on everyone's calendar. Afterwards, begin the same preparations for your next meeting. Keep the process alive and make it part of both your routine and theirs. If you follow the steps outlined in this section, your results will be an efficiently conducted board of advisors meeting that results in increasing your productivity.

CONTACT MANAGEMENT

As a financial advisor, you have many demands on your time. Among them are planning your schedule, prospecting, making phone calls, conducting selling/planning interviews, writing applications, delivering policies, recommending and implementing financial plans with clients, servicing clients, and continuing your education. How to accomplish everything you need to accomplish is a problem that plagues every busy businessperson.

As a professional, you recognize the need for all these activities. You also know that because you can't increase the amount of time available, you must get more out of the time you have. This calls for good business management; it requires a system that will improve your efficiency as a businessperson and increase the time that you have for prospecting and selling. The solution is a contact management system.

> **Because you can't increase the amount of time available, you must get more out of the time you have.**

Contact Management System

contact management
system (CMS)

A *contact management system (CMS)* is a tool to help you contact your prospects and service your clients at key appropriate times, as well as manage the administrative and financial aspects of your business. It was designed to enable you to run your business effectively and keep you in charge of your prospecting activities.

CMS programs are widely available in computerized database management systems. You can also use a paper version. The important point is not which type of CMS you use but that you understand how the system works and why the process of contact management is critical to your success.

> **A CMS will give you a greater sense of control over your time and business activities.**

This section discusses the aspects of a generic manual (or paper version) CMS. If you master the use of a basic 3″ x 5″ card system, you will better comprehend what a computerized version (discussed later) can do for your business. You can then teach an administrative assistant how to use the system to help you expand your prospecting and marketing efforts. A CMS will give you a greater sense of control over your time and business activities.

Features and Benefits of a CMS

A CMS has a wide range of important features to fit various financial services practices. A good CMS should be simple to operate and easy to keep up-to-date and in good working order. It is designed to avoid unnecessary duplication of information. It should also be flexible so that data can be easily added or removed. It must contain all essential information about your prospects and clients.

The benefit to you is that a CMS is a complete and integrated system, devised to improve your work habits and your time control. With a CMS, you should be able to

- keep track of callbacks
- monitor future appointment dates and times
- control service work in progress
- centralize all client and prospect information

Most important, you will have a system that works for you, automatically bringing up sales and service opportunities such as annual financial plan reviews, insurance age changes, term insurance conversions and insurance policy purchase option dates, and mutual fund asset allocations changes. You simply make the proper notation, enter it into the system, and the system will remind you to take the marketing or service actions at the right time. With an efficient contact management system, you will be able to spend less time on detail work and more time for face-to-face contacts with prospects and clients.

Setting up a CMS

A manual CMS is usually kept in a long rectangular file box (metal or sturdy cardboard) that is large enough to hold records for hundreds of prospects and clients. Basically, it is divided into two sections:

prospect file

- *prospect file*—cards containing information about prospects, clients (often your best prospects for repeat sales), and centers of influence, filed alphabetically

tickler file

- *tickler file*—monthly reminders of timely actions to take regarding sales and service opportunities for prospects and clients

(Note: Usually, a CMS also has a separate client file with information about owners of financial products of the company you represent. However, because these client files are company specific, the discussion in this text will be confined to the prospect and tickler files of the typical CMS.)

Managing Contact Information

The prospect file and tickler file are interrelated. Both files need to managed and updated continually.

Prospect File. Recall from chapter 1 that a qualified prospect is someone who meets the following four criteria. He or she

- needs and values your product and/or service
- can afford it
- is suitable for it (if applicable to the product or service)
- can be approached by you on a favorable basis

Until you have had the opportunity to meet with a prospect and perhaps complete a fact finder, however, you will usually not be able to determine whether or not he or she is actually a qualified prospect.

A prospect, on the other hand, is someone who meets at least some of the four criteria above. For instance, a person who is interested in your services and has the potential ability to afford financial products is considered a prospect. Thus, you should complete a prospect card about this individual—as you should for any individual about whom you have sufficient information to classify as a prospect—and keep it in your alphabetical prospect file.

prospect card   Manually or electronically, you will want to develop file records on a computer or on an individual 3″ x 5″ *prospect card* that tracks the same basic elements for each prospect. This qualifying information should include the following information:

- prospect's name
- home, work, and cell phone numbers
- home and e-mail addresses
- date of birth
- family status
- occupation and employer
- annual income
- investable assets
- possible needs the prospect has
- target market or market segment (if applicable)
- source (nominator's name, vendor list, public record, and so on)

- prospecting method (referral, center of influence, seminar, and so forth)
- record of contacts (information about your initial and previous contacts with the prospect)

Also note any additional information that may be of possible interest to the prospect that you could use in your approach and initial conversation with him or her. Indicate on the back of each prospect card the date you want to call that person with a note of the callback date on a tickler card (discussed below). Then return the prospect card to the alphabetical prospect file so you will be able to retrieve the prospect's information at the moment you call him or her. It is helpful to keep records of all dates you speak to the prospect and to summarize those conversations so that you can bring yourself and the prospect up-to-date quickly regarding the progress of your relationship.

Tickler File. The tickler file component of the CMS is designed to direct to your attention the names of particular prospects and clients at their most "sales likely" time(s). These prospects' and clients' names are not organized alphabetically but according to monthly index tabs to "tickle" your memory when you need make contact. Over time, this file will become your most important tool for gaining new, and keeping old, clients. Notations in the file will bring the prospect's or client's name to your attention at a preselected date—you can simply forget about it because your tickler file will remind you. It will enable you to devote time to other important sales and prospecting responsibilities, confident that you will remember to contact a client, send a birthday card, conduct an annual review, and so on.

> **Over time, your tickler file will become your most important tool for gaining new, and keeping old, clients.**

CMS Process at Work

The following demonstrates how you can use the prospect and tickler files in a CMS to manage contacts with a prospect:

- Complete a prospect card when you obtain the name of a prospect (along with sufficient qualifying data) and have enough information to contact the prospect. You may choose not to file this card in the CMS until you have actually made contact with the prospect.
- When you contact the prospect, one of three things will usually happen. You will
  - make an appointment to determine interest, complete a fact finder, and (possibly) close a sale
  - not make an appointment but will obtain permission to place the prospect's name into your prospect file or on your mailing list for future contact
  - not make an appointment and will decide that this person is not a prospect for you

If you decide he or she is *not* a prospect for you, discard the prospect card immediately.

- When you gather facts, you will determine if this person is a prospect for you.
    - If the person is a prospect, prepare a file folder with his or her name and date of birth on the tab; keep the completed fact finder and copies of all correspondence to or from the prospect in this folder.
    - If the person is not a current prospect but you believe he or she will become a prospect within a reasonable time period, add the name to your tickler file and mailing list.

        Again, if you decide the person is not a workable prospect for you, discard the prospect card immediately.

- When you complete a financial product application on the prospect (whether taken during the fact-finding interview or later):
    - Place a copy of any sales illustrations used (if applicable) in the prospect's file folder along with a copy of the application.
    - In the written business section of your monthly business planner (monthly business planners are discussed later in this chapter), write the name of the applicant and the date of application.

- When the financial product is issued, delivered (if applicable), and paid for, clip the upper left corner of the prospect card (which indicates to you that the prospect is now a client), and file the prospect card in the alphabetical file. Current clients represent cross-selling opportunities for the future purchase of additional financial products.

tickler card

- Create a *tickler card* for each day of the calendar year. These cards should be a different color than prospect cards. They should allow room for a prospect's name and the reason for contacting him or her on a specific date in the next year. File these cards for all prospect and client birthdays, anniversaries, insurance age changes, annual reviews, contract option dates, term insurance conversion dates, and any other call-back date that is important to the contact or you. Whenever you add a prospect to your tickler file who has the same reminder date as another prospect, include both names on the same tickler card.

Clients and Prospects. Toward the end of each month, pull all of the tickler cards for the upcoming month; record the birthdays, insurance age changes/annual reviews, and so forth of the prospects and clients for that month in your business planner. Don't evaluate the merits of calling these individuals at this time; if the name is in your tickler file, put it in your planner. After you have recorded all the names in your planner, refile your tickler cards in the correct month.

> **Contact every person in your tickler file each month for the specific reason you included him or her in the file in the first place. Do not postpone the calls for any reason.**

Contact every person in your tickler file each respective month for the specific reason you included him or her in the file in the first place. Do not postpone calls for any reason. If you call only the "best" prospects on your list, you may eventually be left with only your "worst" prospects to call on. Here are some tips for making optimal use of this contact opportunity:

- Call people who have insurance age changes or who require annual reviews well in advance of these events to allow another opportunity for a callback if cannot reach or see them when you first call. Many successful CMS users begin their calls to age-change prospects and clients at least 1 month before the insurance age change or annual review.
- Remember to call your client or prospect to wish him or her a happy birthday or anniversary. Remember, too, that a call in advance of the event may be a timely reminder of a spouse's birthday or a welcome prompt for the client or prospect to remember the couple's wedding anniversary.
- You can also use a birthday or anniversary as an opportunity to buy your prospect or client lunch, breakfast, or coffee. You can then take advantage of the meeting for some referred lead prospecting or other marketing effort.
- Sending cards for special occasions and notes of congratulations to your client's or prospect's family is also an excellent way to keep your name in front of clients and prospects several times each year on a favorable basis.

center of influence contact record

Centers of Influence. You should create a *center of influence contact record,* and file these names alphabetically on cards that are a different color than the prospect or tickler cards. Contact centers of influence regularly—not just when you have run out of client and prospect names to call. Here are a few pointers for contacting your COIs:

- Send your COIs birthday and anniversary cards, just as you would clients and prospects.
- Include COIs on your mailing list, and send them copies of your newsletter to keep your name in front of them on a regular basis.
- Send them compliance-approved information in which they might be interested, along with a short note.
- Call them regularly just to say hello and keep in touch.
- Treat COIs as if they are your most important clients.

Importance of Keeping Records

Keeping meticulous records—whether they refer to clients, prospects, or COIs—allows you to refer to prior conversations; track and insert reminders about upcoming annual reviews, a client's anniversary, a spouse's birthday, and so forth; and account for results of contacts with previous referrals.

> When you consider just how hard you worked to acquire your clients, you certainly don't want to lose even one of them.

Keeping complete and accurate records of all contacts with your clients will enable you to properly service those clients, not just maintain their existing financial plans or products. When you consider just how hard you worked to acquire your clients, you certainly don't want to lose even one of them.

Keeping complete and accurate records of all your COI contact records, including all the referrals the COI gave you, will enable you to report results months or even years after receiving the referral. Without feedback, a COI may not continue to suggest names of people for you to contact.

Automated Database Management System

Next, we will briefly discuss a typical computerized content management system—the automated contact management system (ADMS).

An ADMS is much more than an electronic Rolodex—it is a completely integrated contact and office management tool. Combining insurance/financial product information, activity scheduling, reporting, and word processing functions into one easy-to-use program, an ADMS eliminates the need to purchase and learn multiple programs to run your office.

With a tightly integrated structure and intuitive interface, the ADMS was designed with the end user in mind. An ADMS usually includes at least an electronic database filing system, report generator, word processor, and activity manager.

ADMS Benefits

Because the ADMS automatically shares and integrates contact, business, and insurance/financial product information with other parts of the program, you can set up your prospect and tickler files records with a number of key fields for numerous sorting and retrieving possibilities. The following are a few examples of tasks you can perform to take advantage of these features:

- Create a list based on database criteria you define, and send personalized information in mass mailings.
- Schedule an activity with a contact or business in the calendar function or in a list of activities associated with that contact or business.
- Track time you spent working with a contact or business in a time log.
- Automatically put entries concerning a contact, group of contacts, or business into an electronic notebook that you can view, create, or change.
- Generate reports based on data in fields, and records that you select in a database such as birthdays, income, financial product reports, and so on.

When using any automated database system, whether it is on a desktop or notebook computer, or a personal digital assistant (PDA), it is essential to protect your data. This means regularly backing up your contact information on a remote storage facility, CD, or DVD. It also means using a password to prevent unwanted access to your records if your computer or PDA is lost or stolen.

Summary

Any system, no matter how well it is thought out and set up, is only as good as the day-to-day attention you pay to its upkeep. If the system is current, it will provide you with current information. Therefore, spend a few minutes each day on the initial filing and any necessary changes that will ensure that your system is always up-to-date.

> If your total number of prospects declines or if the number of new prospects does not meet your goals, you must increase your prospecting activities.

It is your responsibility to make your CMS a system that works for you to improve your effectiveness and increases your available selling/planning time. Your prospects are your "inventory." It is critical that you maintain accurate records of the names of all new prospects you obtain and all nonprospects you discard so that each month you will have an accurate count in your inventory. If your total number of prospects declines or if the number of new prospects in a month does not meet your goals, it indicates that you must increase your prospecting activities.

BUSINESS PLANNING AND ORGANIZATION CONCEPTS

You are the entrepreneur of your own business. And as with any business, to succeed, you need to manage your current activities, evaluate your past activities, and plan for your future activities. To achieve these objectives, you need the proper tools, the necessary planning, and the organizational skills.

In this section we will discuss business planning concepts, functions of business management, and methods to achieve a successful financial services business enterprise at the strategic, tactical, and operational levels. (See figure 8-1.)

Business Planning Levels

Major broad-based business ideology and concepts evolve at the strategic planning level—for example, formulating a business plan and composing a mission statement to define the philosophical purpose for being in business. It is at this level that goals are set for at least 1 year with specific plans for achieving them.

At the tactical planning level, there is more emphasis on organizing resources and business practices within a shorter time frame—for example, quarterly and/or monthly. Progress toward achieving annual planning goals is

FIGURE 8-1
Business Planning and Management

Business Planning Levels	Functions of Management	Methods for Achievement
Strategic	Planning Creativity/conceptual Annual planning	Goal setting Mission/vision statement Business plan
Tactical	Organizing Resources, budgeting Monthly planning	Scheduling Monitoring activity Evaluating progress
Operational	Controlling Regulating activity Weekly planning Daily planning	Implementation Record keeping Performance reviews

evaluated and adjustments made to compensate for potential shortfalls. A great deal of scheduling takes place, and activities are carefully monitored to determine their effectiveness.

The operational planning level focuses on controlling and regulating activity on a short-term basis such as daily and weekly. This is the level at which plans are put into action. There is daily record keeping, and performance is reviewed weekly to meet broader activity and income objectives.

The remainder of this section will examine two basic tools—annual and monthly business planners—used in the business planning processes. It will then take a detailed look at how advisors can effectively approach the annual, monthly, weekly, and daily planning processes.

Annual Planning

Annual Business Planner

annual business planner

In addition to the CMS, another tool essential to goal setting and planning is an *annual business planner*. It measures your work activity and marketing efforts from the previous year and uses those results to determine income, activity, and career achievement objectives for the coming year. An annual business planner is usually an 8″ by 11″ booklet that has separate pages for calculating and recording the goals and business information described below.

Income Objectives. Renewal and residual income are based on previous years' activities and are relatively fixed. Therefore, to set your total annual income objective, it is necessary to determine your variable income, which is next year's first-year commissions (FYC) goal. On this page in the planner, you should record a detailed breakdown by category of your estimated expenses and savings goals to help you determine your income objectives for next year.

Activity Objectives. First-year commissions are a result of certain necessary activities through which you produce the business to earn them. These activity requirements are based on your previous year's performance. As discussed in chapter 6, by knowing the average value of each sale you make and the prospecting and sales effectiveness ratios you need to obtain a sale, you can apply this information to your next year's desired income goal. This will enable you to determine the year's activity objectives for closing interviews, fact-finding interviews, initial interviews, number of qualified prospects, appointments seen, and phone contacts required to achieve your goal. On this page you should list the quantity you need in each category of required activity for the year to meet your income goal.

> **A review of last year's objectives will help you pinpoint what achievements you should aim for in the coming year.**

Career Achievement Objectives. This page focuses on specific self-improvement objectives you consider important—for example, industry and company awards you want to earn over the year and/or courses of study you intend to complete. A review of last year's objectives will help you pinpoint what achievements you should aim for in the coming year. A written commitment will help you stay on track.

Written Case Log. This section holds detailed records of written and paid business. You should include demographic and marketing information about a prospect, including age, marital status, occupation and industry, annual income, prospecting source, and the needs addressed by the financial plan/product sold. A careful review of these items will furnish you with pertinent facts to help you analyze the quality of your markets as well as assist you in recognizing the source of your most profitable prospecting sources.

Activity Summary. By transferring the information compiled monthly from your monthly business planner to the pages in this section, you will have an at-a-glance summary of the categories of your prospecting and sales activities. These numbers provide you with the raw data necessary to determine your strengths and weaknesses and to establish the effectiveness ratios used in your year-end activity evaluation.

Year-end Activity Evaluation. This section is where you calculate your prospecting and sales effectiveness ratios. These ratios help you to determine how much of each specific activity you will need to achieve your goals, and they allow you to place a dollar value on each type of activity you perform in the process of making a sale. The evaluations provide a basis for comparing this year with last year to observe your sales effectiveness trends and project your strengths and opportunities for improvement for next year.

Business Expenses. As a businessperson, you will want to record information in this section that will tell you what business expenses are producing results that will help you establish next year's business budget requirements.

Agenda for Annual Planning

annual planning

Annual planning is a strategic business function that advisors should perform over several days at the end of each year and the beginning of the next year. It entails determining income, career achievement, and activity goals for the coming year, using the evaluation of the previous year's activity as a foundation, to establish a working business plan for the new year. You should conduct the planning session in your office privately and without any interruptions, distractions, or phone calls. The discussion below examines the steps involved in this important function.

Complete Your Activity Summary. First, you must tabulate data and total all of your actual numbers for the previous year in the following areas: phone (or face) contacts, appointments seen, number of qualified prospects obtained, fact finders completed, closing interviews conducted, and sales made. (See the top half of section 1 in appendix B.)

Evaluate Year-end Activity. Using the totals from the activity summary section, you must calculate your prospecting and sales effectiveness ratios for the previous year, including phone (or face) contacts to appointments seen, appointments seen to qualified prospects, qualified prospects to fact-finding interviews, fact-finding to closing interviews, closing interviews to sales, and FYCs to sales (to determine the average FYC per case). (See the bottom half of section 1 in appendix B.)

Determine Income Objectives. Next, you must determine your FYC income objective for the year ahead. (See appendix B, section 2.) Begin by estimating next year's personal expenses, business expenses, and savings goals. This involves budgeting amounts for each category of cash outflow you expect and for the savings you want for the upcoming year. Next, determine the new year's income objective to meet your cash outflow and savings goals. Start with the gross income you desire; then subtract anticipated renewal commissions and other business income to arrive at next year's net FYC objective.

Calculate Activity Objectives. Now you can determine the new year's prospecting and sales activity objectives you need to achieve your FYC income goal. (See appendix B, section 3.) This involves calculating next year's required numbers of annual, monthly, and weekly activities. The process begins by *dividing* next year's FYC income goal by the average FYC per sale, to determine the number of sales you need to meet your goal. For example, if your goal is $100,000 in FYCs and each sale has an average value of $1,000, then you need 100 sales to meet your goal.

Next, using last year's prospecting and sales effectiveness ratios and working backwards, start by *multiplying* the number of sales you need by the number of closing interviews it takes to produce one sale. Continue the same process to calculate the total number of fact finders completed, qualified prospects, appointments seen, and phone/face contacts you need to achieve each respective level of prospecting and sales activity that will result in the number of sales required to meet your FYC goal. For example, it may take 3,000 phone calls to produce 750 appointments, to obtain 500 qualified prospects, to complete 400 fact finders, to lead to 200 closes and 100 sales.

Establish Career Achievement Objectives. Establish qualitative career achievement objectives for the new year in the following areas: company award and incentive programs (for example, sales conference meetings), industry awards (for example, MDRT), and self-development activities (company courses, industry courses, continuing education, and so forth).

Monthly Planning

Monthly Business Planner

monthly business planner

A *monthly business planner*, another important tool to complement the CMS, is a pocket planner that advisors use to help keep track of weekly and daily business activities.

The monthly business planner typically includes the following sections:

- monthly activity and production goals
- current and next month's at-a-glance pages
- space to record new sales completed and paid cases
- prospects' and clients' birthdays, insurance age changes, and annual review activity
- lists of prospects to call for initial, fact-finding, and closing interviews
- weekly planning pages
- daily hour-by-hour appointment calendar
- daily and weekly activity summary reports
- space to record business expenses

Agenda for Monthly Planning

monthly planning

> **The purpose of the monthly planning session is to evaluate your progress toward your annual goals and organize your business activities for the month ahead.**

Just as annual planning is critical, so is *monthly planning* an important tactical business function. It occurs at the end of the current month, preferably on the last Friday or Saturday of each month. As with the annual planning session, you should conduct the monthly session with no distractions or interruptions. It is during this time that you will establish your work plan for the coming month. The purpose is to evaluate your progress toward your annual goals and organize your business activities for the month ahead.

You will need the materials you use in your CMS, along with your monthly and annual business planners. The discussion below outlines the monthly planning process.

Review the Current Month. You should take the following four steps to review the current month:

- *Evaluate your goals*. Assess the status of your annual activity and FYC goals for the year-to-date. Factor in sales you made, paid cases, FYCs and renewal income you earned, and any financing payments received.
- *Update activity summary*. Update your log of written and paid business activity in your annual planner. It is essential to keep these records up-to-date so you can evaluate your actual progress against your plans.
- *List prospects you didn't contact*. Take an inventory of prospects you tried to contact this month but were unable to reach. Consider plans to follow up with them next month.
- *List new prospects you obtained*. Take an inventory of new prospects you acquired this month so you can make plans to contact them next month.
- *Review your prospect prioritizer*. Use these forms to evaluate which prospects are the ones it is most important to contact first next month.

Define Work Activities. Attend to all the items that will contribute to prospecting and servicing actions you must schedule in next month's business planner. This includes (but is not limited to) compiling lists of people to whom to send birthday cards and direct mail letters, for example, and lists of individuals to contact regarding insurance age changes and annual reviews, for instance. It also includes reviewing your client file for sales and service opportunities, examining next month's scheduled activities, assessing your center of influence file for possible contact for referrals, studying your tickler file for prospect follow-up, and preparing next month's newsletter list.

Plan Next Month. Plan the following scheduling, prospecting, and organizational activities for next month:

- *Prepare monthly schedule.* Schedule next month's daily planning. Delineate work time, fixed commitments (for example, continuing education classes), telephoning time, canvassing time, study time, and social activities.
- *Prepare next month's business planner.* List your 25 best prospects for initial interviews, fact-finding interviews, and closing interviews. Fill in written and paid business goals for the month. Set up your month-at-a-glance calendar(s).
- *Prepare for telephone calls.* Assemble lists of prospects and clients whom you wish to phone in the next month. (See the section titled "Preparation" in chapter 5 for some tips on how to prepare for phone calls.) Plan to call clients to whom insurance policy deliveries need to be scheduled.

Weekly Planning

Agenda for Weekly Planning

The weekly planning session is also an important business function at the operational planning or implementation level. It occurs at the end of the current week, preferably on Friday afternoon or Saturday. It, too, should be done in your office privately and without any interruptions. The purpose is to establish your work plan for the coming week.

When beginning weekly planning, you should already have a skeleton plan carried over from your previous week's activities, including appointments with people you need to see at certain times or places, and any classroom or study activities you have planned. Your primary goal for a full day's activities is to obtain at least two initial or fact-finding interviews and one new firm commitment for a scheduled appointment for a closing interview.

Materials to Have Available When Planning Your Week. To plan your week, you will need the following materials used in your CMS, along with your monthly business planner:

- all prospect cards for people you had planned to see the previous week and missed contacting
- lists of prospects to whom you sent preapproach letters whom you plan to contact face-to-face or by telephone for follow-up this week
- all referred leads you haven't already called on
- names obtained from publications, newspapers, lists, personal observation, an so on

- names of all clients who have upcoming annual reviews or other events for which you intend to make personal calls
- names from your tickler file for follow-up this month or week
- insurance policies that need delivery
- pending applications and service work to check on this week

Steps in Weekly Planning. The following is a brief summary of the weekly planning process:

- *prospecting activities.* Separate all prospect cards into two categories—names of people to contact personally and those to telephone. Organize personal calls into zones (East, West, North, South, downtown, a particular industry, and so forth). Collect all your prospect cards for telephone contacts, evaluate each one, and divide them according to what you think would be the best day and time of day to call.
- *work and organization.* Complete your direct mail lists—preapproach letters with reply cards for follow-up the next week. Make sure that your study assignments and projects for the previous week are complete, and plan for study time for the upcoming week. Prepare all insurance age change and birthday cards for mailing. Check that your tickler file and CMS are organized and current. Set (review) your personal goals for the week and consider how you'll feel when you reach them.

Things to Remember. During the planning process, always put your "best" prospects in your plan first. (But remember not to neglect your other prospects or you might have no prospects at all.) Also, remember to call each prospect to remind him or her of the time of your appointment. Finally, don't waste your time calling the same people over and over if they aren't interested.

Daily Planning

Another important business function at the operational planning or implementation level is daily planning. Daily planning is a microcosm of weekly planning. That is, essentially, it is planning the same prospecting activities; only now you need to execute the plans as well.

Chapter 6 discussed the organization of time into offensive (green zone) and defensive (red zone) activities and suggested percentages of time that you should devote daily to each type of activity. You should plan and execute each activity carefully every day to manage your time most effectively. Because chapter 6 discussed daily planning techniques in detail, it is not necessary to repeat them here. You may, however, want to review them as you study this chapter. You may also want to take a look at the box titled "An Organized Daily Schedule" for some pointers on how to improve your overall marketing efforts.

An Organized Daily Schedule

- Start work at a set time each morning.
- Have at least a full 8-hour work schedule each day.
- Have a minimum of two definite appointments for interviews.
- Call on at least one person each day for the specific purpose of securing some new prospects.
- Try to arrange a luncheon appointment each day with a prospect or a center of influence.
- During business hours, do only the activities related to the selling/planning process.
- Do service work at odd hours; save the best hours for the selling/planning process.
- Devote some part of your day to case preparation.
- Know the recommendations that you are going to present to each prospect and know what you are going to say.
- Have a definite time for reviewing the day's work and for planning the next day's work.
- Keep a record of what has happened each day.
- Include 1/2 to 1 hour for reading or studying.
- Remember that time is your working capital—that every hour, every minute, has a definite value.

COMPLIANCE, ETHICS, AND PROFESSIONALISM

Your career carries with it a tremendous responsibility. As a financial advisor, you approach both friends and strangers, ask them for the opportunity to help them plan for their future financial security, and then ask them to accept your advice and trust your recommendations. In doing so, you have an absolute obligation to maintain the highest ethical standards of professional behavior.

You assume responsibility for helping prospects and clients meet their financial needs. It is this responsibility, combined with your training, specialized knowledge in areas of financial planning that are often difficult for prospects and clients to understand, and the promise of continued service that elevates your selling/planning activities to the level of career professional.

> **Ethical behavior in all business dealings is the responsibility of every professional advisor. What must you do to live up to this responsibility?**

In today's financial services marketplace, there is a tremendous emphasis on ethical business practices and legal compliance issues. Ethical behavior in all business dealings is the responsibility of every professional advisor. What must you do to live up to this responsibility?

Certainly, an advisor must comply with

- state regulations for the sale of all insurance products
- federal regulations for the sale of securities and registered products
- company rules and procedures for all marketing activities
- codes of ethical conduct
- principles of professionalism

In fulfilling your responsibility to your prospects and clients and conducting yourself as a professional advisor, you must comply with all applicable laws and regulations. State and federal laws regulate the insurance and securities industries to protect consumers from unfair sales practices.

These regulations will be discussed first, followed by company rules and procedures, codes of ethical conduct, and professionalism.

State Regulation

Traditionally, insurance companies and insurance policies have been regulated primarily at the state, not the federal, level. Each of the states has established its own department of insurance to regulate the insurance activities conducted within the particular state. In addition to the state insurance department, the legislative bodies of each state set policy for the regulation of insurance. Each state legislature passes laws to guide the insurance company's activities and products.

The state regulation of insurance companies has several key functions:

- insurance company licensing
- producer licensing
- product regulation
- market conduct
- financial regulation
- consumer services

Making sure their insurance companies stay solvent is of primary importance to state regulators.

State insurance departments regulate the financial aspects of insurance companies. Therefore, states must be concerned with their insurers' financial solvency. They want assurance that the insurance companies have enough money invested to cover their policyowners. Making sure their insurance companies stay solvent to pay claims is of primary importance to state regulators.

There is also state regulation of market conduct. State insurance departments supervise the sales and marketing practices of both insurance companies and advisors. The goal is to make sure consumers are treated fairly. In fact, each state insurance department's number one priority is the protection of its insurance consumers.

Licensing

Insurance Companies. To be able to sell insurance products in a particular state, the insurance company that offers the products must first be licensed by the state department of insurance. The insurance company must apply for a license and meet the specific requirements of the particular state.

The insurance company must file all the products it sells with the state insurance department. Some states require a mere informational filing, depending on the type of product. Other states require that the insurance company receive an official stamp of approval by the insurance department before the insurer can offer its products for sale.

Advisors. The advisor who is selling insurance products must also be licensed in the particular state. Typically, the basic life/health insurance license allows its holder to sell most life insurance, health insurance, and fixed-annuity products. Usually, the state requires that the advisor be licensed in the state where the application is written. This may or may not be the policyowner's state of residence.

Advisors may be licensed in more than one state to sell insurance products. Generally, an advisor will carry a resident license in his or her home state and have additional nonresident licenses in other states.

The advisor must also obtain an appointment from one or more companies. This appointment is what allows the advisor to sell a particular insurance company's products. During the appointment process, the insurance company will verify the existence of the advisor's insurance license and will most likely perform both financial and criminal background checks. If an advisor sells for several insurance companies, he or she will have several appointments.

Variable life insurance, variable annuities, and other investment or equity-based products, however, are generally not covered by the basic life/health insurance license. These products require special licenses and registration with the National Association of Securities Dealers (NASD). The NASD is a self-regulatory organization. It was established under authority granted by the Securities Exchange Act of 1934 to provide voluntary self-regulation of broker/dealers under the oversight of the Securities and Exchange Commission (SEC). Depending on the state regulations, separate state licenses may also be necessary.

Although state laws that regulate insurance company and advisor licensing vary, the state commissioners work together through the National Association of Insurance Commissioners to identify the issues of greatest importance to consumers and to set legislative standards through model legislation. Because of their work, there is general agreement on some practices in insurance sales.

Your Financial Services Practice: Unauthorized Entities

Regulation of insurance products and services varies from state to state. In Florida, for example, regulations prohibit doing business with an unauthorized insurance entity. An unauthorized entity is an insurance company that has not gained approval to place insurance in the jurisdiction where it or a producer wants to sell insurance. These carriers are unlicensed and prohibited from doing business in that state. In most cases where these carriers have operated, they have characterized themselves as one of several types that are exempt from state regulation. It is the financial advisor's responsibility to exercise due diligence to make sure the carriers for whom they are selling are approved by the department of insurance in that state.

Advice

States are also concerned with what advisors call themselves and the kinds of advice they give their prospects and clients. Advisors who call themselves financial planners or financial consultants may be breaking state laws unless they have obtained special licenses. In many states, financial advisors, planners, and consultants are considered separate professional groups, requiring specific licenses.

Even in routine contacts, an insurance or financial advisor's discussions with a prospect or client may touch on legal or tax matters. Giving a client or prospect specific legal or tax advice can be construed as practicing law without a proper license, and that is illegal. Although discussing legal or tax matters with clients in very general terms is allowed, you cannot give specific advice in these areas. When specific advice is requested, recommend that your prospects and clients consult an attorney or tax advisor.

Federal Regulation

Although the states are the primary regulators of the insurance industry, the main responsibility for regulation of securities products rests clearly with the federal government. Securities products include the variable life and variable annuity products many insurance companies now offer. Without the proper registration, an advisor is not allowed to discuss equity-based investment products with a prospective buyer.

Selling Securities Products

To sell securities products, you must be properly registered with the National Association of Securities Dealers. The NASD has the power to require and monitor compliance with standardized rules of fair practice for the industry. NASD regulatory responsibilities include registration and testing of securities professionals, review of members' advertising and sales literature, and services such as arbitration of investor disputes. Registered representatives must provide the NASD with personal information, including prior employment and any history of securities-related disciplinary action.

An NASD registration will not be issued until you are affiliated with a broker/dealer. Most large life insurance companies have broker/dealer subsidiaries and thus make sponsorship and broker/dealer affiliation easy for their advisors. For independent advisors, there are broker/dealers who stand ready to establish an affiliate relationship.

Marketing Securities Products

The regulations for marketing securities products are extensive. The Securities and Exchange Commission and the broker/dealer must approve all

> Failure to follow any of the rules to market securities can lead to the loss of licensing, fines and penalties, and suspension of a company's right to do business.

advertising materials, including letters that mention the sale or availability of securities products, as well as any sales literature. Even stationery and business cards that the registered representatives use must meet certain standards. Failure to follow any of the rules can lead to the loss of licensing, significant fines and penalties, and a suspension of a company's right to do business.

Before securities products can be sold, the law requires that potential buyers be given information about the product in a form known as a prospectus. As with the other materials associated with the sale of securities, the prospectus must follow certain formats and guidelines.

Registered Investment Advisor

The Investment Advisers Act of 1940 defines an investment advisor as any person who, for compensation, engages in the business of advising others as to the value of securities or the advisability of acquiring or disposing of securities. Most persons who fall within the act's definition of an investment advisor and who make use of the mails or any instrumentality of interstate commerce are required to register with the SEC as a registered investment advisor.

> Merely dealing with a security does not, by itself, make someone an investment advisor.

Merely dealing with a security does not, by itself, make someone an investment advisor. In the 1980s, the SEC issued three tests to determine whether or not individuals must register as an investment advisor. If all three tests are answered in the affirmative, registration is required unless specifically excluded or exempted. If any of the tests is answered in the negative, there is no need to become a registered investment advisor. According to the SEC's three tests, the individual or entity must

- give advice or analysis concerning securities (security advice test)
- be engaged in the business of advising others regarding securities (security business test)
- be in receipt of compensation (compensation test)

It is important to remember that the purpose of the SEC in devising the three tests is to protect clients from fraud and other abusive situations. The SEC does not guarantee the competence or investment abilities of any individuals who register under the act; it merely seeks to provide a mechanism for discouraging unethical behavior, principally via full disclosure to clients.

Although the SEC does not require investment advisors who seek registration to successfully complete a qualifying exam, the individual advisor and advisory firm must follow strict rules for registration, record keeping, and compliance. Many states do require successful completion of a qualifying exam before registration as an investment advisor.

Financial Planners

There are no special federal licensing or registration requirements for financial planners. The services of most financial planners generally include analyzing their clients' financial situations, setting achievable financial goals, and developing and implementing sound financial plans. The development of a plan may include recommendations of securities to be purchased or sold, and there may be a fee charged for the plan. Generally, if the financial planner meets the three tests cited above, the individuals involved with developing the plan must be registered as registered investment advisors with the SEC.

Many states have implemented legislation for persons calling themselves financial planners or those charging a fee for financial advice. The legislation requires licensing of those who hold themselves out as financial planners. Other states are considering legislation. Contact your State Division of Securities for information on your state's licensing requirements.

> **You will need to know if your state allows you to charge a fee for planning and to collect commissions for products sold on the same case. This is illegal in some states.**

You will need to know if your state allows you to charge a fee for planning and to collect commissions for products sold on the same case. This is illegal in some states. If you have any questions concerning your requirements, contact your broker/dealer and clarify your position immediately.

Company Rules and Procedures

Financial services company rules are developed to make certain that the company and its advisors meet all state and federal regulations. They are also designed to make sure that the company has complete and accurate information on which to base the issuance of suitable financial products to its customers.

As a financial advisor who sells financial products, you function as an agent of the company you represent. In its simplest terms, an agent represents the company with a limited right to speak for the company he or she represents. What you say and do as an agent may be binding on the company. Clearly, investment and insurance companies must take due care to protect themselves from the possible misbehavior of those representing them.

Company rules also have a value to advisors. They are designed to help advisors make sure that they meet all applicable legal requirements. Keep in mind that most financial products are legal contracts between the owner and the financial services company. If the contract is not carefully and properly executed, it could lead to serious legal complications for the owner, the company, and the advisor. A company's rules are designed to ensure that all legal requirements have been met.

Advertising

One area of marketing in which compliance problems can occur is advertising. Financial advisors must make sure that the materials they use to advertise,

Advertising that
contains untrue,
unclear, incomplete,
or deceptive statements
is unethical.

sell their products, and portray themselves are true and clearly comprehensible. Advertising that contains untrue, unclear, incomplete, or deceptive statements is unethical. Furthermore, state and federal laws specifically prohibit such advertising. The application of these laws not only applies to home office material but also to agency-prepared material or material created by the financial advisor.

If you prepare customized materials for your clients, recognize that these materials are subject to regulatory oversight. This is the major reason that insurers require that the company's home office approve in advance all sales materials an advisor uses. It is a simple way for the insurer to make certain that its products and benefits are not being presented in a manner that goes against federal or state regulations. Be sure to comply with the process.

Ethical Considerations

Suitability

Your professional obligation to prospects is to help them determine and carry out the most suitable solutions to meet their financial planning needs. In identifying the need for financial products that address prospects' concerns, helping prospects understand how certain products meet those concerns, and implementing solutions to prospects' financial planning needs, you have fulfilled your professional obligation.

Clients expect their
advisors not only to
be accurate in their
analysis but also to
recommend suitable
products to satisfy
their needs and goals.

Client Needs. Clients expect their professional financial advisors not only to be accurate in their analysis but also to recommend suitable products to satisfy their needs and goals. To accomplish this, advisors must conduct thorough fact finding. They must uncover client needs for insurance and other financial products, as well as clarify the client's personal goals.

Thorough fact finding will also help the advisor determine the client's risk tolerance and time horizon. Risk tolerance has to do with a person's ability to withstand a financial loss. Some clients will accept far more risk in their investment portfolio than others. Time horizon deals with the period of time involved in a financial need or decision. Determining needs, goals, risk tolerance, and time horizon provides the advisor with information necessary to make a product recommendation. It also gives the advisor guidelines regarding the suitability of a particular product for a particular client.

Compliance and Ethics

compliance

Compliance means following the laws and regulations, including company rules that apply to the sale of all financial products. These are the minimum standards.

The conflict that may arise between the prospect's need for an insurance or investment product and the financial services company's underwriting rules or issuing standards spans both compliance and ethics. You may encounter situations when fair business practices, legal requirements, and company rules seem to conflict with your best efforts. How the apparent conflict is resolved is a matter of ethics, and it goes beyond compliance with the law.

Ethical behavior is doing the right thing. It is doing what is right and putting the prospect's best interests before your own. It is maintaining the highest possible standard of behavior in all your business dealings. It is continuing to develop your skills so you can provide the best possible service to those with whom you work. It is representing the industry, its companies, and its advisors, in the best possible light.

Professionalism and ethical conduct demand more than mere compliance with laws and regulations. Nevertheless, following these regulations and rules is the first step in professional conduct.

Professional Code of Conduct and Ethics

professional ethics

Professional ethics can be defined as behaving according to the principles of right and wrong—a code of ethics—that are accepted by the profession. By adopting and practicing a professional code of ethics, you will achieve the high standard of professionalism that a career in financial services demands.

All professional organizations within the financial services industry publish pledges and ethical codes that apply to their members. These codes rest on ethical common sense and seven common themes:

- Every code calls on professionals to look out for the best interests of the client.
- Most codes in one way or another ask professionals to conduct themselves with fairness, objectivity, honesty, and integrity.
- Each code requires professionals to protect the confidential information of their clients.
- Most codes require that professionals present enough information to allow the client to make an informed decision.
- Each code requires professionals to continue the learning process throughout their careers.
- Each code asks professionals to conduct themselves in a way that brings honor to themselves and to their professions.
- Most codes specify that financial services professionals should comply with the law.

Knowledge of the codes will enable you to better deal with the complexities of today's marketplace. The code of ethics of The American College is an example of standards of professional behavior to which you should constantly adhere. (See figure 8-2.)

FIGURE 8-2
The American College Code of Ethics

To underscore the importance of ethical standards for Huebner School designations, the Board of Trustees of The American College adopted a Code of Ethics in 1984. Embodied in the Code is the Professional Pledge and eight Canons.

The Professional Pledge and the Canons

The Pledge to which all Huebner School designees subscribe is as follows:

In all my professional relationships, I pledge myself to the following rule of ethical conduct: I shall, in light of all conditions surrounding those I serve, which I shall make every conscientious effort to ascertain and understand, render that service which, in the same circumstances, I would apply to myself.

The eight Canons are:

I. Conduct yourself at all times with honor and dignity.
II. Avoid practices that would bring dishonor upon your profession or The American College.
III. Publicize your achievement in ways that enhance the integrity of your profession.
IV. Continue your studies throughout your working life so as to maintain a high level of professional competence.
V. Do your utmost to attain a distinguished record of professional service.
VI. Support the established institutions and organizations concerned with the integrity of your profession.
VII. Participate in building your profession by encouraging and providing appropriate assistance to qualified persons pursuing professional studies.
VIII. Comply with all laws and regulations, particularly as they relate to professional and business activities.

Being a Professional

Professionalism

professionalism

To be successful as a financial advisor, you must be a professional. *Professionalism* means having the technical knowledge necessary to provide meaningful support and accurate advice to your prospects and clients. As a competent financial services professional, you must also be fully aware of the legal and tax ramifications of the recommendations you make. In addition, you must be able to outline the positive and negative implications of the various investment and insurance options available so that your prospects and clients can make informed purchasing decisions. You, therefore, must have a thorough understanding of your products, the problems that your prospects and clients face, and the solutions your products can provide.

Client Focus. To be a professional, you also need to be client-centered. This means you have to put your clients' interests before your own. In financial, estate, and retirement planning, your job goes beyond simply

> **Your job goes beyond making a sale. Your responsibility is to help your prospects and clients identify and implement all the steps that will help them accomplish their financial planning goals,**

making a financial product sale. Your professional responsibility is to help your prospects and clients identify and implement all the steps that will help them accomplish their financial planning goals.

Client Confidentiality. Prospects and clients are entitled to a high level of confidentiality regarding the personal information that they give to the advisor. It is expected that every advisor will keep this information private and not share it with anyone other than those involved in the company's risk evaluation process. It is also necessary that prospects and clients be treated fairly, with honesty and respect, for this is the only way to foster a professional relationship that will benefit all the parties involved.

Professional Responsibility

Financial advisors have an obligation to think of themselves and conduct themselves as professionals. The public expects professional advice, and the courts see financial advisors as professionals. Financial advisors advertise themselves as having the special skills provide investment guidance to clients and to protect their property and human life values. This advertising brings a higher legal standard of performance to the advisor's work.

You have an important role to serve your prospects and clients in the best way you can. Part of your professional responsibility is to encourage your prospects and clients to take actions they can reasonably afford that will enhance their financial security and that of their families. To do this, you need to educate them about the consequences of action or inaction so that they can make informed financial planning decisions. A well-informed prospect is likely to make a good long-term client if your products and services provide solutions to his or her planning needs.

Professional Development through Education

To educate your prospects and clients, you must first educate yourself. Not only do you have to understand the basic concepts involved in financial planning, but you must also keep studying in order to learn advanced concepts that are often part of the responsible planning process. You must stay abreast of new product innovations, legislative trends, and tax rulings that can affect your ability to provide the highest possible level of service.

> **By continually educating yourself, you will become a competent member of your clients' financial planning team.**

By continually educating yourself, you will become a competent member of your clients' financial planning teams. Additional knowledge and skill development can result from pursuing the recognized professional designations of the financial services industry. The LUTCF, FSS, CLU, ChFC, and CFP designations, which are all earned in part by the successful completion of a qualified course of study, indicate your ongoing commitment to professionalism.

Membership and participation in the industry's professional organizations, such as NAIFA and the Society of Financial Service Professionals, offer an opportunity for continuing education. You may also want to explore the various training and educational programs provided by insurance companies, universities, proprietary training organizations and other professional organizations. *Advisor Today* (the magazine of NAIFA) and *Life Insurance Selling* magazine are excellent sources of insurance news and sales ideas.

Formal programs can supplement your personal self-improvement regimen of daily and weekly readings in financial literature. For more information on additional training resources aimed at enhancing the skill and knowledge of the dedicated financial services professional, log on to The American College's website at www.theamericancollege.edu.

CONCLUSION

Freedom Has Its Price

One of the reasons advisors are attracted to the financial services business is the freedom to manage their careers the way they choose (under current compliance regulations, of course). This so-called freedom is also one of the major reasons they leave the business; many of them enter the profession without realizing the price they must pay for success. The successful professional advisors in financial services paid that price with hard work and fortitude. To succeed, you must be willing to pay it too.

This text has given you the prospecting tools, strategies, and knowledge that are the accumulation of hundreds of combined years of experience. Although knowing what to do is important, it takes internal fortitude to stay disciplined and focused on creating the results that you want.

What is fortitude? Merriam-Webster defines it as "strength of mind that enables a person to encounter danger or bear pain or adversity with courage." The same source also describes fortitude as "a quality of character combining courage and staying power."

There are no shortcuts on the road to success. It takes hard work. Ideally, what you have learned in this text will help you to reach your desired destination.

CHAPTER EIGHT REVIEW

Key Terms and Concepts are explained in the Glossary. Answers to the Review Questions and Self-Test Questions are found in the back of the textbook in the Answers to Questions section.

Key Terms and Concepts

mentoring	contact management system
strategic alliance	(CMS)
personal board of advisors	prospect file

tickler file
prospect card
tickler card
center of influence contact
        record
annual business planner

annual planning
monthly business planner
monthly planning
compliance
professional ethics
professionalism

Review Questions

8-1.   Briefly describe the events in each of the four phases of a mentoring relationship.

8-2.   Identify and describe how working with other professionals can help the advisor in his or her prospecting activities.

8-3.   Identify the six key jobs that members of a personal board of advisors should assume.

8-4.   Identify the features, benefits, and components of a contact management system.

8-5.   Identify the sections in a typical annual and monthly business planner.

8-6.   Briefly describe the steps in the annual planning process.

8-7.   Briefly describe the steps involved in the monthly planning process.

8-8.   Identify the key functions that state regulation of insurance companies and advisors addresses.

8-9.   Explain the distinction between compliance and ethical behavior.

8-10. Identify the common themes used widely in professional codes of conduct and ethics among organizations within the financial services industry.

Self-Test Questions

Instructions: Read chapter 8 first, then answer the following questions to test your knowledge. There are 10 questions; circle the correct answer, then check your answers with the answer key in the back of the book.

8-1.   In which phase of mentoring does most of the learning between mentoring partners takes place?

(A)   preparing
(B)   negotiating
(C)   enabling
(D)   coming to closure

8-2.   Which of the statements concerning the typical procedure to manage contacts with a prospect is correct?

(A)   Tickler cards for days of the year should be created only on an as-needed basis.

(B)   A prospect card should be completed only when the advisor has obtained enough information to consider the prospect qualified.

(C)   If the advisor decides that the person contacted is not a prospect, the advisor should keep the prospect card for future call-backs.

(D)   To indicate that a prospect has become a client, the advisor should clip the upper left corner of the prospect card.

8-3.   Which of the following statements concerning an automated database management system is correct?

(A)   It usually includes an electronic database filing system, report generator, word processor, and activity manager.

(B)   It is essentially an electronic Rolodex.

(C)   It separates contact information from other parts of the program.

(D)   It requires the purchase of multiple programs in order to run the office effectively.

8-4.   Which of the following best describes an advisor's professional obligation to the prospect?

(A)   to obtain the necessary insurance coverage issued for the prospect

(B)   to assist the prospect in presenting medical information in the most favorable light

(C)   to help the prospect identify and implement all the steps that will help the prospect accomplish his or her financial planning goals

(D)   to keep information received from the prospect in confidence from the home office

8-5.   Which of the following statements concerning state regulation of insurance products is (are) correct?

I.   All states subscribe to the same governing body to regulate insurance activities conducted in that state.

II.   Traditionally, insurance companies' policies have been regulated primarily at the state, not the federal, level.

(A)   I only

(B)   II only

(C)   Both I and II

(D)   Neither I nor II

8-6.   A financial advisor may be required to register as a registered investment advisor if which of the following is (are) answered in the affirmative?

   I.   Is the financial advisor registered with the National Association of Security Dealers to sell securities?
   II.   Is the financial advisor engaged in the business of advising others regarding securities?

   (A)   I only
   (B)   II only
   (C)   Both I and II
   (D)   Neither I nor II

8-7.   Which of the following statements concerning the monthly business planner is (are) correct?

   I.   The client case log is its most important section.
   II.   It is a pocket planner to help the advisor keep track of weekly and daily business activities.

   (A)   I only
   (B)   II only
   (C)   Both I and II
   (D)   Neither I nor II

8-8.   All the following are key jobs that members of the board of advisors should assume EXCEPT

   (A)   strategist
   (B)   coach
   (C)   cheerleader
   (D)   manager

8-9.   The monthly business planner can be used effectively in all the following planning sessions EXCEPT

   (A)   annual planning
   (B)   monthly planning
   (C)   weekly planning
   (D)   daily planning

8-10.   All the following statements concerning what the advisor should look for in a mentor are correct EXCEPT

(A)   an indivudal whose success the advisor would like to emulate
(B)   a person who possesses the qualities of honesty and integrity
(C)   someone the advisor agrees with about about everything and does everything the same way the advisor does
(D)   an individual who is able to provide guidance to the advisor

NOTES

1.   Lois J. Zachary, *The Mentor's Guide: Facilitating Effective Learning Relationships* (Hoboken, NJ: Jossey-Bass, Inc., 2000).
2.   Whiteley, Richard C., *Love the Work You're With* (New York, NY: Henry Holt and Company, 2001).

Appendix A

Telephone Prospecting Guidelines

SECTION 1: TELEPHONE SCRIPTS

Note to Advisors: The following scripts in this section are grouped according to the category of prospect you are calling. All of them—except the first one below, which consists of all seven script components—assume that you have previously stated your greeting (part A) and identified yourself (part B), or that you are already known to the prospect.

The point at which you identify your company affiliation (part C) depends on whether you have met the person prior to the call.

These scripts consist primarily of parts D (the connector), E (offering of information), and F (the benefit). Alternative-choice closes (part G) are found in section 4 of this appendix.

All the telephone scripts printed in this appendix are copyrighted property of Gail B. Goodman, president of ConsulTel Telephone Skills Training. They are provided here with her permission. Any unauthorized use or reproduction of them is prohibited.

For more comprehensive information on telephone training materials available from ConsulTel Telephone Skills Training, contact Ms. Goodman at www.phoneteacher.com.

Natural and Initial Market

Former Coworkers and Acquaintances

Note to Advisors: Normally, when calling cold to a person who has not met you before, you would state your name and identify yourself as a financial advisor in the first sentence under component C before continuing with the body of the script. However, because this call is to a person who already knows you and may also already know you are an advisor, we have merged the company affiliation in part C with the connector in part D.

Note on Small Talk: It's natural that there be some small talk with someone you know, but keep it brief or it may interfere with the business purpose of your call. To keep small talk brief, it is best to ask a particular catch-up question—for example, "How did your son enjoy camp this summer?"— as opposed to the general "How are you?" or "How's everyone?" A catch-up question tends to keep the chit-chat to a minimum.

Hi, (name). This is (your name). (small talk) [A and B]

(Name), I don't know if you've heard, but I recently changed careers. I've joined (name of your company) as a financial advisor, and I'm really excited about what I'm doing. I've just completed an intensive training program and learned a lot of new and exciting planning concepts and ideas. [C and D]

While I was going through this training, I was thinking about the people I know who might appreciate seeing some of these ideas, and your name naturally came to mind. [E]

What I'd like to do is get together for a quick cup of coffee and share information about some financial approaches that may help you protect your family, prepare for retirement, or just accumulate funds in a tax-deferred vehicle. (If the prospect is better known, you may want to meet for a breakfast or lunch appointment). [F]

I know you're busy, so what is a less hectic time of the day for you—morning or afternoon? [G]

Good. Let's get together at (place) on (day) at (time) o'clock. I'll call you to confirm.

Generic Script

Experienced Advisor

We've known each other for a long time from (mention group/affiliation). In all that time I've never spoken to you on a professional basis, but I feel that it would be irresponsible of me if I didn't offer myself as a resource for financial matters. If at any time you have a question about financial products or services of any kind, or if you find that you need to talk about any of the financial programs you are currently participating in, I'd like to make myself available to you, in whatever way you might find my knowledge and experience helpful. Whatever happens, we'll still be friends at (name of group) and still enjoy that aspect of our friendship.

Personal Observation

(This script assumes a specific conversation took place.)

We met last Tuesday before our spinning class and you mentioned that you need to start investing but don't know where to begin. You may remember that I'm from (company) and I've been thinking about that conversation for a while. I called to ask you to get together to discuss some easy ideas on how to take the first steps in planning for your future. This way, instead of losing sleep over how to start, you can begin to think about the part of the world where you and your husband would like to begin your retirement travels.

Business Owners You Know

For the past ___ years, I've been with (company), and my primary focus has been to assist professionals and business owners in analyzing and then enhancing their personal and business insurance/retirement/estate planning goals. Last week when I was sitting with a client who has a business like yours, I thought of you. I'd like to get together to share some of the same ideas with you that she found so helpful.

Calls to Previous Industry Associates

Recently, I left the _____ industry and joined (your company), which specializes in insurance and financial programs. You would be amazed at the technology available to customize financial ideas and programs for a business owner—especially in the (prospect's industry) business. I'd like to visit with you for about 15 minutes to explore the possibility of helping you and giving you some of the innovative ideas I've learned. They really apply well to (prospect's industry).

Referrals

(When you are calling a referred lead, part C comes after part D.)

(Referring person) suggested that I give you a call. She and I recently met, I did some very good financial work with her, and she wanted me to call and say hello to you. I am an advisor with (name of your company). All I would like to do at this point is position myself as a financial resource to you and schedule a time when I can share with you the scope of the work that I do. That way, you can use me in any way that makes you feel comfortable.

Following Letter of Introduction

Recently, (name of referring person) sent you a letter about some of the work I've done with her. Perhaps you have had a chance to call her because she thought that it might be helpful for you to meet with me and hear about some ideas that were interesting to her. I'd like to set a time for us to meet so that I can share some of the financial programs (referring person) was intrigued by.

Center of Influence Referral

Hello, this is (your name), and I was referred to you by (name of referring person). The reason I'm calling is that I am looking for other professionals who can provide complementary services to my clients. I give my clients only outstanding service, and I am looking for other professionals who pride themselves on doing the same thing. (Referrer) thought you were that type of person and told me to call you. So I'm calling to see if I could set up a time for us to get together and discuss the possibility of pooling our resources for the benefit of our potential mutual clients.

Trade/Home Show Leads

(Filled out a lead card with information requested)

We recently met at the (name of show) last week. I wanted to thank you for stopping at the (name of your company) booth. You had indicated an interest in (read from the card what the individual is interested in), and I would like to get together with you to share information and options regarding that. Usually, it is more productive to share information on a face-to-face basis so that I get a better understanding of your specific needs.

Seminars

(After seminar, expressed interest in an appointment)

At our seminar on (topic), you indicated that you were interested in a private consultation so we can personalize the information for you. What is a better time for us to meet—days or evenings?

Chamber of Commerce

We recently met at the Chamber of Commerce business luncheon last week, but we didn't really get a chance to talk. I'd like to get together with you—perhaps for an early morning cup of coffee—so I can learn more about your business and tell you a bit about mine. That way, we can refer business to each other.

Financial Discovery Process

The reason I'm calling is that my work focuses on helping people to get an overall picture of their current financial position. I accomplish this by using an interesting process that helps you (1) to discover just where your family finances are and (2) to make intelligent decisions on financial programs that will help you move more effectively toward your financial goals.

Retirement

I specialize in working with professional women to help them establish retirement programs. So often, women neglect to prepare themselves for this important area of financial planning, and I would like to meet with you to explain some of the options available to you.

Calling Owners of Small Businesses

Canvas Follow-up

I visited your office (store) the other day and dropped off my business card. The reason for my call is that I've worked with small business owners like you, and I understand your financial needs and concerns. I can help you address these concerns and find financial programs that are the best fit for you.

Small Business Pension Plans

The reason for my call is that (your company) offers ways to fund private pension plans for business owners like you. The unique feature of these plans is that they allow you to supplement what you currently have without having to include your employees. What I'd like to do is spend a few minutes with you to fully explain this innovative idea.

Retirement

The reason for my call is that I specialize in helping small business owners like you to fund individual retirement plans to supplement their retirement income. I have a variety of ideas, which many of my clients like to tailor to their needs to increase their personal retirement income. I'd like an opportunity to share these ideas with you and see if I can give you a feeling of comfort when you think ahead to those years on the golf course.

Responders to a Mailing or Purchased Leads

Mortgage Protection

You recently expressed an interested in protecting your mortgage with (mortgage company/bank). The focus of my business is in helping people to protect their most valuable assets. Since this is best done in person, I would like to get together with you over a cup of coffee to discuss your needs and explore some ideas I have that may be of some use to you.

New Baby

You recently responded to a marketing program and indicated that you would like more information on family protection programs. My work focuses on helping new parents to select the right program for themselves and their new baby. I know this is a hectic time for you. However, I would like to arrange a brief meeting when we can sit and talk about your life insurance options.

College Funding

You may remember me from the letter I sent you regarding our booklet on college costs. You indicated that you would like a copy. I work with a lot of people who want their children to be able to attend college but do not know where the money is coming from 15 or so years down the road. Along with giving you our booklet, I'd like to get together and share some ideas that may help you find the means necessary to ensure your child's education.

SECTION 2: QUESTIONS . . . AND GREAT ANSWERS

Note to Advisors: Many times, a prospect will pose a question to you in response to your offering of an appointment. This is a potentially hazardous situation for the advisor. You want to avoid providing the prospect with too much information on the phone so that you do not obtain a face-to-face meeting.

It is imperative that you be brief in responding to questions, and remember to ask for an appointment after answering them. In addition, by the third question from the prospect, you should be able to set up the appointment because three questions are a sign of interest.

Below are some of the most common questions you will hear. Each question is followed by a brief answer that has worked successfully for advisors in the field.

Can you mail me a brochure?

Of course I can. There are hundreds of brochures published by this company, but that isn't the way I do business. I'd like to meet you first, see if we want to have a business relationship, and then, after I find out about your financial concerns, I promise to send you whichever brochures you prefer.

Can I call you back?

There is never a definite time that I will be at my desk, so rather than playing phone tag, if you will give me a time to call when you know you'll be more available to talk, I will make sure to call *you* back at that specific time.

Can you explain it over the phone?

No, because the best thing I could do for you is to meet with you so that I can be there to answer all your questions in person.

Where did you get my name?

From time to time, our company will compile names of people who fit the profile of our clients with whom we do our best work.

Does my spouse have to be there?

It is important that all individuals involved in the planning process be there.

Is this about insurance? Is this about life insurance?

Well, that depends. Of course, (your company) offers insurance products, but we also offer a full portfolio of other financial programs. But we're getting ahead of ourselves. I'd rather get together with you and assess your financial needs before recommending anything.

Can't you just send me a quote?

No, that would be professionally irresponsible of me. The way I do business, I would much rather meet with you personally and collect the information necessary to do a proper quote.

Can you call me back next week?

Yes, but let's schedule a tentative appointment for 2 weeks from now so I can get you on my calendar and I can get on yours. If you need to, we can reschedule when I call to confirm.

What are you selling?

Right now, I'm not selling anything. I'm talking about getting together to determine if we want to have a professional relationship.

What is it going to cost?

The initial appointment and the information are both free and without obligation.

How much do you charge?

Nothing. The initial meeting is without fee or obligation. If you choose to do business with me, I am compensated by the companies we use. At no time do you pay me a fee for the services I provide.

How long will this take?

My initial professional presentation usually takes 20 to 30 minutes, but it will take longer if you have questions you want me to answer.

SECTION 3: PROBLEMS (MORE COMMONLY KNOWN AS OBJECTIONS)

Note to Advisors: *For phone prospecting, the feel-felt-found technique is powerful in closing for the appointment.*

Say these three statements immediately after the prospect states his or her objection or problem:

<u>Standard Response</u>	<u>Alternative Response</u>
• *I can appreciate that you feel that way.*	• *I can appreciate you telling me that.*
• *Other clients of mine initially felt that way too.*	• *Other people initially told me the same thing.*
• *But they found, after meeting with me that. . . .*	• *But they found, after meeting with me that. . . .*

With either type of response, following the third sentence, you must insert a benefit that relates to the stated problem.

After the three feel-felt-found sentences and benefit statement, you then add:

- *I'd like to see if I can do the same for you.*
- *Sound fair enough?*
- *Close. (See section 4 of this appendix for closes.)*

The section below gives you excellent benefit statements to answer the most commonly heard problems.

I have an advisor. I like my advisor. I have been doing business with this financial advisor for 10 years.

. . . no one person has a monopoly on all the good ideas. Any suggestions I make will only be those that complement your existing portfolio. I will not undo any good work that your other advisor has done.

Send me literature.

. . . brochures are generic, but their financial lives are specific. The work I do is tailor-made to my clients, and a brochure is unable to do that.

I'm getting money in the future; call me in 6 months. The time isn't right.

. . . they were glad they had all the facts and figures in advance so that when they *were* ready, they were able to make an intelligent decision.

. . . looking at different ideas for what to do with the money—without pressure—allows them to make better decisions.

I am busy.

. . . they were pleased that the appointment was for only 15 minutes and that I was direct and to the point. I'd be happy to meet you during your coffee break.

I have no money.

. . . my work is about organizing people's money, not necessarily spending it.

I'm covered by my employer/my spouse's employment.

. . . looking closely at the coverage they have, alongside the financial profile we develop, gave them a better idea of how that benefit program takes care of their family's needs.

Call me after the holidays.

. . . they were glad that they did since this is a time to be thinking about family. Our conversation was, in fact, about providing for the security of their family.

I do my own investing/I invest on-line.

. . . many times two heads are better than one.

. . . having a financial advisor who can serve as a resource was helpful.

I don't want to waste your time.

. . . building a professional relationship is never a waste of time. If we meet now, then when you are ready, you'll think to call me first.

My (wife/husband) makes all the decisions. Meet with him/her alone.

. . . they each had very different questions. As a courtesy to both parties, I like to set the appointment when it is convenient for both of you.

My (accountant, lawyer, CFP, whoever) takes care of that.

. . . the work I do complements the work they do with their other professionals.

My brother-in-law (relative) is an advisor.

. . . they were actually more comfortable disclosing private financial information precisely because I was not their (relative).

I have enough insurance.

. . . my assessment of their entire financial portfolio reassured them that anything that was in it was performing an important financial function for their overall goals.

I just did it months ago.

. . . only one of two things could happen: one, I'd look at their new plan and agree that it was excellent, or two, I'd come up with a way to enhance their plan. In either case, you will be happy that we met.

I'm already saving money.

. . . my ideas can complement their existing program.

I'm getting divorced/changing jobs/buying a new house.

. . . this was a particularly good time to get a handle on their financial portfolio, both for now and the future.

SECTION 4: ALTERNATIVE-CHOICE CLOSES

__Note to Advisors:__ Whenever you finish a phone presentation, it is important to remember to ask for the appointment. Always finish your presentation with a request to see the prospect.

Avoid being specific about the day of the week or the time of day when initially closing for the appointment.

All people divide time into certain kinds of larger concepts such as

- *day or evening*
- *beginning or end of the week*
- *weekdays or weekends*
- *before or after lunch*
- *early in the day or late in the day*

Use these in your first attempt to close. As soon as a prospect chooses one of the offered times, you've got an appointment. Then be specific about day and time, trying to keep the amount of time you spend driving around town to a minimum.

Select the alternative-choice closes you like best from the ones below, and add them to any of the scripts in this book.

Business Closes

I know you're a very busy person, and with that in mind, when would be a less hectic part of your day—the morning or afternoon?

Which do you prefer—to meet before or after lunch?

With respect for your very busy schedule, would you prefer to meet at the very beginning of the day or at the end of your day, say, at around 5 o'clock?

With that in mind, are mornings or afternoons better for you?

(When you have another appointment near this prospect's office)

I have another appointment in (name the town, or better, the building you're going to) and I'd love to see you the same day. Can you meet me before or after my 1:00 appointment?

Personal Market

With that in mind, what is better for both of you—weekday evenings or a weekend time?

Which do you prefer—to meet at my office or that I come to visit you in the comfort of your own home?

Which is better for you—earlier in the evening before dinner, or after dinner when the kids are settled down?

I'm all booked up for the rest of this week, but next week is open. Is earlier or later in the week better for you?

SECTION 5: TRACKING FORMS

Counting Sheet

Name _____ Actual Dialing Time _____ to _____ Date _____

Total Dialing Time

Dials

1	2	3	4	5	6	7	8	9	10	11	12	13	14	15	16	17	18	19	20

Total Dials

Contacts

1	2	3	4	5	6	7	8	9	10	11	12	13	14	15	16	17	18	19	20

Total Contacts

Presentations

1	2	3	4	5	6	7	8	9	10	11	12	13	14	15	16	17	18	19	20

Total Presentations

Appointments

Name	Date of Appt.	Time of Appt.	Source of Lead
1.			
2.			
3.			
4.			
5.			
6.			
7.			
8.			

Dialing Session Form

Date: _____

Day of Week: _____

Time of Day: _____

Dials	Contacts	Presents.	Appts.	Time

Effectiveness Percentages

Contact % _____

$$(Contacts \div Dials)$$

Presentation % _____

$$(Presentations \div Contacts)$$

Appointment % _____

$$(Appointments \div Presentations)$$

Appointments per Hour _____

$$(Appointments\ Made \div Hours\ on\ the\ Phone)$$

Appendix B

Annual Planning Forms

SECTION 1: ANNUAL ACTIVITY EVALUATION

ACTIVITY SUMMARY

During this year, _____, my total annual activities were:

_____	Phone (face) contacts for appointments
_____	Appointments seen
_____	New qualified prospects obtained (in initial interviews)
_____	Fact-finding interviews
_____	Closing interviews
_____	Completed sales

PROSPECTING AND SALES EFFECTIVENESS RATIOS

These activities created the following ratios:

_____	Total contacts to produce one appointment (phone [face] contacts ÷ appointments seen)
_____	Appointments seen to produce one qualified prospect (appointments seen ÷ qualified prospects)
_____	Qualified prospects to produce one fact-finding interview (qualified prospects ÷ fact-finding interviews)
_____	Fact-finding interviews to produce one closing interview (fact-finding interviews ÷ closing interviews)
_____	Closing interviews to produce one completed sale (closing interviews ÷ sales)
$_____	Average size case (first-year commissions ÷ number of sales)

SECTION 2: EXPENSE ESTIMATES AND INCOME OBJECTIVES

		Monthly Total	x 12	Annual Total
Fixed Expenses	Rent or mortgage payments	$		
(Nonbusiness)	Utilities			
	Property, auto, and life insurance			
	Property taxes			
	All health insurances			
	Income taxes			
	Social Security taxes			
	Other			
	Total Fixed Nonbusiness Expenses	$	X 12	$
Variable Living	Food			
Expenses	Clothing			
	Laundry and dry cleaning			
	Outside meals (nonbusiness)			
	Transportation (nonbusiness)			
	Medical—doctor, dentist, drugs			
	Other			
	Total Living Expenses	$	x 12	$
Business Expenses	Rent			
	Sales promotion, advertising, and mail			
	Telephone			
	Transportation			
	Entertainment and meals			
	Administrative assistance			
	Self-development			
	Association dues			
	Other			
	Total Business Expenses	$	x 12	$
Miscellaneous	Charity, gifts, church			
Expenses	Recreation, vacation			
	Hobbies, clubs			
	Other			
	Total Miscellaneous Expenses	$	x 12	$
Savings	Savings			
Accumulations	Investments			
	Other			
	Total Savings Accumulations	$	x 12	$

Total Annual Income Objective $ _____

    Less: Estimated Renewal Commissions $ _____

        Other Income $ _____

Annual First-Year Commission Objective $ _____

SECTION 3: INCOME AND ACTIVITY OBJECTIVES

Income Objective	Weekly	Monthly	Annual
FYC Income Goal	$	$	$
Average FYC per Sale	$	$	$
Required Prospecting and Sales Activity Objectives			
Sales required to reach your goal Based on dividing your FYC goal of $ _____ by your average FYC of $ _____ per sale			
Closing interviews required Using your ratio of closing interviews to sales: _____ to 1, multiply sales required by the number of closing interviews per sale			
Fact-finding interviews required Using your ratio of fact-finding interviews to closing interviews: _____ to 1, multiply closing interviews required by the number of fact-finding interviews required to get one closing interview			
Qualified prospects required Using your ratio of qualified prospects to fact- finding interviews: _____ to 1, multiply fact- finding interviews required by the number of qualified prospects required to get one fact- finding interview			
Appointments required Using your ratio of appointments to qualified prospects: _____ to 1, multiply qualified prospects required by the number of appoint- ments required to generate one qualified prospect			
Total contacts required Using your ratio of total contacts to appoint- ments: _____ to 1, multiply appointments required by the number of total contacts (phone and face) required to see one appointment			

Appendix C

Profile Spotlights

Jody J. Ackerman
Consultant
Transamerica Life Insurance and Annuity Company

> You have to get on the phone and call people. That's the only way you're going to succeed.

What It Takes to Succeed in the Financial Services Industry

If I could go back and do it all over again, I would have been a wholesaler first, and then I would have become a financial advisor, an insurance agent, or a broker. When you start out, if it's your first job out of college, it's extremely difficult. It's a difficult job even when you don't start right out of college. The challenges I faced were not unusual. First of all, you don't know anybody who is going to listen to a 22-year-old kid tell them what to do with their money, or how to protect their assets, and also how to buy insurance.

You have to pick up the phone and call people. That's the only way you're going to succeed. As a wholesaler, I still constantly have to prospect to my brokers. You always have to face rejection when you are continually being told "no." Once you get over that initial fear and that initial rejection, you'll be fine.

Confidence is a huge part of it, as well as personality. Neither is taught in a textbook. But you really do need the right personality, and you must have drive. You can't avoid the rejection. I'll tell you, there are times when I'll go into a presentation or I'll pick up the phone thinking it's a slam dunk, and then it isn't. You never know what might happen.

On the other hand, the easiest way to avoid rejection is to call the right people. But you have to find out who the right people are. On the brokerage side of financial services, the right people for me are brokers who have a need—for example, brokers who are not experts in the 401(k) arena. A lot of times they say, "Thanks, but I don't prospect for 401(k) business." Eventually, however, if something falls in their lap, they're going to call me for help.

To be successful in this business you must have patience and persistence. For instance, I prospected the same broker for 2 1/2 years and couldn't get anywhere with him. He wouldn't even take my phone calls for the first year. Then I did a presentation for a large group of brokers, which he attended. (With prospecting, getting yourself in front of the right groups is essential to your success; you have to market yourself.) Now he's my top producer. He says it's because I was persistent. He is also my number one advocate who opens many doors for me.

You also have to ask for referrals. I don't just sell to these brokers. They take me into companies that either have a 401(k) or are looking to set up a 401(k). And I ask those people for referrals as well—other business owners whom they know who either want to change their plan or want to set up a plan. You've got nothing to lose; all you have to do is ask.

Nevertheless, you have to overcome the psychological hang-up in asking somebody for a referral because the individual doesn't want to refer friends and family to you until he or she knows you can be trusted. Too many advisors make the mistake of asking for a referral right at the point of sale. You need to first build the relationship, make sure that everything is running smoothly, and then you ask for the referral.

All advisors should have a long-distance-race mindset, challenging themselves to get past that 2-mile mark and then the 4-mile mark, and so on. You have to constantly want to do better, to find a better way to do it, to find a smarter way to do it.

Just as there is training involved in a distance race, so is there training in this business. And that's where patience comes in, because your first couple of years as an advisor (or as a wholesaler) will be difficult. There are only certain things that you can control: picking up the phone, making the dials, making the contacts, and seeing the people. By working hard, eventually business will flow. But it takes time.

Remember, you can't sell a product if you can't sell yourself on the value of your services. Once you are convinced of your self-worth, then the prospecting will take care of itself.

Robert Berz, CLU
Insurance Broker
Berz, White, and Cooper Insurance and Financial Services

> How did I get over my call reluctance? Well, believe or not, I never have—at least not completely. But I have compensated for it.

Overcoming Prospecting Hurdles

I always had a real interest in getting into a marketing-oriented business. I majored in finance in college, and I started taking all my courses in insurance toward the end of my senior year. I was influenced by some people who were successful in the life insurance business, and I think that, coupled with my curiosity about how this would fit, really got me going. Of course, I struggled in the first 5 years, but I stayed with it.

The sources of my struggling were call reluctance and fear of rejection. But they were overshadowed by my persistence—my belief that there had to be a way that I could succeed. Fear of rejection was a serious problem. There were all these people around me I knew would be glad to see me, but I couldn't make it to their doorstep.

How did I overcome my reluctance? Well, believe it or not, I never have—at least not completely. But I have compensated for it. I admit, I would still almost rather take a beating than pick up the phone and make a cold phone call. But you have to figure a way to compensate for whatever your deficiencies are. We all have them.

Prospecting is the ability to walk with kings without losing the common touch, and the ability to be able to converse with people, no matter what station of life they're in, or who they are or what they are. That's communication and it is probably one of the compensating factors that enabled me to overcome whatever deficiencies I had, whether they were reluctance or fear.

What causes fear in a lot of people is pressure to make a sale. There is the mentality that says "I'm trying to get something from this other person." But if you really keep your focus on just connecting with people, you can find out about their needs, interests, and goals. There are some people for whom doing this is easy. If they get rejected, it doesn't bring them down. That's a marvelous personality characteristic. I've seen people come into this business with that trait and do really well.

As time goes by, some other key traits emerge that have a further impact on prospecting. One is self-confidence and a belief in what you do. The other is reputation. The fact that I've built a reputation in my market and my community is a big plus, but it took several years.

In the beginning, almost everybody you talk to is going to be somebody you already know. You have to go through that first. After all, if you have something that is of real value to people, you should want your friends to know about it.

Prospecting is tremendously enhanced by knowledge. The more knowledge a person has, the easier the job of prospecting is.

Product knowledge is essential, but more important is how the product affects people's lives. That understanding can stimulate conversations with people. There's an old saying: "If you don't learn, you won't know; and if you don't know, you won't do." Education and knowledge, therefore, do help to create a lot of activity. Understanding and believing in your product just wipe out a lot of resistance to prospecting. They can be quite motivating.

Community involvement is also essential, but not if you are doing it only with the idea that it will get you some customers. Participation must be for the purpose of truly wanting to be involved and contributing. If you are not doing this, you're not a whole person in this business. At the same time, if you are contributing and involved, it will help your business.

Thomas W. Curry, CLU, ChFC
President
Curry & Company, LLC

> **All of a sudden we got smart! Asking for and getting referrals made a tremendous difference.**

The Value of Referrals

Today, the financial services industry, which includes insurance and financial planning and a whole spectrum of products, is probably the most lucrative, wonderful opportunity for a young person who is willing to work for it. There's no place where you'll earn as much money based on what you can produce. You don't receive an hourly paycheck.

When I began in this business, I made cold calls, which is really not a fun thing to do. I also called on the people whom I knew and had met one way or another. In addition, I used preapproach letters. They were nondescript, asking for the prospect's name and date of birth in exchange for an idea or some useful information. If I sent out 100 letters in a week I would get about three responses. Out of those three responses I would sell two. We had specific target areas to which we would mail the letters, such as new births, new mortgages, or new marriages. We even mailed letters to people who received promotions as announced in the local paper. And that was how we worked.

After a few sales, all of a sudden we got smart! That's when we began to ask for referrals. Asking for and getting referrals made a tremendous difference.

The other thing that I did early on was to make a lot a business calls and to arrange my schedule around regular business hours. I started working at 8:00 a.m., and at dinner time I was done. I began to call on businesses during the day, and to prospect for what we call business insurance, which was anything someone in business would buy. I became proficient in selling group insurance. The corporate need for life insurance was a big thing.

When you work the small to medium-sized businesses, you'll discover that there's not much difference between business life insurance needs and personal life insurance needs. They overlap. When I was working on the business aspect of selling key person insurance, group insurance, or long-term disability income insurance, I also found myself discussing the family's needs and the personal estate. And that just kept blossoming and blossoming.

As the years went by, I decided I would specialize, do one thing very well, and become known as the best person around for doing that. I started focusing on nonqualified deferred compensation and personal retirement plans, working in the corporate marketplace.

It's a matter of referrals. Over the years, I've done a lot of different things in the insurance business; one of them was to become well known in the legal community. I have numerous clients who are attorneys and CPAs. We refer business to each other. These referrals really made a difference.

William Harrison, CLU, CSA
President
General Insurance Consultants Agency

> Unless you have conditioned yourself to just pick up the phone and call, you tend to procrastinate. That's what drives people out of the business.

Conquering Call Reluctance

There is a great phantom in our industry known as "call reluctance." Unless you have conditioned yourself to just pick up the phone and call, you tend to procrastinate. That's what drives people out of the business.

When I was managing other agents, everybody came in on Monday nights, got on the phone, and called potential clients to make appointments. It was mandatory. They came in at 6:00 p.m. and stayed until 9:00 p.m. It was humorous to see what they would do to get out of making those phone calls. It was call reluctance at its best! And that mandate probably drove 90 percent of the agents out of the business. It wasn't only my company. It was the same with every company. Every now and then you had a general agent who was a little smarter than most, and would teach his people how to develop prospecting leads. And that's what is needed in our industry.

You have to believe that your services are valuable. Why is it that when agents deliver a claim check to a widow, they become much better agents? That's an incredible experience. When the check is delivered, the agent's work isn't finished. First of all, she's got money right now. Second, if she doesn't get appropriate advice, every brother-in-law she has—along with everybody else—will tell her how to invest the money, but nobody's going to tell her how to preserve that money because the breadwinner is gone. That's the legacy of his life. But once you deliver the claim check and see how it's going to enable her two kids to go to college, and how it's going to permit her to live without having to work 24 hours a day, then that call reluctance diminishes substantially. You no longer have a product; you now have a mission.

Over the years, my passion has increased tremendously. My first claim check delivery was to the neighbor across the street whose husband had just died. He had been in tremendous shape; he worked out everyday. One day he came home with an alarmingly high temperature, and I took him to the hospital. The doctors submerged him in a bath of ice to try to lower his temperature. It didn't work. Three days later, he was gone. He had died of a virus. Five years earlier, I had sold him $150,000 worth of life insurance. His wife was angry with me at the time, claiming that I had sold it for the commissions.

Now, 5 years later, upon her husband's death, I brought a $150,000 claim check to her. She put her arms around me and hugged me for what seemed like an hour. She cried. And she said, "You know what? All my relatives have been very good to me through this. But you're the only one who's bringing me a check, so my children will be able to get through college and I won't have to go to work." Wow.

When it comes to service, though, the first thing you need to do is ask pertinent questions of your client's situation. In other words, you need to listen and get to know your client. You certainly need to know how much money he or she makes and how the money is spent. I talk to all my clients about the need for budgeting their money.

A lot of times the spending is just about "stuff"—whether it's for themselves, their kid, or their grandkids. But when you talk about budgeting, you have to get down to the bread and butter of things. This is particularly vital to younger families. Often, one of the first things families cut down on is their insurance. It always seems too expensive—except when they have a claim. Nobody ever asks about premiums when they've just totaled their car. And the widow to whom I took the $150,000 check after her husband died never once mentioned premiums.

I have to ask very personal questions. Most of us don't want to talk about budgeting. But people who don't budget are more likely to experience credit card abuse. Even people who make good money are having a hard time budgeting. I know I've got a client for life when I've helped that client see the need for and then actually create a budget.

Michael W. Lee, CLU, ChFC, CSA, CEBA
Vice President, Senior Estate Planner
Estate Planning Professionals

Focus beats brilliance
every time. So pick a
couple of prospecting
strategies, and don't
give up right away if
they don't work.

Prospecting and Building Relationships

I'm reminded of what a good friend of mine says about marketing and sales: Marketing has to do with figuring out who and where the potential customer is. Prospecting has to do with how you specifically approach him or her. And sales are what you do when you get there. There are smart people who know lots of good planning solutions, but they don't have the ability to talk to clients or prospective clients about those planning solutions. Therefore, developing a stream of potential customers and clients is key to any business—not just ours.

One of the biggest hurdles is that prospecting is hard work. It is not always fun—although it can be. There are many ways to prospect and to try to find those people to whom you can talk to about your products and services.

Someone new to the business needs to develop a few automatic root systems for prospecting. Remember that focus beats brilliance every time. So pick a couple of prospecting strategies, and don't give up right away if they don't work. Just about any system is a good system as long as you give it enough time to work. Just do it, and make it an everyday part of what you do.

Prospecting is difficult in any kind of business. By forming business strategic alliances with large property and casualty agencies, we are provided with a steady stream of people. I have a lot of relationships with CPAs. We take our planning techniques and educate our CPA partners so that they become our eyes and ears. They say, "Gee, I have a client who fits that particular profile for that planning technique." They'll introduce us and we'll work the case and, assuming they're all duly licensed, as most of them are for insurance or annuity products, we share everything. Everybody wins—especially the client.

It's a long relationship-building process. The first strategic partnership that you build is probably the most difficult and takes the longest. You have to choose a couple of people—whether they are a property and casualty agents or CPAs or attorneys—and start cultivating relationships.

I've been successful over the years developing networks with CPAs and attorneys, and I keep in touch with those people on a regular basis to discuss a new planning technique or just to say hello. Often, they'll broker their insurance cases through us, so I speak to them frequently. It took a long time to develop those relationships. But now our business partners refer us to other potential business partners. Prospecting is just something that goes on subliminally. You have to do it—a little bit every day.

We have some things we do automatically, like send out an investment newsletter to clients once a month. We also send Thanksgiving cards, St. Patrick's Day cards, and birthday cards (with a birthday cookie) to our business partners and clients. When we do business with a client for the first time, we send him or her 10 cookies.

We probably get in touch with our clients, aside from a follow-up telephone call or a review-meeting-type telephone call, eight to 10 times a year. If I haven't heard from a strategic business partner for a while, I'll call him or her. When you have a core group of people—which our business partners and our centers of influence tend to be—that's not a big network. You can certainly contact them on a regular basis.

You have to be really excited about what you're doing. That's another key piece. It's easier to prospect, develop networks, and build relationships with potential sources of business partners or clients if you really believe in and are excited about what you're doing. We have some unique planning techniques that differentiate us from what other people in the insurance or investment business are providing their clients. I'm excited about that.

And that's another big point: You've got to differentiate yourself from everybody who looks like you. You've got to find the prospect and then, in only 5 to 10 minutes, be able to define yourself in a way that distinguishes your product and service from what the prospect hears from everybody else.

When you meet somebody for the first time, ask yourself this question: If I were that prospect, what would I want to hear in only 5 minutes that would make me want to listen and hear more? It had better be something that quickly differentiates you from the crowd.

Ronald F. Mayer, CLU, ChFC
Financial Consultant
AXA Advisors

> As a new entrant to this business, you have to be totally committed to it.

An Important Trait for New Financial Advisors

When I was interviewing prospective new associates, I always asked this question: What was your greatest success and your greatest failure prior to age 14?" Quite a few candidates told me that it was an interesting question. I got a lot of comments and a lot of angry interviewees. Most people didn't even know how to respond, but you have to understand the guy who was asking it. My dad died when I was 5. How do you survive something like that? Luckily, we had a great family network. There was never any pity or sorrow, and I've known people whose mom or dad died at a young age and their whole life was, "Poor me."

I was also involved with Boy Scouts, and at age 13 1/2, I had my first Eagle board review. This is the one that no one ever fails—until me. A father/Scoutmaster asked me, "What do you want to do when you grow up?" I said, "I want to be a pilot." He looked at my merit badges—you needed 21, and I had 28.

He said, "But you don't have an aviation merit badge."

I said, "That's right, I've read the requirements, and they're outdated—created in the 1920s and 1930s. This is the 1960s. They are totally out-of-date and totally irrelevant. I will never get the aviation merit badge, but I will become a pilot."

It was definitely the wrong answer. The father was very upset; he voted no. The other fathers argued but they were intimidated by him.

Six months later, at age 14, I came back and met with a different group of fathers, except for one—the one who had voted against me the last time. He asked the same question, and he got the same answer. I was told later that the discussion became quite heated because he voted no again. The other fathers said, "You're changing your vote. You can't say no. You may not like his answer, but it's an honest answer."

I went on to become an Eagle Scout. I went on to air force flight school, learned to love flying, and I am still a flight instructor today.

What does this have to do with financial services?

As a new entrant to this business, you have to be totally committed to it. You also need training, and you must learn to prospect. When I started out in Detroit, with family back in New York, family in Ohio, and knowing only five people in Detroit, I had to learn how to survive.

This book on prospecting will definitely be helpful.

Jim McCarty, CLU, RFC, RHU, LUTCF
National Vice President
American Express Financial Advisors

> People do business with people they like. So don't prospect; visit.

Everybody Prospects

Everybody has to prospect. Even if you're looking for a date, or for a spouse, you're prospecting. Advisors are afraid of the word "prospecting." You know, "They're not prospects; they're human beings." As far as I'm concerned, when you walk down the street, everybody you come in contact with is a "prospect."

In the life insurance business, a prospect is a person who is in good health, has a need, and has the money to solve the problem. We have a tendency, as advisors to say, "Well, what do I say to these prospects? How do I approach them? How can I gain their attention? Should I mail something to them? Should I ...?"

Slow down. You're making it too complicated. And if it's too complicated, you probably won't do it.

Suppose you met a management consultant at your church or Rotary Club, for example. How would the conversation go? Here's how:

"Hi, I'm Jane Smith, and you are ... ?"

"Oh, I'm Jim McCarty."

"Jim, are you from here? I mean, are you a native of Columbus?"

"Well, no, as a matter of fact, I live in Florida."

"Really? So what exactly is it that you do?"

"Well, I'm the national spokesperson for American Express Financial Advisors. What is it that you do?" I would ask out of politeness. It's called engaging somebody. And we are reluctant, as human beings, to engage people.

We are especially reluctant to engage the person if he or she has a title such as president or CEO of the company. That's intimidating. Even if the person is a welder, it's intimidating because most of us know nothing about welding. So we say nothing. Also, whenever you put a label on a person, you become more reluctant to approach that person. And prospecting simply is engaging people on a regular, ongoing basis, looking for common ground.

By engaging people, we can prospect without losing that human connection. We must always be in the mode of engaging and showing an interest in other people.

Try to find people whom you genuinely like. Too many times, advisors say, "What can I do to make you like *me*?" That's not the answer. People will want to do business with people who like *them*.

The mind is a funny thing. I'm not a psychologist, but I do strongly believe that sometimes our minds really don't know right from wrong, up from down, left from right—unless we've told our minds whatever it is, over and over again, for a long period of time. And our minds associates actions with the words we say—that's self-talk. For example, suppose we say to a new advisor, "You're going to have to prospect or perish." The word prospect turns a lot of people off. They think, "I'm not any good at prospecting. That sounds like something pushy. That's not me. I'm not good at that."

I had an agent one time who had been really productive, but he was in a bad slump; for about 3 months, he had sold hardly anything. I was about to terminate him, but he was a great guy. I said to him one day, "Look, for the next 2 weeks, I want you to do me a favor. Don't come into the office. Go over to the mall and just hang out. I want you there when the people are walking around, and I want you there only until noon. Will you do that for me?" He agreed.

Guess what happened? He started writing a lot of business—not at the mall—but because the activity at the mall got him excited again so that he could get back on track. It saved his career. When he was sitting on a bench or in the food court, people would say to him, "I have seen you here for the last 3 days. What's going on? What do you do?" They would engage him in conversations about what it was that he actually did. He enjoyed talking to them, and they enjoyed talking to him. Remember, people do business with people they like. You can't get to that stage unless you engage individuals in a conversation and visit with them. Don't prospect; visit.

I have a friend who is a plastic surgeon. Guess what he's got to do on a regular basis? He has to prospect.

Now, if you asked him if he did any prospecting, I'm not sure what he'd say. But everybody does it, in every walk of life, through websites, running ads in the newspaper, sponsoring fund raisers . . . it's called prospecting.

Robert M. Roach, CLU, ChFC
Business and Estate Planning Specialist
Northwestern Mutual Financial Network

> The key to successful prospecting is telephoning. You've got to telephone every day.

The Role of Prospecting

Prospecting is critical. After you've been in the financial services business for a long time, it really becomes a matter of prospecting not for more clients, but for higher net-worth clients.

I decided in my first year in the business that this was going to be my career. I didn't approach it from the standpoint of just earning a paycheck every 2 weeks, but more from the standpoint of building relationships, knowing that a lot of people aren't going to buy immediately, but realizing that sooner or later, they will buy. I knew that if I built a strong enough relationship, when they were in a position to do something, they'd do it with me.

My practice in the financial services business has evolved. For 20 years, I sold life insurance and risk-based products. One of the things early in my career that helped me, and continues to help me today, is that I tried to make as many attorneys my clients as I could. I set up "nests" of clients in law firms. I sold one or two of them and then, every year, got a list of new attorneys who joined the firm. I'd go to the office, pass out calendars, socialize with the attorneys, and get to know them as people. Eventually, they became clients. I managed to get into four large law firms and acquire sizable numbers of clients using this method. Now those clients are in positions in their tax, business, or estate planning departments, and they send me quite a few referrals. They know the kind of service I give and that's important to both them and me.

Today, I'm at the opposite end of the spectrum, and I am looking to use younger financial representatives to service some of my clients so that I can work with fewer people and do more work for a higher level of prospective buyer. I'll be focusing on relationships with higher leverage ability.

I enjoy helping people achieve their goals and giving financial advice. Right now I don't see myself saying goodbye to the business and packing it all in. I envision myself working actively for quite a long time. Even though I picture a time when I will have somebody to take care of my clientele, I will still be available on a case-by-case basis.

If I could sum up the key to successful prospecting up in simple terms, it would be that telephoning is the biggest thing. You've got to telephone everyday. Even though I have assistants to call for me, I still do a lot of phoning myself. I continually look at the calendar to make sure I have enough meetings booked.

On a daily basis, the most important thing is to get up early. I'm convinced that getting up early, exercising, and doing some meditation really helps the day go much, much better.

Glossary

10-second commercial • a variation of the elevator speech, designed to stimulate a prospect's awareness of and interest in the advisor's work and usually in response to the question, "What do you do for a living?"

80/20 rule • the principle that states that 20 percent of causes are responsible for 80 percent of effects, and that 20 percent of efforts produce 80 percent of the results. Also known as the Pareto principle.

A-B-C method (client classification) • a method of prioritizing and segmenting prospects and clients into "C" (those who are not interested in developing a relationship with the advisor, nor the advisor with them), "B" (potential customers whom the advisor wishes to turn into clients), and "A" (those with whom the advisor wants to do business and who see the advisor as a trusted professional)

access • the ability to approach a group and gain entry with greater ease because of having something in common. This translates into a personal understanding of their needs and the way they think.

advertising • the use of persuasive messages communicated through the mass media, promoting a level of familiarity with the advisor. The ultimate goal is to create new clients.

affinity • the natural gravitation toward certain people because of shared interests, hobbies, background, lifestyle, or stage of life. Natural markets based on affinity include such groups as friends, family, and acquaintances.

annual business planner • usually an 8″ x 11″ booklet that has separate pages on which to calculate and record an advisor's annual activity, income, and career objectives. It also contains pages to log cases written, business expenses, and an annual activity summary and evaluation.

annual planning • a strategic business function for advisors that involves the formulation of income, career, and activity goals in the coming year, using the evaluation of the previous year's activity as a foundation. In this process, the advisor establishes a working business plan for the year.

appointments • in using a tracking system for telephone activities, a record of each person's name, the date, time of appointment, and source of lead

appointment-setting rate • the total number of appointments the advisor obtains divided by the total number of presentations he or she makes in a telephoning session

appointments per hour • the number of appointments the advisor makes divided by the total time he or she spends on the phone in a telephoning session

behavioristic variables • a segmentation variable that characterizes people by their buying and usage behaviors

blog • *See* weblog

brand • a name, term, design, symbol, or any other feature that identifies one seller's goods or service as distinct from those of other sellers. For financial advisors, the identifying features are typically the prospect's perception of the advisor's knowledge, skills, ideas, and personality that differentiate him or her from competing advisors.

buying process • the five stages of the sale from the prospect's perspective: problem recognition, information search, evaluation of alternatives, purchase decision, and postpurchase behavior

cancellation • an appointment the prospect refuses to attend even though the advisor has offered to reschedule

center of influence • an influential person the advisor knows, who knows the advisor favorably, and agrees to introduce him or her to others

center of influence contact record • a 3″ x 5″ card in a contact management system used to record personal and business information about a center of influence, along with every incident of contact an advisor has with him or her

client • a person who has bought from an advisor, but also someone with whom the advisor had developed, or is developing, a strong interpersonal business relationship

client-focused approach • an approach to selling/planning that emphasizes helping clients recognize and understand their financial needs, providing necessary information, removing obstacles to financial success, and assisting them in taking positive action to achieve their goals

client loyalty • a level in the client-advisor relationship in which a client tells others about the advisor, refers the advisor to others, and buys again; a level of relationship above client satisfaction

client satisfaction • a level in the client-advisor relationship in which a client receives what he or she expects from the advisor and the advisor does what is right for the client; level of relationship that lacks client loyalty

CMS • *See* contact management system

COI • *See* center of influence

cold canvassing • a method of prospecting that involves going out and knocking on doors to approach people

community involvement • the advisor's involvement in social, civic, or charitable organizations and causes in the community that are important to both the advisor and his or her target markets

compliance • following the laws and regulations, including company rules, that apply to the sales of insurance and financial products

confirmed appointment • an appointment that is scheduled with a prospect for a specific date, time, and place, to which the prospect (within 24 hours before it is set to begin) reiterates his or her commitment to attend

contact management system • a business management tool that will help the advisor handle the paperwork necessary to contact prospects and service clients, as well as manage the administrative and financial aspects of business activities

contact percentage • ratio of the total number of contacts the advisor obtains divided by the total number of dials he or she makes in a telephoning session

contacts • in tracking telephone activities, a record of the times the advisor reaches the proper party

creating • a learning process in which creators teach themselves how to bridge the gap between their goals and their current capacity, leading to tangible, recognizable results

creative tension • during the process of creating, the discrepancy or gap between vision and reality; a pragmatic view of where the individual is as opposed to where he or she wants to be

customer • someone who has bought a product from the advisor but with whom the advisor has not developed or is not interested in developing a client-advisor relationship

defense time • an advisor's low-priority time during the workday in which he or she needs to hone selling/planning, product knowledge, and service skills. Activities include ongoing service to clients and building existing client-advisor relationships.

deliverables • the products and service experiences the advisor promises and provides to prospects and clients

demographic variables • the most commonly used segmentation technique, which includes such variables as gender, education, ethnicity, occupation, income, size of family, marital status, religion, generational cohort, and family situation. Differences in prospects' needs and wants are often linked to these variables.

dialing time • actual time the advisor is on the phone in hours and fractions of hours

dials • every single time the advisor dials the phone in an attempt to reach a prospect

direct mail • sending letters with reply cards that prospects can return to the advisor if they are interested in making an appointment or obtaining more information

Do Not Call Law • legislation that gives consumers the opportunity to limit unsolicited telemarketing calls made without the consumer's prior consent or sales calls to persons with whom the caller does not have an established business relationship. *See also* Do Not Call Registry

Do Not Call Registry • a list of phone numbers from consumers who have indicated their preference to limit the telemarketing calls they receive. The registry is managed by the Federal Trade Commission. The Federal Communication Commission has adopted rules that make use of the Do Not Call Registry and sets forth the ways that telemarketers must check the list to ensure compliance. *See also* Do Not Call Law

elevator speech • a short (15 to 30 seconds, 150 words or fewer) sound bite that succinctly introduces an advisor and helps him or her to connect with a person the advisor has just met. It is designed to create interest, give the prospect a quick synopsis of what the advisor does, and advertise the results that the advisor helps clients achieve.

extraordinary service • customer service that goes beyond satisfaction to develop client loyalty. This service exceeds client expectations and leads to repeat business and referrals.

fact finding • gathering all relevant prospect data regarding financial resources and obligations; personal information, needs, concerns and goals; and values, attitudes, and expectations in order to determine financial strategies to meet the prospect's personal and financial goals

"fit" goal • the problem-solving approach to goal setting in which people fit their goals into the same structure that helped create the problems in the first place

general market • everyone within the advisor's territory as defined by geography, whether voluntarily selected by the advisor or assigned by a company, and restricted to the jurisdictions in which the advisor is licensed to practice

geographic variables • variables that segment a market by using political divisions such as states, counties, cities, boroughs, and so forth, or by territories delineated by neighborhoods, regions, miles, and so on. For most financial advisors, geography is used to define their territory, mainly because of licensing requirements.

goal • a desired end point in personal or organizational development. The intention is usually to reach the goal in a finite period of time.

goal prioritization • the process of ranking goals according to their importance to determine the order for their achievement

goal setting • the process of establishing clearly definable, measurable, achievable, and challenging objectives toward which an individual can target his or her effort, resources, plans, and actions

ideal client profile • the type of prospects with whom the advisor prefers to work. It concentrates more on personality traits and attitudes than on demographics.

marketing • the process of identifying needs that people have, clarifying those needs, and creating and supplying the products and services appropriate to satisfy those needs

marketing coverage strategies • approaches that an advisor can use to select and target markets. There are five basic marketing coverage strategies: single-market concentration, selective specialization, product specialization, market specialization, and full-market coverage.

market segment • a potential target market determined by applying relevant segmentation variables to the advisor's natural markets

Maslow's hierarchy of needs • Abraham Maslow's theory that five primary needs motivate human beings and that there is an order in which these needs are satisfied. In ascending order of importance, these needs are physiological, safety, affiliation, esteem and self-actualization.

mass marketing • a strategy that aims products and services to a general market without accounting for differences in the characteristics or needs of the various groups within it

mentoring • a four-phase process that involves the transfer of knowledge and information from a mentor and knowledge acquisition, application of that knowledge, and critical reflection on the part of the aspirant

moment of truth • a time of customer contact, during which the advisor must make a choice about how he or she will respond to the client. Each of these moments is divided into three parts: the moment (client contact), the choice (advisor behavior), and the truth revealed (demonstration of advisor's attitude toward the client).

monthly business planner • a pocket planner used to help advisors keep track of their weekly and daily business activities and to manage their time

monthly planning • a session in which an advisor establishes a work plan for the coming month. The purpose is to evaluate progress toward annual goals and organize business activities for the month ahead.

natural market • a group of people to whom the advisor has a natural affinity or access because of similar values, lifestyles, experiences, and attitudes

nests (of prospects) • groups of prospects to whom the advisor has access and/or affinity and are therefore easier to contact and establish rapport with

networking • the process of ongoing communication and sharing of ideas and prospect names with others whose work does not compete with the advisor's but whose clients might also be eligible to become the advisor's clients

niche market • *See* target market

nominator • someone who provides referrals by giving names to the advisor and/or by giving the advisor's name to others

no-show • a scheduled and confirmed appointment that the prospect fails to attend, often referred to as a stand-up

objection • a fundamental concern that the advisor's recommendation is not needed (now), that the solution is not right, that it is too expensive, or that it is too much trouble now. These concerns or questions must be resolved before the sale can proceed.

offense time • an advisor's high-priority time during the workday in which he or she must be proactive about acquiring new business and finding new prospects

orphan • a customer whose advisor is no longer available to provide client service

personal board of advisors • a group of three to five (or more) centers of influence who want to see the advisor succeed and are willing to meet with the advisor periodically to give advice, counseling, encouragement, and motivation to follow through on career commitments

personal brochure • typically a one-page document that introduces and is the prospect's first impression of the advisor. Therefore, it should impress, inform, and create interest.

personal observation • a prospecting method based on watching, listening, and engaging in which the advisor gathers information about prospects and target markets and begins to develop a relationship

personal traits • personality, experiences, hobbies, and so forth that make an advisor unique; assets for creating marketing opportunities

Platinum Rule—a variation of the Golden Rule that counsels the advisor to treat the other person the way that person wants to be treated as part of providing extraordinary service

positioning • designing the company's offering and image to occupy a distinctive place, and therefore create a favorable perception of the advisor, in the minds of members in the target market(s)

preapproach • the first step toward making a sale, which involves identifying the prospect and planning how to contact him or her for an appointment by stimulating his or her interest in what the advisor has to offer

preapproach letter • a letter or postcard mailed (or e-mailed) to a prospect to introduce the advisor and arouse the prospect's interest in meeting with the advisor

preconditioning • creating interest so that the prospect knows what to expect and will be receptive to the advisor's request for an appointment

presentation rate • ratio of the total number of presentations the advisor makes divided by the total number of contacts he or she reached in a telephoning session

presentations • in tracking telephone activities, the number of times that an advisor completes his or her script from beginning to end without interruption

prestige building • the advisor's public relations campaign to position his or her personal brand favorably in the advisor's target markets

proactive service • service requests that the advisor cultivates in order to create client contact and opportunities to demonstrate the advisor's reliability and responsiveness. Proactive service sets the advisor apart and is the key to additional sales and referrals.

professional ethics • behavior according to the principles of right and wrong—a code of ethics—that are accepted by the profession

professionalism • the knowledge, conduct, responsibility, and client-focused attitude that characterize a professional financial advisor

prospect • a potential buyer who has been identified by the advisor

prospect card • a 3″ x 5″ index card containing various types of basic data and additional qualifying information about a prospect

prospect file • an alphabetical file within a contact management system that contains the names of prospects, clients, and centers of influence

prospecting • the advisor's continual activity of exploring for new people to meet and talk to about the products he or she sells

prospecting and sales effectiveness ratios • quantitative measurements of an advisor's effectiveness in performing key sequential prospecting and sales activities as they relate to each other and to attaining desired production goals

prospecting methods • different ways to access the prospecting sources for specific names and contact information such as referrals, direct mail, personal observation, networking, Tips Clubs, centers of influence, social mobility, cold canvassing, public speaking, and seminars

prospecting sources • where the advisor will find prospects. There are three types of prospecting sources: people who know the advisor favorably, people recommended by those who know the advisor favorably (referred leads), and people who do not know the advisor at all.

psychographic variables • variables that divide a market by lifestyle and attitudes—for example, leisure activities, values, personality, interests, and hobbies

public image • all the characteristics that make the advisor recognizable and memorable; how people—the advisor's target market in particular—perceive the advisor

qualified prospect • a person who has a need for a financial product, an ability to pay, is suitable for it, and can be approached by the advisor on a favorable basis

qualifying prospects • as the advisor collects information during fact finding, the process of determining whether a prospect is a possible (and eligible) buyer, a probable buyer, or an eager buyer; what the prospect is likely to buy; and when and why he or she will buy

rapport • a relationship marked by harmony and accord, aided by the advisor's friendly, interested concern in the prospect or client

reactive service • service that the advisor provides at the client's initiative

referral • a person to whom the advisor is introduced by someone who knows and values the advisor's work

rescheduled appointment • an appointment that the prospect changes, either at the time the advisor calls to confirm or shortly thereafter

rule of two • the prospect's negative response two times during a phone call to the advisor's request for an appointment. If the prospect shows such a lack of interest, it is best for the advisor to let go of the call and politely excuse himself or herself.

selling • giving up something (property) to another in exchange for something of value (money). This is a transactional view. Client-focused selling has more to do with helping people get what they want and need through a process of discovery. By developing a relationship the long-term benefits to both parties are served and selling takes a back seat to service.

selling/planning process • the eight-step procedure for the advisor to follow from selecting the prospect to completing the sale and servicing the client. It is similar to the financial planning process.

seminar • a prospecting method in which the advisor, alone or as a part of a team of professionals, conducts an educational and motivational meeting for a group of people who are interested in the topic presented. Seminar selling permits advisors to meet a large number of potential clients in one location and at one time.

service package • the level of service the advisor offers to clients, based on the type of client and the relationship that has developed or is expected to develop with the advisor

social mobility • the advisor's involvement in his or her community and recognition as a person who is willing to work for common community interests. Social mobility is a way of building prestige and a good reputation in order to develop prospects and centers of influence.

speaking to groups • giving presentations to organizations or businesses as a way to build prestige, as opposed to generating immediate appointments and sales

strategic alliance • a relationship that a financial advisor develops with professionals in complementary disciplines to whom the advisor can refer clients when their financial situations require the other professionals' expertise and services. These alliances are often reciprocal, involving an exchange of referrals to each other's clients.

stress • mental tension resulting from the gap between an individual's expectations and reality

stretch and leverage strategy • an approach to goal setting used by businesses in which they set what appear to be impossible goals for their size and capacity. Then, relying on resourcefulness and innovation, they leverage available resources into outstanding results.

structure • the arrangement of the key elements in a person's life in relation to each other

suitability • the appropriateness or the proper fit of the characteristics of a product to the needs and goals of a prospect

target market • a group of prospects with common needs and characteristics that make them distinct from nonmembers of the group. The group is large enough to provide a continual flow of prospects and, ideally, has a communication network through which to share information.

target marketing • the marketing strategy that aims products and services at well-defined target or niche markets

telephone approach • the process of the advisor's obtaining the prospect's consent to see the advisor about the services he or she provides, as well as the date, time, and place of the appointment

telephone effectiveness ratios • using statistics compiled for dials, contacts, presentations, appointments, and dialing time, ratios that measure how productive the advisor's time is during each telephone session. The ratios help the advisor determine how much time to schedule each week to accomplish appointment-setting goals when prospecting on the telephone.

telephone techniques • methods used to improve prospecting results, including beginning with a pleasant greeting, asking if the prospect has time to speak, creating interest in the call, asking for an appointment, and confirming the appointment

tickler card • a 3″ x 5″ index card used in a contact management system containing the name of a prospect, client, or center of influence; the reason for contacting him or her; and a specific future date on which to contact the person in the coming year

tickler file • a monthly file that provides reminders of timely actions for the advisor to take regarding sales and service opportunities for prospects and clients

Tips Club • a group of local salespersons who represent diverse industries, formed solely for the regular exchange of information on prospects. Each member of the Tips Club shares his or her own expertise, business connections, and social contacts with the group.

vacillation • bouncing back and forth between good habits and bad ones

value proposition • a compelling reason the advisor offers to prospects to conduct business with him or her; the tangible results the advisor commits to deliver

weblog (blog) • an online journal that allows interaction between the advisor and a web audience

Answers to Questions

Chapter 1

Answers to Review Questions

1-1. Selling and planning are much easier when the advisor is dealing with someone who needs and wants his or her help, can afford it, is suitable for it, and is approachable. In other words, if the advisor sees the right prospects, he or she will have more selling and planning success.

1-2. Through service, the advisor cements the relationship with each new buyer. This gives the advisor the opportunity to make additional sales and obtain referrals. Some service is reactive—the buyer or client initiates it when a change, such as a coverage amount or a dividend option, is needed. In these situations, of course, the buyer or client should expect to receive excellent service. What differentiates one advisor from another, however, is the proactive element of his or her service strategy. Many people buy a product and never hear from the advisor again. Proactive servicing strategies, such as monitoring the plan through periodic reviews, mailing newsletters, and sending birthday cards, enable an advisor to stay in touch with clients. It is this high-contact service that builds clientele.

1-3. Marketing is the planning and implementation of a process that involves
   1. identifying specific consumer needs and wants that the advisor's products and services can meet
   2. defining groups of consumers who have those needs and wants
   3. creating or customizing a solution that meets those needs and wants effectively
   4. creating and articulating messages that raise consumer awareness of these needs and wants
   5. positioning the advisor's ability in the consumers' minds to meet those needs and wants
   6. identifying and contacting specific consumers with whom to meet
   7. motivating consumers to buy the product or service that will satisfy those needs and wants from the advisor
   8. exchanging the product or service for something of value (money)
   9. managing postsale relationships

   Both selling and prospecting are marketing activities. Specifically, selling is the process of motivating, guiding, and asking the prospect to buy—to become a buyer. (Selling involves activities 7 and 8 in the above list.) Prospecting is the advisor's continual activity of identifying, approaching, and prequalifying new people to meet with and talk to concerning the advisor's business. It is the marketing activity that narrows an advisor's focus from the broader market to the selling of products and services to individual prospects.

1-4. The first characteristic of a qualified prospect is that the prospect needs and values the advisor's products and services. If a person does not need the advisor's products or services, no matter how much he or she wants them, that person is not a qualified prospect. A sale made in this situation has a great chance of being noncompliant and unethical. Need is important, but value is critical. Value determines want, and the reality is that people will most often buy what they want before what they need.

   The second characteristic is that the prospect can afford the products or services. The issue of affordability often can be resolved by implementing a spending plan (a budget). However, this works only if the person has an adequate amount of discretionary expenditures that can be reduced and wants what the advisor offers badly enough to make the necessary sacrifice.

   The third characteristic of a qualified prospect is that he or she is suitable for the products or services. Unfortunately, there are people who need and want the products and services the advisor provides but cannot meet the necessary financial and/or health requirements. Working with prospects who will probably satisfy any financial and/or health conditions will increase the advisor's effectiveness and efficiency.

   Finally, a qualified prospect is someone the advisor can approach on a favorable basis. In lines of business where relationships drive most activity, this criterion is critical. A person may meet all the other requirements, but

Answers.1

if the advisor cannot approach him or her on a favorable basis, then he or she is not a qualified prospect right now. Rapport is crucial, especially in a business that gets personal. Financial advisors get about as personal with their clients as anyone can. In fact, the only people who may get any closer are usually health care professionals. However, the advisor should not throw the unapproachable prospect's name away—there may be a way to approach him or her on a favorable basis at some point in the future.

Working with people who have a high probability of being qualified prospects is the key to success for financial advisors.

1-5.   An advisor can overcome a lack of conviction in the following ways:
- Analyze his or her portfolio of products and/or services and identify what excites him or her about these products or services.
- Ask clients how his or her products and/or services are meeting their needs or enriching their lives.
- Ask a mentor to define why he or she picked this business and not something else.
- Identify brand "you"—what the advisor brings to the client-advisor relationship that is unique and valuable.

1-6.   Two ways that an advisor can overcome rejection are as follows:
- Ask, "What is the worst that could happen?"
- Compare prospecting with other activities that were difficult at first, using the same strategy the advisor used to overcome his or her fears involving them: Believe you can do it; do it; and keep on doing it.

1-7.   Ten characteristics of doers:
- They know where they are going.
- They are passionate.
- They believe.
- They create plans.
- They do.
- They are people of strong character.
- They have energy.
- They communicate.
- They take responsibility.
- They are courageous.

1-8.   The five stages of the buying process are
- problem recognition. The prospect becomes aware of a problem or need.
- information search. The prospect searches for relevant information to quantify the need and the possible solutions.
- evaluation of alternatives. The prospect compares the alternatives using various criteria.
- purchase decision. The prospect selects a product or service.
- postpurchase behavior. The buyer is either satisfied or dissatisfied. Satisfaction leads to future purchases and referrals.

1-9.   The following are the different needs in Maslow's hierarchy:
- Physiological needs are for food, water, air, shelter, and clothing. People who want any of these things are first preoccupied with attaining them. Only after these basic needs are met can the individual consider satisfying higher-level needs.
- Security needs are for safety, stability, and the absence of pain or illness. Again, people with these needs become preoccupied with satisfying them. Many workers require medical, unemployment, and retirement benefits to help satisfy needs in this level of the hierarchy.
- Affiliation needs are for love, affection, and a feeling of friendship and belonging. When their physiological and security needs are satisfied, social or affiliation needs arise to motivate individual behavior. These needs are most evident in a person's conduct as it relates to job satisfaction, work ethic, team participation, family ties, and general well-being.
- Esteem needs include personal feelings of self-worth, recognition, and respect from others. People with esteem needs want others to accept them for what they are and to view them as capable. The needs are fulfilled when the individual receives recognition and feedback from others regarding his or her competence and ability.

- Self-actualization needs are the highest in Maslow's ranking. They are the needs for self-fulfillment and the realization of personal potential. People who are striving for self-actualization usually accept themselves and others, are superior at problem solving, are more detached, and have a desire for privacy. One irony is that people driven by ambition often willingly sacrifice everything to achieve their quest.

1-10. An advisor can apply information on prospect behavior to the four marketing activities associated with prospecting in the following ways:
- By identifying the problem-recognition triggers for the advisor's products and services (to the degree they are observable), the advisor can create a profile for prospects.
- People need to know who the advisor is before they are willing to trust him or her. Awareness creation strategies that feature more personal interaction with advisors and prospects are typically more effective than advertisements.
- Not everyone who needs a particular financial product experiences a life event that causes him or her to acknowledge a possible financial need. Even when life events do occur, they often do not motivate prospects to begin searching for information. Thus, advisors need to create and implement strategies that help trigger prospects' recognition of their needs.
- Appointment-setting scripts should reflect the prospects' possible financial and emotional needs.
- Service strategies that cultivate relationships will result in repeat buyers who will refer the advisor—clients.

Answers to Self-Test Questions
1-1. C
1-2. B
1-3. D
1-4. A
1-5. D
1-6. C
1-7. A
1-8. B
1-9. C
1-10. B

Chapter 2

Answers to Review Questions
2-1. An advisor who uses a mass marketing strategy aims his or her products and services to a general market without accounting for differences in the characteristics or needs of the various groups within it. In contrast to mass marketing, target marketing aims products and services at well-defined target or niche markets. A target market (niche market) is a group of prospects with common needs and characteristics that make them distinct from nonmembers of the group. The group is large enough to provide a continual flow of prospects and, ideally, has a communication network through which to share information. Furthermore, the common characteristics are actionable, meaning that they provide a practical way to distinguish members from nonmembers and to create unique marketing messages and approaches that produce sales.

2-2. Five advantages to advisors that result from target marketing are
- enhanced referability
- increased customer loyalty
- less competition
- higher profits
- greater sense of satisfaction in their work

2-3. It is important that an advisor define his or her products in terms of the prospect's financial needs because people buy a product for various reasons, financial and emotional. One person may buy a mutual fund to save for a child's education. Another may do so to accrue a sizable walk-away fund. Identifying financial needs can open the advisor's eyes to the real needs that his or her products can address. In addition, having an understanding of the financial needs allows the advisor to spot triggers, typically life events, that are likely to motivate prospects to take action. Knowledge of these triggers enables the advisor to create specific marketing messages and perhaps send them at appropriate times (the birth of a child, a marriage, and so on).

In addition, the advisor needs to define his or her product in terms of the advisor's personal traits because of the importance of approaching prospects on a favorable basis. Personal traits have much to do with likability. Prospects want to work with advisors they can trust. All things being equal, they will trust someone whom they like. Thus, it is important for the advisor to identify his or her personal traits not only for selecting appropriate target markets and prospects, but also for the personal branding that will help the advisor connect with them.

2-4. a. An example of affinity is an advisor's friends. These are people the advisor knows and with whom he or she shares a common interest, stage of life, hobby, and so on.

b. An example of access is members of an advisor's former profession or his or her spouse's current profession. Natural markets created by access are groups of people that the advisor does not know directly (yet) but with whom he or she has enough in common to open the door.

2-5. The four types of segmentation variables are as follows:

- geographic variables. These variables segment a market by using political divisions such as states, counties, cities, boroughs, and so forth, or by territories delineated by neighborhoods, regions, miles, and so on. For most financial advisors, geography is used to define their territory, mainly because of licensing requirements (advisors must be licensed in each state in which they practice).

- demographic variables. Partially because they are easier to measure (through a census or marketing data companies), demographic variables are the most commonly used segmentation variable. Even more important is that the differences in prospects' needs and wants are often linked to these variables. Along with age, demographic segmentation includes variables such as gender, education, ethnicity, occupation, income, size of family, marital status, religion, generational cohort, (baby boom, generation X, silent generation, and so forth), and family situation (single, married with kids and a single income, married with no kids and two incomes, single parent, empty nester(s), divorced, and so on).

- psychographic variables. Advisors use psychographic variables to divide a market by lifestyle and attitudes. These variables include things like leisure activities, values, personality, interests, and hobbies. For example, the advisor may have a few clients who are running enthusiasts, play in a softball league, participate in a bowling league, and so forth.

- behavioristic variables. Finally, behavioristic variables group people by their buying and usage behaviors. This type of segmentation categorizes people according to when they buy—life events (birth of a child, marriage, divorce), type of user (do-it-yourselfer, collaborator, delegator), brand loyalty, benefits sought (convenience, price, quality), and so forth.

2-6. Advisors can create their own criteria for selecting a target market. Three general categories are often used by marketers in general:

- fit of resources to segment's needs. How well do the advisor's products match the segment's needs? The advisor should think in terms of the priority each market segment places on his or her products as well as their affordability. In addition, the advisor should evaluate how well his or her personal traits match with each segment. Will the advisor's personality and experiences give him or her an advantage in a particular segment?

- level of income. The advisor needs to evaluate each segment in terms of the likely product purchases members will make over a lifetime. Also, the advisor should consider how referable he or she will be within a market segment. Do members often communicate with one another? Are they known for giving referrals? Furthermore, the advisor should estimate the size of each market segment and whether or not the segment is growing enough to warrant the advisor's attention.

- level of competitiveness. Finally, the advisor should assess how competitive he or she will be in each market, looking at how many competitors there are, how are they positioned, and whether a market segment has a high or low regard for them.

2-7. The five basic market coverage strategies are

- single-segment concentration. This strategy is the narrowest. It involves marketing one product to a single market segment. It may be either a rather large market segment or a small but very lucrative one (doctors, dentists, professional athletes, and so forth).

- selective specialization. This coverage strategy involves marketing a few different products to multiple target markets. It diversifies an advisor's market portfolio in terms of both its products and target markets. Advisors who sell more than one product can pursue this coverage strategy effectively.
- product specialization. Advisors who use the product specialization strategy market a product to multiple target markets.
- market specialization. Market specialization is like the single-segment concentration strategy in that the advisor specializes in one market's needs. It allows the advisor to devote his or her undivided attention to a particular target market and to become an expert in that market's financial needs. The difference is that the advisor sells multiple products to that market.
- full-market coverage. In this strategy, all products are marketed to all segments. This is the least efficient strategy for individual financial advisors but is a viable strategy for multi-advisor practices.

2-8.   At the heart of positioning is persuasion. Per Aristotle, there are three elements to persuasion: ethos, logos, and pathos. Ethos refers to the person's character. Is the person credible and trustworthy? Logos refers to the facts of an argument. Is the argument logical? Finally, pathos refers to the emotional appeal of the argument. How does it make the listener feel?

All persuasion must be rooted in the truthful presentation or reflection of the facts. It is important to identify the relevant facts that pertain to the prominent financial needs of the target market. Recall that a buyer responds to external stimuli that trigger a sense of need. Triggers are usually based on an emotional response. Therefore, the positioning of the advisor's products must also appeal to prospects' emotions.

2-9.   Positioning a product should accomplish three things:
- First, it should help prospects see their financial needs clearly. This takes a few well-placed facts that stir up prospects' emotions. The advisor should consider creative ways to pique prospects' curiosity.
- Second, prospects should have a general sense of the cost of doing nothing compared to the cost of addressing their financial needs. In other words, the advisor must help them see that "a stitch in time saves nine"—that it is more costly to do nothing than it is to tackle the problem now while they can.
- Third, the positioning of products should motivate prospects to take action. It should be a trigger to move them from awareness of their need to searching for information and being willing to solve their problems.

Answers to Self-Test Questions

2-1.   D
2-2.   B
2-3.   B
2-4.   C
2-5.   C
2-6.   C
2-7.   D
2-8.   C
2-9.   A
2-10. C

Chapter 3

Answers to Review Questions

3-1.   Prospects are people advisors are looking for, prospecting sources are where advisors will find prospects, and prospecting methods are the different ways to access the prospecting sources for specific names and contact information.

3-2.   The three types of prospects are as follows:
- people who know the advisor favorably. These are prospects who already have a favorable impression of the advisor, essentially the advisor's natural markets. Examples include current clients and buyers, friends and family, businesses the advisor patronizes, current organizations to which the advisor belongs, and any community contacts.

- people recommended by those who know the advisor favorably. These are prospects the advisor obtains by taking people from the first group and asking them to provide referrals. Examples include nominators (referrals and centers of influence) and orphan accounts.
- people who do not know the advisor at all. These are prospects who are total strangers and for whom the advisor does not know someone who could nominate him or her. Examples include directories and membership lists, local papers and publications, public records, business information sources, and direct marketing lists.

3-3.   Principles for selecting prospecting methods are as follows:
- The type of prospecting source will influence the type of method the advisor uses to obtain a constant stream of names. Some methods are more appropriate for one type of source than another. But prospecting requires some creativity. By not pigeonholing a method to one type of source, the advisor can evaluate its merit without any bias.
- The personal nature of the business means that personal methods for prospecting are generally better.
- Some prospecting methods are also effective for creating awareness and building prestige, as well as functioning as a preapproach and even an approach for an appointment. Seminars fall into this category.

3-4.   Advisors should follow three basic steps to get started with building a network:
- Start with people they know. Advisors should identify people they already know who are members of a target market, work with members of a target market, or know people who can help them. Through friends and family, advisors probably already have a strong network in place.
- Extend their contact base. They should collect new contacts each day, and utilize current contacts to meet new ones.
- Build relationships. It is the much deeper side to networking, and it is a little more challenging. But building relationships is also better and more powerful than simply making contacts because relationships are ongoing and more sincere.

3-5.   The following are the eight principles for asking for referrals:
- Be referable. Successful advisors earn referrals by cultivating relationships with clients and centers of influence.
- Pave the way. Advisors should not let the request for referrals surprise their prospects. They should mention it early in the process.
- Recognize opportunities. There are some key situations in the selling process where people are often open to providing referrals: after the interview is completed, at policy delivery, and after service has been provided. In addition, good times to ask are when advisors hear prospects say something about the value they are receiving from meeting together.
- Know what to say. If advisors plan fail to plan, they will miss the opportunity. They should write a script and practice it.
- Prepare to handle concerns. Advisors need to plan how they will approach prospects who are reluctant to give them referrals. They must know how they will probe to understand prospects' reluctance and what they will say to exit gracefully if the reluctance is deep-seated.
- Qualify the names. Information beyond the standard name and contact information is necessary for qualifying prospects, and advisors need to know how to ask for it and to recognize what information is important.
- Obtain a personal introduction. Advisors should know what they will say to obtain one. Depending on the situation or relationship, the referrer could arrange for a face-to-face meeting with a prospect and an advisor. If that is not possible, an introductory phone call, letter, and e-mail are other options.
- Follow up with the referrer. Finally, the referral must become a priority, and the advisor should follow up with a note or phone call to let the referrer know the result. Some advisors have used token thank-you gifts. Others eschew such measures.

3-6.   Many times advisors fail to obtain referrals not because the referrer does not want to help the advisor, but because the advisor asks in a manner that does not give the referrer a clear understanding of the type of prospects the advisor seeks. Effective approaches for asking for referrals:

- list approach. The advisor makes a list of prospects that the referrer probably knows (neighbors, other members of an organization, other employees, and so forth) and asks the referrer if he or she knows any of them.
- the whom-do-you-know approach. The advisor creates a list of criteria (ideal client profile, target market profile, or observable triggers) and asks the prospect if he or she knows anyone who meets the criteria.
- association approach. Another way is for the advisor to ask clients or friends about their involvement with local professional or industry associations. The objective is to be introduced to an officer of the association or organization.

3-7. Advisors can cultivate a relationship with a center of influence by
- focusing on the center of influence's interest and what the advisor can do for the COI, specifically finding ways to help
- volunteering to complete a task, whether business-related or even community-oriented. By completing the task well, the advisor demonstrates his or her competence.

3-8. Three aspects of a prospecting system:
- It manages contact information.
- It allows for quick access to information.
- It enables advisors to identify priority prospects.

3-9. Three record keeping-tasks and their importance:
- First, advisors should summarize information about their prospects, specifically prospecting sources and methods. This will help advisors eliminate sources and methods that are not producing prospects and focus on ones that are.
- Second, advisors should budget and track their time to ensure that they are spending the time they need to complete the activities that lay the groundwork for successful prospecting in the future—activities such as networking.
- Third, advisors should monitor their prospecting and sales numbers. Reviewing these numbers daily and weekly will help advisors remain on target to meet their goals.

Answers to Self-Test Questions

3-1. A
3-2. B
3-3. D
3-4. C
3-5. A
3-6. C
3-7. D
3-8. B
3-9. A
3-10. C

Chapter 4

Answers to Review Questions

4-1. The four levels of participation related to community involvement are
- sponsoring and giving. This is the easiest level in terms of time and energy. All it involves is donating money, sponsoring an event, or buying a much-needed item for a local organization.
- volunteering. At the volunteer level, the advisor has a low level of commitment but is still afforded a great opportunity to build relationships with other volunteers and members. In addition, it is an excellent way to explore whether or not to join the organization.
- joining. An advisor should join organizations only if he or she is serious about supporting the charter and goals of the organization. The advisor should carefully assess the amount of time needed to honor commitments made to the organization before he or she makes a decision to join. The benefits of

membership are that it gives other members (and the community) an opportunity to discover the advisor's talents, values, and character.

- leading. At the highest level of commitment is leading, either by holding an office or chairing an initiative the organization sponsors. This level of involvement provides the greatest level of visibility and requires a great deal of time and attention. Not only will the advisor be able to showcase his or her talents and abilities, but he or she will also rub shoulders with the leaders of the organization who are usually the more successful and influential people in the community, and thus are potential centers of influence.

4-2.   An advisor who wants to use writing as a means to increase social mobility should consider doing the following:
- Research publications that prospects and clients read.
- Explore these publications. Contact the editors and ask about their needs.
- Be aware that the article should not pitch the advisor's product or service. Instead, it should offer useful information that is relevant to the target market the publication (and the advisor) is aiming to reach.
- If the advisor gets stuck, he or she can relate a client's experience (without the name, of course), outlining the client's situation and need as well as a brief synopsis of the solution.
- Another approach is to present solutions or a message in numbered points, such as "The Ten Secrets of Estate Taxation."
- Leverage the article for all it is worth. If the advisor cannot reprint it for free or a nominal cost, he or she can quote from it or obtain permission to link to it from his or her website.
- Give credit where it is due. This includes attribution to sources as well as a few brief sentences about the advisor.
- Obtain approval from the compliance department.

4-3.   To get started with speaking to groups, the advisor can take a class on public speaking and join a speech club like Toastmasters International. In addition, the advisor can begin saving articles and information on various topics that his or her target market would find interesting. Once the advisor has a certain level of confidence, he or she can begin to give a few speeches on topics with which he or she is familiar.

4-4.   Marcie should describe how the Chartered Advisor in Philanthropy will help her clients and prospective donors achieve their philanthropic goals. For example: "Marcie earned her Chartered Advisor in Philanthropy in 2006. This designation gave her exposure to both tried-and-true and innovative strategies that she can use to help you make a difference in carrying out your philanthropic goals."

4-5.   The five objectives for the website are as follows:
- Create an emotional connection. This is the most basic and universal objective for a website (and is the same as for a personal brochure). The website should help prospects feel comfortable with the advisor before contact is made with them. To do this, the advisor should provide an overview of what the advisor does, communicating his or her value proposition from the prospects' perspective. In addition, the advisor should personalize the website to the degree that the target market desires. This could mean including a photograph, a brief overview of the advisor's personal history, and a description as to why he or she is in this business.
- Establish credibility. Prospects want to do business with an advisor they trust. A website is an opportunity to establish some credibility. The advisor can do this by listing his or her credentials and explaining how they benefit prospects. In addition, a website can be a great place to demonstrate expertise through articles, case histories, and testimonials from satisfied clients—emphasizing results rather than enumerating platitudes.
- Sell. Although this has not been a very successful use of a website in general, there are a few advisors who have done well.
- Educate and motivate. Most people go online to find information about the product or service they are interested in purchasing. This is especially true for members of generations X or Y. Thus, advisors should strongly consider devoting a good portion of their websites to education, including articles about financial topics, financial calculators, worksheets, and spreadsheets to help prospects make decisions, and links to other helpful websites.
- Service. Many clients will appreciate the convenience of online service for simple transactions. It is advisable to have instructions for what the client needs to do and the information he or she needs to effect a transaction. This will expedite servicing and increase office efficiency.

4-6.   There are four key advantages related to seminars:
- They increase efficiency. The multitasking nature of seminars makes them a very efficient marketing tool. If conducted properly, seminars enable the advisor to qualify several prospects all at once, resulting in a time savings, because the amount of time needed to qualify prospects in the initial interview is reduced.
- They are an effective prestige builder and preapproach method. Assuming that the advisor makes educational and motivational presentations, seminars build the advisor's credibility as an expert. In addition, seminars create an awareness and a sense of the financial needs prospects should address.
- They maximize the advisor's public speaking skills. For the advisor who naturally gravitates toward public speaking, seminars are an excellent way to leverage this asset.
- They provide a natural segue into an appointment. If the seminar has effectively presented the need and created a sense of importance and urgency, offering an appointment is the natural next step. The advisor can either ask for the appointment at the conclusion of the seminar or mention that he or she will be calling in the next week to set up an appointment.

4-7.   Important tasks for planning a seminar:
- Set an objective or goal.
- Establish a budget and work to stay within it.
- Define the target market. Successful seminars involve target marketing. The prospects invited should have a common need or interest. That allows the advisor to address specific needs with greater ease.
- Determine seminar content. The content of the seminar should be a blend of technical information and motivational material. The topic should reflect the needs and interests of the target market. (The advisor also needs to consider the constraints he or she may face to satisfy compliance requirements. In fact, compliance approval should be built into the timeline if the advisor is going to develop his or her own materials.)
- Consider the benefits of a team approach. The use of other professionals, such as an attorney or accountant, communicates more objectivity and can potentially provide more value, depending on the topic presented.
- Select a date and time. The choice of a date and time may seem mundane, but it can mean the difference between having a large audience or a lot of empty seats. The advisor should be aware of other events (concerts, sporting events, television shows) that could potentially compete with a seminar.
- Choose a location. The site should be convenient for the members of the targeted group. Parking may be a prime consideration in urban and suburban areas. Also, the location should be neutral. Thus, the advisor's office conference room may not be a suitable choice. Consider a high school or college classroom or the conference room at the local library. These are inexpensive and neutral locations.
- Create the invitation. Once a target audience is selected, it is time to create the invitation. The invitation should clearly inform prospects that the seminar will be educational in nature. It should provide the topic, date, time, and length of each seminar as well as any fees to be paid. The seminar title should be clear and relate to the perceived needs of the targeted audience.
- Check and recheck the facilities. It is a good idea for the advisor to visit the facility while another meeting is in progress to evaluate the lighting, the sound system, the projection system, and so on.
- Create a mechanism to evaluate seminar attendees' feedback regarding the quality and usefulness of the presentation.

4-8.   Writing an effective preapproach letter takes a lot of work. Below are a few guidelines:
- Write something that grabs the prospect's attention.
- If possible and necessary, establish a basis for contact by referring to how the advisor heard of the prospect.
- Describe the most probable and acute financial need of the prospect that the advisor's product addresses.
- Identify the link between the financial need and an appropriate emotional need.
- Do not overstate the need; that is manipulative.
- State the value proposition clearly and concisely, using one to two sentences. Avoid platitudes and clichés.
- Confirm the credibility of any statistics. Use statistics responsibly and appropriately (do not overuse them).
- Pay strict attention to wording, grammar, spelling, and punctuation.
- Use quality stationery and typeface to ensure that the letter conveys an image of professionalism.
- Consider asking a current client from the target market to review the letter's message and appearance.

Answers to Self-Test Questions

4-1.	C
4-2.	A
4-3.	A
4-4.	C
4-5.	B
4-6.	D
4-7.	C
4-8.	D
4-9.	A
4-10.	B

Chapter 5

Answers to Review Questions

5-1. An elevator speech is a short (15 to 30 seconds, 150 words or fewer) sound bite that succinctly introduces the advisor and helps him or her connect with someone the advisor has just met. It is designed to create interest and give the prospect a quick synopsis of what the advisor does and how he or she can help people. It is used in such impromptu prospecting settings as when the advisor is in an elevator, shopping at the supermarket, waiting in line at an ATM, or purchasing a morning latte. The elevator speech spotlights the advisor's uniqueness, focuses on the benefits he or she provides, and creates a lasting first impression that showcases the advisor's professionalism. The objective is for the advisor to position himself or herself so the he or she can obtain permission to talk to the prospect again, either on the phone or face-to-face in an appointment setting.

      The 10-second commercial is a variation of the elevator speech, usually in response to the question, "What do you do for a living?" Especially in social situations, it provides an excellent way for the advisor to stimulate both awareness and interest in what he or she does, which can lead to an appointment. The key is for the advisor to customize the 10-second commercial to the prospect and refer to a need that he or she might have that tells the person specifically what the advisor can do for him or her.

5-2. The exceptions under the current Do Not Call (DNC) Law are as follows:

- If there is an established business relationship, the DNC restrictions do not apply. An established business relationship is present when there is a product or service in place, and it continues for 18 months after that product or service is no longer in effect or active. Several states have stricter requirements. If a consumer contacts an advisor, whether by phone, mail, or in person, to inquire about a product or service, an existing business relationship exists for 3 months after that inquiry (referrals do not satisfy the established business relationship exceptions and are not a basis to call someone on the DNC list).
- Advisors may make calls to a person on the DNC lists if they have prior express permission signed by the consumer whereby the consumer agrees to be contacted by telephone. If received before the call, written permission is valid indefinitely unless the consumer revokes it.
- The DNC regulations do not apply to business-to-business calls.
- Calls may be made to people with whom an advisor has a personal relationship such as family members, friends, and acquaintances.
- Any charitable organization is exempt from the Do Not Call Law.
- Individuals can be solicited at their place of employment because business phone numbers are also exempt from the law.

5-3. The following are the four principles of phoning:

- The advisor should not sell on the phone.
- People are irritated by telephone solicitations.
- Not all leads are created equal.
- There is no such thing as call reluctance.

5-4. The three main elements in preparing for a phoning session are that the advisor be

- mentally prepared—in a good mood and enthusiastic about the call
- physically prepared—having the following tools ready:
  - leads with correct phone numbers
  - counting sheets
  - appointment book or palm pilot
  - pen or pencil
  - uncluttered desk
  - clear, working phone
- verbally prepared—knowing what to say. The advisor must memorize the script and various ways to handle prospects' responses.

5-5. The seven basic components in a telephone script are
- part A—the greeting
- part B—the introduction (the advisor's name)
  - part C—company affiliation
  - part D—the connector
  - part E—the offer of information via a face-to-face appointment
  - part F—the benefit of the appointment to the prospect
  - part G—alternative-choice close

5-6. There are three variations of part D (the connector) of the telephone script:
- memory jog (D1)—This connector is used with somebody the advisor has met at least once prior to making the phone call such as acquaintances from personal, social, employment, school, organization, networking, or business affiliations, or a someone he or she met in a casual encounter. Clients, family, and friends are also part of this group. For all prospects in the D1 category except clients, the sequence of the phone script is parts A, B, D, then C.
- referral or orphan lead (D2)—Someone else told the advisor to contact prospects in this category. With referrals, a prospect, client, friend, or center of influence suggested that the advisor contact the individual. With orphan leads, the company for which the advisor works asked him or her to contact the person. The phone script sequence for referrals follows the same sequence as with the memory jog approach: parts A, B, D, then part C. With an orphan lead, the order of the phone script is parts A, B, C, then D. This is because the company name (not the advisor's name) is recognizable to the prospect, and the company asked the advisor to make the call.
- credential or specialty (D3)—This approach establishes a connector to a person the advisor does not know. It is most effective in a direct mail program where the advisor sends a preapproach letter that describes a product or service and follows up with a phone call. The third party here is the letter the advisor sent to the prospect, which serves as the credential that justifies the call in which the advisor responds to the prospect's request for more information. If there was no preapproach letter, the advisor is making a cold call. For this type of call, the connector is the advisor's specialty—that is, his or her area of expertise. The phone script sequence in a D3 connector is parts A, B, C, and then D.

5-7. The way for the advisor to avoid sounding canned on the phone when using a written script is to compose it carefully, making sure to include all the necessary seven components. After reading the script out loud and approving the exact words, the advisor should rewrite the script in short, bulleted phrases. That way, he or she will "talk" the key points in the proper order and remember the important words, which will help the advisor to avoid sounding like he or she is reading a script.

5-8. The procedure for handling a prospect's questions on the telelphone is to respond briefly, using some variation of "Yes," "No," or "That depends." The advisor should avoid answering too many questions; questions answered on the telephone can eliminate the need for an appointment. It is imperative that the advisor is brief in responding to questions and remembers to ask the prospect for an appointment after answering each question. After the prospect's third question, the advisor should be able to set up an appointment because three questions are a definite sign of interest.

5-9. The prospect typically has one of the following four problems with or objections to meeting with an advisor:
- "I don't have any money."
- "I already have an advisor."

- "I'm too busy. Call me sometime in the future."
- "I'm all taken care of at my job."

The advisor's best response is to acknowledge the problem, show that he or she respects the prospect's opinion, and then address the problem so that it does not become an obstacle to the appointment. The focus must be on giving the prospect a reason to meet with the advisor, not on selling any financial product over the telephone.

The "feel, felt, found" technique is an effective method for handling objections. These three statements follow immediately after the propsect states a problem or raises an objection:

- "I can appreciate you feel that way."
- "Other clients of mine initially felt that way too."
- "But they found, after meeting with me that. . . ."

Following the phrase "But they found . . . that" in the third sentence, the advisor must add a benefit that relates to the stated problem (See section 3 of appendix A.) The advisor should then add:

- "I'd like to see if I can do the same for you."
- "Sound fair enough?" (This is a rhetorical question; the advisor should not wait for an answer.)
- "With that in mind. . . ." (The advisor should close the call by making an appointment.)

5-10. It is necessary to use a counting system to track the following six elements: dials, contacts, presentations, appointments, dialing time, and phone session time of day and day of the week. Using statistics compiled for for these elements, the advisor can quantify the following four telephone effectiveness ratios:

- contact percentage (contacts ÷ dials)
- presentation rate (presentations ÷ contacts)
- appointment-setting rate (appointments ÷ presentations)
- appointments per hour (appointments made ÷ dialing time)

These ratios enable the advisor to measure how productive his or her time on the phone is during each session. Effectiveness will vary based on geographic area, the time of day the calls are made, markets called, and the source of the leads. The ratios will help the advisor determine how much time to schedule each week to accomplish his or her appointment-setting goals when prospecting on the telephone.

5-11. There are four possible outcomes regarding appointments made over the phone:

- confirmed appointment—one that is scheduled with a prospect for a specific date, time, and place, to which the prospect (within 24 hours before it is set to begin) reiterates his or her commitment to attend. The advisor should always call to confirm 24 hours before the appointment. If he or she gets the prospect's voice mail, the advisor should leave a brief reminder message, stating the time the advisor will be there for the appointment. If the advisor is calling to confirm an appointment at the prospect's house, the advisor should be sure to get directions. For business owners, on the other hand, it is better to use an Internet service to get directions.

- rescheduled appointment—one that the prospect wishes to change. Rescheduling can occur either at the time the advisor calls to confirm or shortly thereafter. The advisor should try to get a new appointment for the same time in the next week. Sometimes reschedules can be challenging, particularly if the advisor keeps reaching the prospect's voice mail and winds up calling him or her repeatedly to set a new time.

- cancellation—an appointment the prospect refuses to attend even though the advisor has offered to reschedule. Cancellations are relatively rare. When one does occur, the advisor should not take it personally but move on to another prospect.

- no-show—a scheduled and confirmed appointment that the prospect fails to attend, often referred to as a stand-up. If the advisor arrives at an appointment and the prospect is not there, the advisor should call on his or her cell phone, leave a message showing concern for the prospect, and ask him or her to let the advisor know that the prospect is okay. Despite the tendency to get angry, there is a greater likelihood of hearing back from a no-show prospect if the advisor remains calm. A maximum of 20 minutes is a long enough time to wait for the prospect to arrive before leaving a no-show appointment.

5-12. Whether the advisor prequalifies prospects prior to meeting with them depends on the type of practice and personal views. If the advisor markets multiple financial products, prequalification before the initial interview may not be crucial because there are other products to which to turn to satisfy other needs prospects may have. Prequalifying, however, will enable the advisor to prepare a smooth transition to these other needs and products.

If the advisor wishes to prequalify before the initial interview, the advisor must decide what information he or she needs to know and build a script. Choosing what questions to ask is based on the advisor's philosophy of prequalification.

There must be a smooth transition from getting the appointment to asking the prequalifying questions, which is why a script is important. Prequalifying questions should be short and general. After the prospect answers the questions, the advisor should refer to his or her company's underwriting rules to determine how and whether to proceed. (The advisor should prepare a short script to use to let the person down gently if the advisor knows the prospect will not qualify.)

Answers to Self-Test Questions

5-1.   C
5-2.   C
5-3.   B
5-4.   A
5-5.   A
5-6.   C
5-7.   D
5-8.   C
5-9.   C
5-10. D

Chapter 6

Answers to Review Questions

6-1.   Goal setting is the process of establishing clearly definable, measurable, achievable, and challenging objectives toward which an individual can target his or her effort, resources, plans, and actions. On a personal level, goal setting is a process that allows people to identify and then work toward realizing their own objectives such as financial or career-based goals.

Having goals enables an individual to maintain focus. Focus can help drive the individual to greater personal achievement.

6-2.   The three keys advisors can use to strengthen discipline are as follows:
- Delay gratification.
- Make choices before the "moment of choice" arrives.
- Know what they want.

6-3.   Advisors should use the following five techniques to break through walls they may encounter:
- Have a goal.
- Divide each goal into manageable parts.
- Anticipate obstacles and overcome them.
- Stay objective and focus on the facts.
- Surround themselves with winners.

6-4.   Structure refers to the arrangement of the key elements in a person's life in relation to each other. The structure of an individual's thinking affects his or her actions, attitudes, and emotions. Changing behavior without changing the underlying structure of how the individual works will result in failure.

Structure is a means to help a person achieve a goal. Although the person must adhere to certain structures, he or she also has the power to create new structures that lead toward his or her goals. The challenge is for the individual consciously to establish structures that consistently support his or her highest and truest goals.

6-5.   There are four fatal flaws in the "fit" approach to goal setting:
- First, a focus on getting rid of what someone does not want, or problem solving, rarely results in creating what he or she truly does want.
- Second, small goals fail to bring out the highest and best in anyone. They hardly ever inspire sustained action, let alone consistently outstanding results.
- Third, problem solving seldom accounts for unpredictable changes, which are an inevitable reality.

- Fourth, conventional problem-solving approaches are based on what someone knows is feasible today. This limits the individual from the very beginning. Essentially, the approach prevents the person from stretching for goals for which no conventional approach is currently available.

6-6. Being creative is traditionally viewed as applying existing methods a little differently to a problem for the purpose of becoming more productive.

Creating, on the other hand, goes far beyond what is conventionally regarded as being creative. To create means to bring something new into existence. The creating process is a fundamentally different way for a person to approach what he or she does. It is a reliable, step-by-step process that leads to tangible, recognizable results. The end result of the creating process is the same—a new creation that did not exist previously.

6-7. The four-step goal-setting process is as follows. The individual should

- start with a clear, compelling picture of what he or she wants (the person's vision)
- assess the current situation
- prepare for creative tension
- use feedback and make adjustments

6-8. Time wasters can be divided into two broad categories—major and minor:

- Major time wasters stand between someone and what he or she wants to accomplish. They include problems with the person's attitude, goals, objectives, priorities, plans, and ability to make basic decisions. These problems can surface in the form of procrastination, which is a fairly universal time management challenge that can be difficult to overcome if left unchecked.
- Minor time wasters are distractions that hinder someone once the person is on the way to accomplishing what he or she wants to achieve. They include daily interruptions by coworkers, extended telephone calls, unexpected visitors, lengthy meetings of minor importance, unnecessary reports, excessive e-mails, annoying junk mail, and prolonged phone messages. Others include coffee breaks, surfing the Internet, and searching for or re-creating misplaced materials.

6-9. The green, red, and yellow time management zones can help financial advisors use their time more effectively as follows:

- In the green zone (offense), the advisor must be proactive about acquiring new business and finding new prospects.
- In the red zone (defense), the advisor needs to hone his or her selling/planning, product knowledge, and service skills. Generally, advisors tend to spend too much time in the red zone. It is important for the advisor to devote time and energy to activities that only he or she can do and delegate the remainder to others.
- In the yellow zone (neutral), advisors perform activities that are not necessarily directed toward business goals but that give their life a sense of completeness.

The green, red, and yellow time management zones allow the advisor to accomplish the following:

- Build structure. The main point of a structured day is to enable the advisor to manage his or her time, rather than letting time manage the advisor.
- Stay focused. By not allowing other types of activities to intrude during the advisor's planned time zones, he or she will be able to stay focused on the scheduled activity.
- Maintain balance. This system will help the advisor to see on what type of activities he or she is spending his or her time.

6-10. Some techniques advisors can use to schedule their time, prioritize their activities, and manage interruptions are as follows:

- To better schedule their time, advisors should make an effort to combat procrastination, use the 80/20 rule to help concentrate their efforts on the 20 percent of the tasks that will generate 80 percent of the benefits, and delegate wisely.
- To prioritize their time, advisors should keep a daily to-do list; spend 10 minutes in a priority-setting conference at the beginning of the day; and use a daily calendar to keep track of activities, to-do lists, relaxation cues, and meeting agendas.
- To help manage interruptions, advisors should control phone interruptions by being unavailable during selected parts of the day, and schedule time for interruptions and unforeseen problems and quiet periods for relaxing and unwinding.

Answers to Self-Test Questions

6-1. B
6-2. C
6-3. B
6-4. C
6-5. A
6-6. C
6-7. B
6-8. A
6-9. B
6-10. D

Chapter 7

Answers to Review Questions

7-1. A financial services advisor's job is to service clients, not just sell them products. People like to be served. A client-focused sales approach emphasizes helping clients by providing solutions to their needs and goals. The client's satisfaction with the process and the results is crucial to maintaining existing business and procuring new business.

Providing extraordinary service is essential to building client loyalty—and client loyalty is absolutely necessary for repeat business and introductions to new prospects. This is where service and prospecting intersect. The sale is made when the application is completed and the check is signed, but it is the service that leads to repeat sales and referrals to new prospects. Some advisors incorrectly view client service as something that takes time away from their selling activities. The successful advisor, however, recognizes that each service opportunity is a marketing opportunity—a chance to demonstrate to clients that he or she is reliable, responsive, trustworthy, ethical, and empathetic to clients' needs. When servicing opportunities are seen as marketing opportunities, referrals and more sales follow.

7-2. Every advisor should take a moment and imagine that the advisor is his or her own client. The advisor should ask this question, "Would I hire me? The advisor should place himself or herself in the client's position. How would the advisor rate his or her own service? Is the advisor providing the kind of service that he or she would want to receive? Is the advisor fulfilling the "Golden Rule"? This is a tough question; advisors need to answer it honestly.

Perhaps an even better way to look at the service and client-advisor relationship question is to consider the Platinum Rule "Treat others the way they want to be treated." By focusing on the feelings of others, the relationship shifts from "This is what I want, so I'll give everyone the same thing" to "Let me first understand what the other person wants and then I'll give it to that person."

7-3. Extraordinary service has the following characteristics:
- People go above and beyond what someone would normally expect—just to please that person.
- People are polite, respectful, and it is obvious that they care.
- Service is outstanding, delivering what the customer wants, and then some.
- Products are top quality.

7-4. A moment of truth is a time of client contact, during which the advisor is confronted with a situation for which he or she must make a choice about how he or she will respond to the client. The choice that the advisor makes reveals the truth about how deeply the advisor cares about his or her clients. In the client relationship cycle, there are many different moments of truth. Each of these moments—and their outcomes—can be divided into three parts: the moment (the stimulus), the choice (the response), and the truth revealed (the message the advisor sends to the client).

7-5. Orphans are clients who have no advisor. Because they have been given no service for some time—perhaps several years—they feel that they have been abandoned and are therefore not often receptive to speaking with any advisor of that particular company.

7-6. The eight laws of extraordinary service are as follows:
- Law # 1: Serve for the joy of it. The advisor should serve because he or she wants to, for the enjoyment of helping others.

- Law # 2: Never substitute convenience for service. Service should never be an inconvenience; it should always be an honor and a privilege.
- Law # 3: Every complaint is a cry for service, so listen to the client and understand what he or she needs.
- Law # 4: Seek out moments of truth, and become legendary. They allow the advisor to show his or her ability, willingness to serve, and level of care about his or her clients.
- Law # 5: As an advisor, remember whom you serve. The needs of the customer, not the organization, must come first.
- Law # 6: No person is an island. An advisor cannot deliver extraordinary service alone. The advisor is part of a larger company and possibly a team of associates within his or her own organization.
- Law # 7: Serve more; sell more. Focusing on benefits to clients will make the advisor more valuable.
- Law # 8: What goes around, comes around. By serving and concentrating attention on others, the advisor will receive help in return, in the form of additional business and referrals.

7-7. Realistically, not everyone who purchases a product from the advisor will want to become his or her client. Likewise, there will be some people whom the advisor would prefer not become clients. This means that the advisor needs to segment customers into different categories, with each level receiving a different service package from the advisor. The A-B-C method to categorize customers divides them into three levels:

- "A"—individuals who believe in the advisor, the products he or she sells, and the services the advisor provides. "A" clients are a source for repeat business and referrals. The advisor's long-term goal is to deal only with these clients, who merit the highest level of service.
- "B"—customers whom the advisor wishes to turn into clients or customers who have purchased a product from the advisor but have not yet committed to a full client-advisor relationship. The advisor's goal is eventually to lower them to the "C" group or raise them to client status.
- "C"— someone who has bought a product from the advisor, but that's all he or she wants (or all the advisor wants). "C" customers do not wish to enter into an ongoing client-advisor relationship, and the advisor offers these individuals only basic services.

7-8. There are many—almost limitless—ways that advisors can provide extraordinary service to clients. The following is a brief recap of those discussed in the textbook: Treat clients to hobby-related events and gifts, provide complimentary services, remember special occasions such as birthdays and anniversaries, keep in touch with handwritten notes, clip and mail newspaper articles of interest, send newsletters, phone contacts regularly, hold client appreciation events, give clients tickets to sporting events or concerts, send them food and beverage baskets, invite them to educational meetings and seminars, and deliver customized gifts to them.

Answers to Self-Test Questions

7-1. B
7-2. C
7-3. C
7-4. B
7-5. A
7-6. C
7-7. C
7-8. D
7-9. D
7-10. A

Chapter 8

Answers to Review Questions

8-1. The events in each of the four phases of a mentoring relationship are as follows:

- During the preparing phase of a mentoring relationship, mentors explore their personal motivation and readiness to be a mentor. They assess their individual mentoring skills to identify areas in which they need additional learning and development.

- Negotiating is the contracting phase of the mentoring relationship. This is when details of the relationship get spelled out: when and how to meet, who has what responsibilities, what the criteria are for success, how accountability is determined, and how and when to bring the relationship to closure.
- The enabling phase is when most of the learning between mentoring partners takes place. Each mentoring relationship is different and finds its own path during this phase. This phase is the longest of the four, and it offers a tremendous opportunity to nurture learning and development. It is also the phase during which mentoring partners are most vulnerable to relationship derailment.
- Coming to closure is the time when mentors and aspirants process their learning and move on, regardless of whether or not the mentoring relationship has been positive. It is a time to assess the experience, acknowledge progress, and celebrate achievement.

8-2. Working with other professionals, including attorneys, accountants, insurance specialists, trust officers, bankers, investment advisors, and fee-only financial planners, can help the advisor in his or her prospecting activities in the following ways:

- When funding is needed for the legal instruments prepared by attorneys, the advisor can be the resource to provide attorneys' clients with the appropriate financial products.
- The advisor can work with accountants to provide their clients with the financial products that will accomplish their planning needs and objectives.
- Someone who works and is licensed exclusively in either the life or P & C insurance fields may be able to exchange leads with an insurance specialist who does not directly compete; they can provide each other's clients with expertise in complementary fields of insurance specialty.
- Trust officers can help financial advisors' clients execute their estate planning objectives, and financial advisors can help them provide funding for the trust instruments they administer.
- Bankers can be great allies for an exchange of prospects when the bank has no investment or insurance representative. This can give the advisor the opportunity to receive referrals for his or her services.
- Many investment advisors have no interest in or knowledge of insurance products, which leaves the potential to provide their clients with these asset-protecting financial products.
- Fee-only financial planners recommend solutions for the client to implement on his or her own. This can represent a tremendous opportunity to offer insurance and investment products to the planner's clients.

8-3. The six key jobs that members of a personal board of advisors should assume are mentor, strategist, problem solver, coach, motivator, and cheerleader.

8-4. A CMS has a wide range of important features to fit various financial services practices. A good CMS

- is simple to operate and easy to keep up-to-date and in good working order
- avoids unnecessary duplication of information
- is flexible so that data can be easily added or removed
- contains all essential information about prospects and clients

The benefit to the advisor is that a CMS is a complete and integrated system, devised to improve work habits and time control. With a CMS, the advisor should be able to

- keep track of callbacks
- monitor future appointment dates and times
- control service work in progress
- centralize all client and prospect information
- automatically bring up sales and service opportunities

A manual CMS is usually kept in a long rectangular file box (metal or sturdy cardboard) that is large enough to hold records for hundreds of prospects and clients. Basically, it is divided into two sections:

- prospect file—cards containing information about prospects, clients, and centers of influence, filed alphabetically
- tickler file—monthly reminders of timely actions to take regarding sales and service opportunities for prospects and clients

8-5. The sections in a typical annual business planner are income objectives, activity objectives, career achievement objectives, client written case log, activity summary, year-end activity evaluation, and business expenses.

The monthly business planner typically includes the following sections:

- monthly activity and production goals
- current and next month's at-a-glance pages
- space to record new sales completed and paid cases
- prospects' and clients' birthdays, insurance age changes, and annual review activity, and so on
- lists of prospects to call for initial, fact-finding, and closing interviews
- weekly planning pages
- daily hour-by-hour appointment calendar
- daily and weekly activity summary reports
- space to record business expenses

8-6.   The steps in the annual planning process are as follows:

- Perform activity summary. First, the advisor must tabulate data and total all of his or her actual numbers for the previous year in the following areas: phone (or face) contacts, appointments seen, number of qualified prospects obtained, initial interviews conducted, fact finders completed, closing interviews conducted, sales made, and referred leads obtained.
- Evaluate year-end activity. Using the totals from the activity summary sections, the advisor must calculate his or her prospecting and sales effectiveness ratios, including phone (or face) contacts to appointments seen, appointments to qualified prospects, qualified prospects to fact-finding interviews, fact-finding to closing interviews, closing interviews to sales, and FYCs to sales (to determine the average FYC per case).
- Determine income objectives. To determine the FYC income objective for the year ahead, the advisor must estimate next year's personal expenses, business expenses, and savings goals. This involves budgeting amounts for each category of cash outflow the advisor expects and for the savings he or she wants for the upcoming year. Next, the advisor must calculate the year's income objective to meet his or her cash outflow and savings goals. This is determined by starting with the gross income the advisor desires, then subtracting anticipated renewal commissions and other business income to arrive at next year's net FYC objective.
- Calculate activity objectives. The advisor must determine next year's prospecting and sales activity objectives he or she needs to achieve his or her FYC goal. The process begins by *dividing* next year's FYC goal by the average FYC per sale, to determine the number of sales needed to meet the goal. For example, if the goal is $100,000 in FYCs and each sale has an average value of $1,000, then the advisor needs 100 sales to meet his or her goal. Next, using last year's prospecting and sales effectiveness ratios and working backwards, the advisor must *multiply* the number of sales he or she needs by the number of closing interviews it takes to produce one sale. The advisor continues the same process to calculate the total number of fact finders completed, qualified prospects, appointments seen, and phone/face contacts needed to achieve each respective level of prospecting and sales activity that will result in the number of sales required to meet the advisor's FYC goal. For example, it may take 3,000 phone calls to produce 750 appointments, to obtain 500 qualified prospects, to complete 400 fact finders, to lead to 200 closes, and 100 sales.
- Establish career achievement objectives. The advisor should establish qualitative career achievement objectives in the following areas: company award and incentive programs (for example, sales conference meetings), industry awards (for example, MDRT), and self-development activities (company courses, industry courses, continuing education, and so forth).

8-7.   The steps in the monthly planning process are as explained below.

- To review the current month, the advisor should do the following:
    - Assess the status of annual activity and FYC goals for the year-to-date.
    - Update activity summary. Update the log of written and paid business activity in the annual planner.
    - List prospects not contacted and consider plans to follow up with them next month.
    - List new prospects obtained in order to make plans to contact them next month.
    - Review the prospect prioritizer to evaluate which prospects to contact first next month.
- Define work activities. The advisor should attend to all the items that will contribute to prospecting and servicing actions that he or she must schedule in next month's business planner. This includes the following:
    - Compile lists of people to whom to send birthday cards and direct mail letters.
    - Review the client file for sales and service opportunities.

       – Examine next month's scheduled activities.
       — Assess the center of influence file for possible contact for referrals.
       — Study the tickler file for prospect follow-up, and prepare next month's newsletter list.
- Plan next month. The advisor should plan the following scheduling, prospecting, and organizational activities for the next month:
  - Prepare a monthly schedule. Schedule next month's daily planning. Delineate work time, fixed commitments, telephoning time, canvassing time, study time, and social activities.
  - Prepare next month's business planner. List the 25 best prospects for initial interviews, fact-finding interviews, and closing interviews. Fill in written and paid business goals for the month.
  - Set up the month-at–a-glance calendar(s).
  - Prepare for telephone calls. Assemble lists of prospects and clients to phone next month.

8-8. The key functions that the state regulation of insurance companies and advisors address are

- insurance company licensing
- producer licensing
- product regulation
- market conduct
- financial regulation
- consumer services

8-9. Compliance means following the laws and regulations, including company rules that apply to the sale of all financial products. These are the minimum standards. Ethical behavior for the advisor involves

- doing the right thing
- putting the prospect's best interests before the advisor's own
- maintaining the highest possible standard of behavior in all business dealings
- continuing to develop professional skills in order to provide the best possible service
- representing the industry, its companies, and its advisors in the best possible light

8-10. The common themes used widely in all professional codes of ethics are as follows:

- Every code calls on professionals to look out for the best interests of the client.
- Most codes in one way or another, ask professionals to conduct themselves with fairness, objectivity, honesty, and integrity.
- Each code requires professionals to protect the confidential information of clients.
- Most codes require that professionals present enough information to allow the client to make an informed decision.
- Each code requires professionals to continue the learning process throughout their careers.
- Each code asks professionals to conduct themselves in a way that brings honor to themselves and to their professions.
- Most codes specify that financial services professionals should comply with the law.

Answers to Self-Test Questions

8-1. C
8-2. D
8-3. A
8-4. C
8-5. B
8-6. B
8-7. B
8-8. D
8-9. A
8-10. C

Index